Springer Tracts on Transportation and Traffic

Volume 7

Series editor

Roger P. Roess, New York University Polytechnic School of Engineering,
New York, USA
email: rpr246@nyu.edu

About this Series

The book series "Springer Tracts on Transportation and Traffic" (STTT) publishes current and historical insights and new developments in the fields of Transportation and Traffic research. The intent is to cover all the technical contents, applications, and multidisciplinary aspects of Transportation and Traffic, as well as the methodologies behind them. The objective of the book series is to publish monographs, handbooks, selected contributions from specialized conferences and workshops, and textbooks, rapidly and informally but with a high quality. The STTT book series is intended to cover both the state-of-the-art and recent developments, hence leading to deeper insight and understanding in Transportation and Traffic Engineering. The series provides valuable references for researchers, engineering practitioners, graduate students and communicates new findings to a large interdisciplinary audience.

More information about this series at http://www.springer.com/series/11059

John C. Falcocchio · Herbert S. Levinson

Road Traffic Congestion: A Concise Guide

Springer

John C. Falcocchio
NYU Polytechnic School of Engineering
New York
USA

Herbert S. Levinson
Transportation Consultant
Wallingford, CT
USA

and

Region 2 Urban Transportation Research Center
City College
New York
USA

ISSN 2194-8119 ISSN 2194-8127 (electronic)
Springer Tracts on Transportation and Traffic
ISBN 978-3-319-15164-9 ISBN 978-3-319-15165-6 (eBook)
DOI 10.1007/978-3-319-15165-6

Library of Congress Control Number: 2015930028

Springer Cham Heidelberg New York Dordrecht London

Printed on acid-free paper

Springer International Publishing AG Switzerland is part of Springer Science+Business Media (www.springer.com)

*We dedicate this book to our families—
especially the grandchildren—who have
endured endless hours of abandon while we
were laboring to bring this book to a closure.*

Preface

This book on road traffic congestion in cities and suburbs describes congestion problems and shows how they can be relieved. The first part (Chaps. 1–3) shows how congestion reflects transportation technologies and settlement patterns. The second part (Chaps. 4–13) describes the causes, characteristics, and consequences of congestion. The third part (Chaps. 14–23) presents various relief strategies—including supply adaptation and demand mitigation—for nonrecurring and recurring congestion. The last part (Chap. 24) gives general guidelines for congestion relief and provides a general outlook for the future.

The book will be useful for a wide audience—including students, practitioners and researchers in a variety of professional endeavors: traffic engineers, transportation planners, public transport specialists, city planners, public administrators, and private enterprises that depend on transportation for their activities.

Acknowledgments

This book is the product of our many years' experience in teaching, research, and practice in urban transportation system planning, transportation engineering, and transportation system management. Our book benefits from the work of many public agencies and research groups for the availability of their public data, and of many transportation professionals—especially those who are acknowledged in the individual chapters—who have greatly contributed to our understanding of urban transportation. Special thanks are due to our closest colleagues with whom we collaborated over the years through many venues, and who have enriched our understanding of urban transportation as part of the larger urban system that establishes quality of life parameters. Colleagues that deserve special mention include: Professors Robert (Buzz) Paaswell at City College and the Director Emeritus of the University Transportation Research Center (II), Camille Kamga (current Director of UTRC), Rae Zimmerman at NYU Wagner School, Sig Grava, whose untimely death diminished the transportation community, George List at North Carolina State University, Roger Roess and Ilan Juran at NYU Polytechnic School of Engineering, Dr. William R. McShane of KLD Associates, Dr. Michael Horodniceanu at the NY Metropolitan Transportation Authority, Sam Schwartz of Sam Schwartz Engineering, Wayne Berman of the Federal Highway Administration, Bill Eisele and Tim Lomax of the Texas Transportation Institute, Robert Skinner of the Transportation Research Board, and Lisa Tierney of the Institute of Transportation Engineers, Richard Pratt (whose material we used), and Sam Zimmerman.

Special thanks is due to Zeng Xu, a Ph.D. student in the Department of Civil and Urban Engineering at the NYU Polytechnic School of Engineering, who has diligently assisted in preparing the manuscript by taking care of crucial matters including keeping track of references and the preparation of graphics.

Contents

Part I Background

Part II Traffic Congestion Characteristics, Causes, and Consequences

About the Authors

Dr. John C. Falcocchio is Professor of Transportation Planning and Engineering at the NYU Polytechnic School of Engineering since 1981, Director of the NYU-Poly's Urban Intelligent Transportation system (ITS) Center since 1995, and was the Head of the Department of Civil Engineering (1998-2001).

He co-established the Urban Intelligent Transportation System Center (UITSC) at NYU–Poly. Since 1995, the UITSC has provided the New York City Department of Transportation with an effective framework to assess ITS technology deployment strategies in upgrading its transportation system through research, professional training, demonstration projects, and international outreach.

Strongly committed to applying theoryto-practice in transportation problem solving, Dr. Falcocchio in 1973 was a founding Principal of Urbitran, a New York City\planning/engineering/architectural firm that was ranked by Engineering News Record as one of New York's 20 top firms thirty years later.

His current research concentrates on the development of user-oriented\transportation performance metrics, and on the management of nonrecurring congestion.

Dr. Falcocchio received his BCE and Ph.D. from the Polytechnic Institutes of Brooklyn and New York, respectively (now NYU Polytechnic School of Engineering), and a Certificate in Highway Traffic from Yale University where he attended on a 9-month fellowship. He is a Professional Engineer in Pennsylvania, New York, and California, and a life member of the American Society of Civil Engineers and the Institute of Transportation Engineers. He serves at the Polytechnic School of Engineering representative on the Transportation Research Board, and is Chairman of the Board of the University Transportation Research Center (Federal Region 2).

Herbert S. Levinson is a practitioner, teacher, and researcher who has advised public agencies in the United States and abroad. He is a recognized authority on transportation planning, traffic engineering and public transport, who has authored several books, many research reports, and more than 250 publications. He is a member of the Connecticut Academy of Science and Engineering, the National Academy of Engineering, and the recipient of many awards. He has been an independent consultant since 1980, and is currently an Urban Transportation Research Center Icon Mentor at City College, New York. Mr. Levinson has a B.S in Civil Engineering from Illinois Institute of Technology, a Certificate in Highway Traffic from Yale University, and an honorary Doctor of Engineering degree from Illinois Institute of Technology. He is a member of the American Planning Association, a Fellow in the ASCE, an Honorary Member of the Institute of Transportation Engineers, and has been member of the Transportation Research Board's Executive Committee.

Part I
Background

Chapter 1
Introduction

1.1 The Nature of the Problem

Congestion in transportation facilities—walkways, stairways, roads, busways, railways, etc.—happens when demand for their use exceeds their capacity.

Travelers tend to complain about traffic congestion because it adds to their travel time and takes away from the time they can dedicate to other activities. Truck drivers complain because it reduces their productivity and increases their operating costs. Transit service providers complain about roadway traffic congestion because it increases the number of buses and drivers needed to provide the service. Congestion increases business costs, air pollutant emissions and fuel consumed.

Congestion also can influence investment decisions, and therefore it becomes a major economic concern. It influences where people live, work and how they travel. Therefore reducing congestion benefits a wide constituency.

Traffic congestion has been a fact of city life from ancient times when movement was by walking and animal-drawn coaches to today's cities that rely on various means of mechanized travel. It is a byproduct of economic activities that grow faster than the growth in transportation infrastructure.

Traffic congestion is now found in cities throughout the world. It continues to increase as the cities' population and motorization grow and as travel growth outpaces investments in roads and public transportation. The beginning of congestion is generally perceived by drivers when their trip time increases by approximately 0.4–0.5 min/mile, and they become acutely aware of congestion when it increases by 0.8–1.0 min/mile.

Traffic congestion may also be the hallmark of a vibrant economy: a city without a traffic congestion problem is likely to experience an economic recession, or a declining population. But where congestion is too pervasive and trip time reliability is a problem, the city may become a less desirable attraction for economic growth.

J.C. Falcocchio and H.S. Levinson, *Road Traffic Congestion: A Concise Guide*,
Springer Tracts on Transportation and Traffic 7,
DOI 10.1007/978-3-319-15165-6_1

People who live a large metropolitan area are concerned about traffic congestion because it affects most of their daily activities—arriving on time to work or at a business meeting, to meet a friend, catching a plane, etc.

Below are examples of different traffic congestion experiences and the type of responses that each engenders:

- If you have moved your young family in the suburbs where you could afford the house and your commute has become longer and more stressful, you will favor the construction of more road capacity, or an affordable, faster transit service.
- If you can afford to buy or rent in the central city, roadway traffic congestion may not bother you too much, but crowded buses or trains, or station platforms will. If you live in the city, therefore, you would favor improving transit service and bicycle routes for your mobility needs.
- If you are an urban economist, you are concerned with marginal cost pricing and are likely to favor reducing traffic demand through congestion pricing. You will be supported by environmentalists and those living near congested roadways because less motor vehicle traffic improves air quality. But congestion pricing is likely to be opposed by suburban commuters because it will increase their commuting cost—upsetting the cost balance of their housing and commuting that they were counting on when they decided on the housing location choice. In addition, low-income commuters will tend to oppose congestion pricing preferring "free" roads that require waiting on traffic queues to toll roads that reduce congested travel.
- If you are an environmental advocate you will support higher land density developments such as "smart growth" because you want to reduce the growth of vehicle miles of travel (VMT). But if you are a developer, you are concerned about the demand for high density housing in suburban areas.
- Transportation planners and environmental groups advocate more transit capacity to encourage travelers to use transit service and they are typically joined by economists in promoting the idea of using revenues from congestion pricing to finance transit improvements.
- If you are a traffic engineer, you will seek to reduce traffic congestion by removing capacity bottlenecks through capacity expansion, and you will favor the application of advanced technologies to improve the efficiency of the road network.

These examples show that the sources and perspectives of traffic congestion are many and diverse. In these examples there is no single overall solution to the congestion problem that meets every situation because the contexts are different. And where these contexts do not overlap it is usually impossible to find a solution strategy that satisfies every need.

1.2 Why this Book

This book has been prepared to fill the need for a clear and comprehensive look at the many dimensions of traffic congestion. It defines and describes congestion, explains its causes, describes its consequences, and identifies ways to provide congestion relief.

Traffic congestion has been extensively explored for many years in various articles and books. But these documents have usually treated congestion from specific perspectives (person travel or goods movement) or discipline (e.g., traffic engineering, transit operations, economics, land use planning and zoning).

In fact, there is no lack of interest or knowledge to reduce or manage urban traffic congestion to meet one's expectations. However, to implement solutions to the traffic congestion problem that are acceptable requires agreement among the diverse stakeholders involved. But these diverse stakeholders—including the various disciplines—are unlikely to find convergence on what needs to be done about the growing traffic congestion problem without a shared language and common objectives.

Although they may all use the same words—congestion, mobility, accessibility—in debating the congestion issue, they do not necessarily share the same meaning that these words convey. To discuss and debate the congestion problem in a public forum it is necessary to use definitions and metrics that allow for clear and unambiguous exchange of ideas among interest groups. Traffic congestion solution strategies need to be described in terms that impacted stakeholders find relevant to their daily lives.

This book, therefore, has been prepared in response to the many needs for a comprehensive, clear, and objective look at the many dimensions and impacts of traffic congestion in metropolitan areas.

The book gives practitioners and researchers, local elected officials, and community leaders, information on urban traffic congestion—its causes, characteristics and consequences—they can use to create a framework that allows diverse interest groups to debate the issue of traffic congestion by using the joint platform of mobility and accessibility. To develop rational policies for managing the urban traffic congestion problem, a focus on mobility <u>and</u> accessibility is needed. Not just mobility as traffic engineers are inclined to favor; and not only accessibility, as "smart growth" advocates favor.

The book lays the foundations for achieving a common understanding among the various stakeholders and disciplines and presents simple quantitative methods for estimating the effects of congestion on mobility, accessibility, travel time reliability, and other quality of life indicators.

Building on this understanding the book presents a rational analysis framework that a city, suburb, or a metropolitan area can use when managing growing traffic congestion problems. Thus the book is useful not only to transportation students and transportation professionals, but also to urban planners, and transportation policy analysts and policy makers.

In summary, the book focuses on four key objectives:

1. To understand and address the factors that contribute to traffic congestion,
2. To understand the issues involved in quantifying urban traffic congestion,
3. To assess the impacts of congestion on urban and suburban mobility, access to activities, network productivity, and environmental quality, and,
4. To provide congestion relief strategies that increase traffic efficiency and increase the use of alternative modes of transportation.

Each of these objectives is examined from a concise multi-disciplinary perspective using illustrations and techniques that provide for a broad, yet clear, understanding of traffic congestion and its impacts, and that will describe adaptation and mitigation strategies that are likely to provide congestion relief.

1.3 Overview of the Book

The book is organized into 24 chapters grouped into four parts:

1. Part I—Background (Chaps. 1–3):

In addition to this chapter Part I includes Chap. 2, "How Transportation Technology has Shaped Urban Travel Patterns," and Chap. 3, "Historical Perspective of Urban Traffic Congestion." Chapter 2 examines urban development and traffic congestion from historical and contemporary viewpoints. It shows how technology, transportation technology in particular, has extended urbanized areas and traffic congestion. The chapter shows that urban traffic congestion is not only a current phenomenon, but has existed in cities since ancient times. Chapter 3 shows how growth in population, employment, motorization and vehicle miles since World War II has contributed to the spread of congestion from the city center to the entire metropolitan area.

2. Part II—Traffic Congestion Characteristics, Causes, and Consequences (Chaps. 4–13):

Chapter 4 describes the underlying causes of traffic congestion. They include the concentration of travel demand in time and space (Chap. 5); the effect of growth in population employment and car use, population density, and the lag between roadway capacity growth and travel growth (Chap. 6); and the effect of bottlenecks (Chap. 7). Chapter 8 describes the criteria and metrics used to describe and quantify congestion; and Chaps. 9–13 address the impacts of congestion on trip time and reliability, mobility, accessibility, traffic productivity, transportation costs, and quality of life issues.

3. Part III—Congestion Relief Strategies (Chaps. 14–23):

Chapter 14 provides an overview of possible adaptive and mitigation strategies for managing congestion. The chapter provides a framework for the various capacity expansion and demand mitigation strategies for managing nonrecurring congestion (Chap. 15) and recurring congestion (Chaps. 16–23). Capacity-oriented (adaptive) strategies (Chaps. 16–17) aim at increasing roadway capacity to keep up with traffic demand. Traffic reduction (mitigation) strategies focus on reducing the use of automobile travel (VMT) by relying on changes in travel behavior motivated by pricing, regulatory, or employer-based strategies (Chaps. 18–23). While some capacity oriented strategies are relatively easy to implement (for example removing a physical bottleneck or improving the timing and coordination of traffic signals), strategies aimed at reducing automobile use (VMT) require behavioral changes. Modifying travel behavior of individuals by restricting their travel choices for the larger societal good is more difficult to implement.

4. Part IV—Conclusions

This concluding chapter summarizes the book's key points, sets forth suggested congestion relief strategies for typical problem locations, and provides a future outlook to the congestion problem in light of expected changes in socio-demographics, and in transportation technology.

1.4 Who Can Benefit from this Book

This book is intended for a wide audience. It will be especially useful to transportation students, practitioners and researchers. But it will also be helpful to urban planners, policy analysts, and transportation policy makers by providing a broad discussion of the issues framing the traffic congestion problem.

Transportation students will benefit from an integrated understanding of the core issues framing the traffic congestion problem, as opposed from what they can get from books that focus on specific aspects of the congestion problem.

Transportation practitioners are provided with a quick reference framework to evaluate the contextual impacts of individual projects.

Transportation policy analysts will benefit from a better understanding of the factors influencing transportation performance.

Policy makers and the general public will benefit from the book because it is organized to cover topics the public cares about, and because it provides knowledge tools needed to better understand and evaluate alternative solution strategies.

Chapter 2
How Transportation Technology Has Shaped Urban Travel Patterns

2.1 Introduction

The primary functions of transportation are to facilitate the movement of people and goods and to provide access to land use activities located within the service area.

This chapter shows how advances in transportation technology have helped to determine the size, shape and density of urban areas and associated traffic congestion patterns. It provides a brief historical review—from ancient times to the present—of how transportation technology has shaped the size of urban areas over time, and highlights the connection between transportation technology and land use. Each advance in transportation technology (e.g., electric streetcars, subways, automobiles) has produced higher travel speeds; and each time travel speed has increased, the amount of land used for urban growth has increased and population density has decreased. The resulting travel patterns followed the population and employment gradients.

This transition in living conditions from high population density (where activities are located very close to one another) to low population density (where activities are located far from each other) has changed how people travel to work, shop, and pursuit of other endeavors—from a high dependence on walking and transit in high density cities, to an almost exclusive reliance on cars in low-density suburbs.

The underlying theme is that traffic congestion is a product of vibrant urban areas and that people with the means to do so have tried to escape congestion when technological advances provided the opportunity to do so.

It took the transportation advances of the Industrial Revolution (electric streetcars and subways) to enable people to act on their desire to escape the congested industrial city. The automobile accelerated and sustained this desire especially since the end of WWII.

J.C. Falcocchio and H.S. Levinson, *Road Traffic Congestion: A Concise Guide*,
Springer Tracts on Transportation and Traffic 7,
DOI 10.1007/978-3-319-15165-6_2

However, just as city streets before the car era were crowded and congested, the popularity of the suburbs has attracted many people and jobs over time creating traffic congestion on many freeways and arterial roadways.

Understanding how transportation technology influences the character of land development is fundamental to establishing policies aimed at sustaining desirable levels of mobility and accessibility in light of increasing travel growth and traffic congestion. Addressing these concerns is a major challenge especially in the US where the zoning of land use is typically controlled by local governments whose decisions are often made separately from decisions that States make about major transportation investments.

This chapter sets forth some key issues that should be considered when formulating policies and programs addressing urban and suburban traffic congestion, and it shows that traffic congestion has usually followed urban development.

2.2 Transportation Technology, Urbanization, and Travel

The predominant type of transportation available at a particular time in history (non-motorized, fixed route transit, or motor vehicles) has influenced the location and density of residential and non-residential activities.

Transportation and land use are two interconnected elements of the urban system and structure. The locations of activities reflect the daily need for access to jobs, shopping, educational or social needs of the population. Access to these people-oriented activities is determined by the prevailing transportation technology, and by the time people budget for travel.

Traveler and goods mobility was provided by walking and animal power for thousands of years until the dawn of the industrial revolution.

Land travel was by foot (2–3 mph) or by the use of animal power (horse speed of 4–6 mph). At these travel speeds the distance one could cover within acceptable travel times was very short and for this reason land use activities were located close together.

With the introduction of mechanized travel, speeds increased substantially allowing people to travel farther within the same travel time budgets. This increased mobility encouraged the separation of various activities, expanded the amount of urbanized land, and reduced population density in central areas.

The transition from high density urban developments to lower density ones is closely related to the transportation technology prevailing at various times in history.

Lay [1] in his remarkable book "The Ways of the World" provides many examples of how transportation technology influenced the character of cities and urban development. Salient highlights are as follows.

2.2.1 Ancient Time

Ancient cities were compact places with buildings located close to one another and connected by narrow streets. Most people lived within a 15 min walk of their work places, and their streets were predominantly used for pedestrian movement as well as for many commercial and social activities.

Population Densities [1]

Examples of ancient population densities are:

(1) Babylon and Rome with peak populations of over ½ million, were contained within an area of 14 Km^2. or less, and had an effective radius under 2 km.
(2) The population of Baghdad in about 900 AD, was 900,000—the largest that could be practically accommodated within a walking city. Its population density peaking at 600 persons per hectare (243 per acre, or 155,500 persons per square mile).

Ancient cities suffered from street congestion. In Rome, 'Julius Caesar found it necessary to issue an order prohibiting the passage of wagons through the central district for 10 h after sunset' [1]—a more stringent regulation than is found in any modern city.

Mobility in medieval cities—hemmed in by their defensive walls—was provided by walking on narrow and crooked lanes/alleys unsuitable for wheeled traffic.

2.2.2 The Industrial Revolution (ca. 1825–1900)

In the years of the Industrial Revolution, land development in cities continued to locate around the walking mode. During this period cities had high population density; streets were narrow, congested, and often polluted with horse manure and dead animals.

The growth of cities around the beginning of the 20th century was made possible by the steel-framed building construction that allowed taller buildings at the city center, and by electric traction that provided speeds of 8–12 miles per hour. At the same time, mechanization of agriculture enabled many people on the farms to migrate to the cities—a trend that continued through the 20th century.

The rise and spread of cities has paralleled the growth and speed of transportation. Improved transportation has played a crucial role in the transition from a rural to and urban society.

People looking for employment and a more promising economic future migrated from the countryside to the industrial city contributing to its extremely crowded living and travel conditions. By 1900, "population densities in London and Paris peaked at over 700 people/ha. (283 per acre, or 181,000 per square mile), and in New York City they reached 1,350/ha" (546 per acre, or 350,000 per square mile in several neighborhoods) [2]. Overcrowded living conditions became a major social and environmental concern in New York City.

The appearance of streetcars, subways, elevated rail, and commuter rail lines, with their higher operating speeds, replaced the horse drawn cars by extending the distance that people could travel within acceptable travel times. This technological development reduced population densities and increased employment densities in city centers and it transformed the urban landscape by enabling settlements to expand into new territories previously inaccessible by the slower modes of transportation.

New rail transit lines were laid out to connect the population to jobs and shopping locations in the central business district (CBD)—which became the most accessible place in the city.

The steam railroads that appeared in the latter half of the 19th century improved access between cities. Over the years, many small communities that had access to train stations, became suburbs of nearby cities.

The commuter railroad operating at higher speeds (30–35 mph), enabled commuters to work in the city and live farther out from the city limits (away from the dirty air) where living space was more affordable, and the environment more desirable for raising a family. With an average commuter rail speed of 30 mph, one could cover a door-to-door distance of approximately 12 miles in 45 min.[1] This rail-based urban expansion, created new towns and villages whose residential and other land use activities were located within walking distance of the transit stations.

The rail lines allowed (1) increased employment concentration in city centers, and (2) fostered residential developments in outlying areas.

2.2.3 The Private Motor Vehicle Era (1925–Present)

With the coming of the motor vehicle, the land between rail lines and beyond became accessible for development and the distance between land use activities was no longer limited by the rail lines and the walking distance to their stops or stations.

The technology of the automobile provides people with access to one almost total freedom to travel when and where they want. Its use is not constrained by service routes or schedules. It offers reliable door-to-door transportation without the need to change travel modes. It operates at high door-to-door travel speeds relative to most urban travel modes. It ensures seating and privacy as well as weather protection. And, last but not least, it offers pride of ownership.

Its higher operating speed (up to 30–40 mph) makes possible traveling longer distances within acceptable commuting times. Consider a 45 min trip from home to a job location: if the trip is by car one can reach a job located 30 miles away; if the

---

[1] (45 min) − (20 min spent to reach vehicle, wait, and reach destination) = 25 min riding time; 25 min/(60 min/hr) × (30 mph) = 12.5 miles.

trip is by commuter rail, however, only jobs within 12 miles can be reached. Thus the higher door-to-door travel speed of the automobile and its unlimited choice of destination opportunities, make it possible for a commuter to expand her/his area of residential location and job choices.

The motor vehicle allowed urban activities to spread-out by removing the need to locate buildings within walking distance of rail stations, and reduced the reliance on transit for accessing more distant destination opportunities. In the US, the superior mobility provided by the automobile was quickly recognized and its popularity steadily increased. In 1916 there were over 2 million automobiles owned and that increased to 8 million in 1920—a fourfold increase in 4 years. Before the beginning of WWII (1940), there were 32.45 million motor vehicles in the US.

After WWII, the private motor vehicle further accelerated the urbanization of agricultural and developable land beyond the city's limits. This was made possible by the convergence of a number of factors. The construction of high-speed (65 mph) limited access highways made possible by a vast federal road building program that peaked with the Federal-Aid Highway act of 1956 authorizing 41,000 miles of high speed freeways and expressways which by 1972, were to link 90 % of the cities with population of 50,000 population or greater, along with many smaller cities and towns [3]. When combined with affordable prices of automobiles, cheap gasoline, an abundance of FHA low-cost housing mortgages, and a favorable tax code for home owners, these events set in motion a large suburban expansion of the population into low-density housing developments that could only be served by car, and were followed by the spreading of jobs from center cities into suburban areas [4]. Schools, retail stores, industries also became more numerous in suburban settings.

The popularity of the car as a mobility provider enabled vast number of families to escape the city—with its crowded housing, poor public schools, high crime, and racial problems of the 1960s—by moving to the open spaces and affordable larger living quarters offered by the suburbs made accessible by new highways connecting the new residential developments to the jobs in center cities. Modes of Travel in US Metropolitan Areas.

Tables 2.1 and 2.2 show commuter trips within the US metropolitan areas and the major travel modes used in commuting to and from work.

The significance of Tables 2.1 and 2.2 can be summarized as follows:

(1) In the suburbs, where 64 % of metro area commuters live and about 54 % work (Table 2.3), the car is used for 94 % of suburban trip destinations that originate in center cities; 91 % of suburban trip destinations originating in the suburbs; and 93 % of center city destinations originating in the suburbs.
(2) In center cities, where 36 % of commuters live and approximately 46 % of the commuters work, transit is used for 15 % of center city trip destinations originating in center city; 6 % of center city trip destinations originating in the suburbs; and 5 % of suburban trip destinations originating in center cities.

Table 2.1 Intra metropolitan origin/destination of commuter travel, 2,000 (million of trips)

	Central city employment destinations	Suburban employment destinations	Total trip origins
Commuter trips originating in central city	24.5 27.40 %	7.5 8.40 %	32 35.80 %
Commuter trips originating in the suburbs	16.6 18.50 %	40.9 45.70 %	57.5 64.20 %
Total trip destinations	41.1 45.90 %	48.4 54.10 %	89.5 100 %

Source Reference [4], p 49, Fig. 3.3, 2,000 data

Table 2.2 Mode share of metropolitan commuters (2,000)

	Destined to central city (%)		Destined to suburbs (%)	
Trips originating in central city	Drive alone	62	Drive alone	76
	Carpool	12	Carpool	18
	Subtotal car	**74**	**Subtotal car**	**94**
	Transit	**15**	**Transit**	**5**
	Bike	1	Bike	0
	Walk	5	Walk	0
	Work at home	3	Work at home	na
	Other	2	Other	1
Trips originating in the suburbs	Drive alone	82	Drive alone	79
	Carpool	11	Carpool	12
	Subtotal car	**93**	**Subtotal car**	**91**
	Transit	**6**	**Transit**	**1**
	Bike	0	Bike	0
	Walk	0	Walk	3
	Work at home	na	Work at home	5
	Other	1	Other	1

Source Reference [4], p 81, Fig. 3.40 and 3.42, 2,000 data

It should be noted, however, that the above values are averages for all metropolitan areas—from the largest to the smallest. There is a large difference, however, in transit share between the largest and smallest metro areas, as shown in Table 2.3.

The transit share of downtown trips of the 15 metro areas in Table 2.3, ranges from 76.5 % for New York with a downtown worker density of over 351,000 commuters per square mile, to 3.8 % for Austin with a downtown density of 80,000 commuters per square mile.

Table 2.3 Transit commuting to downtowns (as defined)

Area	Total commutes to entire metro area, 2,000	Total commuters to central city	Total transit commuters to central city	Transit share to central city (%)	Total downtown commuters 2,000	Total transit commuters to downtown, 2,000	Downtown land area (square miles)	Transit share of downtown commuters (%)	Worker density (commuters per square mile)
New York	9,429,080	4,545,645	2,065,120	45.43	379,380	290,390	1.08	76.5	351,277.78
Chicago	4,263,430	1,686,150	420,975	24.97	341,014	210,490	1.13	61.7	301,782.30
San Fran cisco	3,523,465	1,809,120	273,430	15.11	320,170	156,764	2.55	49.0	125,556.86
Washington, DC	3,876,675	1,296,840	269,685	20.80	409,505	154,658	3.99	37.8	102,632.83
Boston	2,977,665	1,143,960	222,500	19.45	270,315	137,701	2.32	50.9	116,515.09
Philadelphia	2,790,705	875,785	190,310	21.73	230,358	105,387	2.40	45.7	95,982.50
Seattle	1,785,935	841,560	100,500	11.94	147,905	54,435	2.99	36.8	49,466.56
Los Angeles	6,744,860	2,776,585	185,515	6.68	215,340	43,656	3.78	20.3	56,968.25
Portland	1,107,080	549,160	49,940	9.09	104,810	28,839	2.11	27.5	49,672.99
Houston	2,078,465	1,354,610	62,665	4.63	155,050	25,874	1.68	16.7	92,291.67
Dallas	2,569,405	1,430,395	37,475	2.62	91,786	12,493	0.85	13.6	107,983.53
San Diego	1,293,940	801,530	29,830	3.72	75,850	8,675	2.16	11.4	35,115.74
Sacramento	802,455	308,235	14,855	4.82	64,830	7,959	1.26	12.3	51,452.38
San Antonio	708,445	582,675	18,045	3.10	53,440	3,842	1.15	7.2	46,469.57
Austin	657,455	506,750	15,514	3.06	76,150	2,913	0.95	3.8	80,157.89

Source Adopted from Reference [4], p 94, Table 3.42

Table 2.4 Transport mode and urban form

Item	Type of city		
	Pedestrian	Electric transit	Automobile
Population	3,000,000	3,000,000	3,000,000
Area (square mile)	30	200	500±
Density (persons/sq.mi)	100,000	15,000	6,000
Jobs in city center	200,000	300,000	150,000
Development pattern	Compact	Radial with major corridor	Dispersed
Example	Paris pre 1900	Chicago 1930	Dallas 1990

Assuming an average of 225 square feet of floor space per commuter, the office floor space needed to hold New York's downtown commuters would amount to approximately 79 million square feet, and to accommodate Austin's downtown commuters, 18 million square feet.

2.3 Conclusion

Urban development and congestion patterns reflect the available transportation technologies. Each advance in the speed of travel has increased mobility, influenced land development, the form of cities, and patterns of congestion.

- Walking limited the radius of cities to the distance one could cover in 30–40 min (an average of about 2 miles).
- The electric street car extended the radius of the city, focused development along street car lines, reduced residential density in city centers and spread congestion outward. Large cities such as Boston and Philadelphia placed their street car lines underground to avoid congestion in city centers.
- A handful of cities built rapid transit lines that complemented suburban rail lines in improving mobility. These facilities had the dual effects of further concentrating development in the city center and extending urban development outward along the rapid transit lines. In a few cases, parallel rapid transit lines were built to accommodate the increased demand.
- Automobiles and the roadways that were built to serve them further decentralized development and traffic congestion.
- The changes in transport technology progressively flattened the population density gradient—the decline in population density with increasing distances from the city center. These changes are illustrated in Table 2.4 that gives illustrative population and employment densities for pedestrian, electric transit and automobile cities [5].

References

1. Lay GM (1992) Ways of the world—a history of the world's roads and of the vehicles that used them. Rutgers University Press, New Brunswick
2. The New York Times. s.l. (1922) Population density (in NYC) reduced by subway—comparison with London. The New York Times, 1 May 1922
3. Wener E (1992) Urban transporation planning in the United States—an historical overview. Technology Sharing Program, US Department of Transportation, Washington DC
4. Pikasrki AE (2006) Commuting in America III. NCHRP report 550 and TCRP report 110
5. Levinson HS, Falcocchio JC (2011) Urban development and traffic congestion. In: Proceedings of first transportation and development institute (TDI) congress. American Society of Civil Engineers, Chicago, pp 948–956

Chapter 3
Historical Perspective of Urban Traffic Congestion

3.1 Introduction

Congestion is not new. It predates the industrial revolution, the motor vehicle, and the modern city. It was common in ancient time, in 17th century London, and in 19th century New York. The produce markets, port areas, and downtowns of yesteryear all were overcrowded and congested.

The industrial city that grew before the automobile era experienced traffic congestion caused by high population and employment densities that produced travel demands in excess of road capacity. The emergence of skyscrapers in late 19th century, coupled with the mix of horse-drawn vehicles created severe congestion during the latter years of the 19th century.

Before the automobile and electric railways, congestion was mainly limited to city centers and their immediate environs. Traffic congestion now permeates the metropolitan area [1].

The geographic spread of congestion over the past century is illustrated in Fig. 3.1. This dispersion reflects improved transportation mobility and its impact on suburban development patterns.

3.2 Historical Examples

Traffic congestion has consistently reflected city size and structure. Modes, locations, and intensities have changed over the years, but the common themes remain—the concentration of people and vehicles in major employment centers, and the inability to manage the conflicts among competing travelers and road users, and to eliminate physical obstructions to movement [2].

Before the automobile era, congestion was characterized by stagecoaches, wagons, and pedestrians vying for downtown space. The street railway, introduced

J.C. Falcocchio and H.S. Levinson, *Road Traffic Congestion: A Concise Guide*,
Springer Tracts on Transportation and Traffic 7,
DOI 10.1007/978-3-319-15165-6_3

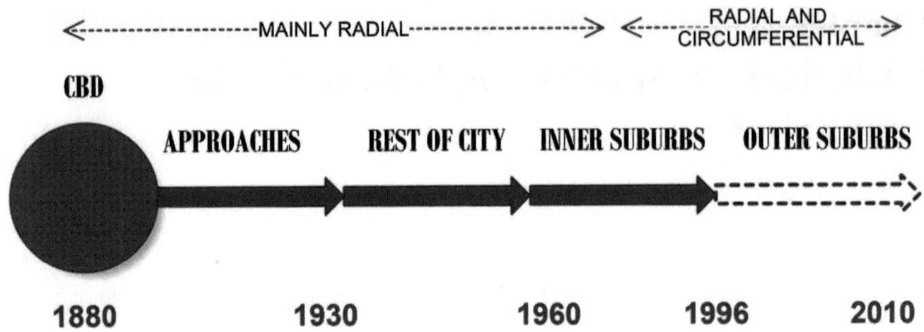

Fig. 3.1 The spreading of traffic congestion

another disparate element into the traffic stream, adding conflicts and congestion. The street cars often operated at very close spacing and often congested each other.

- In 1891, on Tremont Street in Boston: "during the afternoon rush hour, cars were packed so close together that one could walk from Scully Square to Boylston Street on the car roofs" [3].
- In Philadelphia, street cars, horse drawn vehicles comingled and nothing moved (Fig. 3.2).
- In Chicago, Dearborn Street and Randolph Street experienced gridlock conditions as a result of rush hours vehicle-vehicle and vehicle-pedestrian conflicts (Fig. 3.3). Street cars and horse-drawn vehicles were major contributors to congestion.
- In Tokyo's Honjo Quarter, Fig. 3.4 shows the intensity of traffic congestion experienced in 1924, caused by the demand volume of vehicles and pedestrians approaching a bridge with insufficient capacity.

The growth of motorization in the 1920s often preceded effective traffic control and management. This led to both chaotic confusion and congestion in many business centers. Figure 3.5 shows the situation in downtown Los Angeles during the mid 1920s.

These conditions led many cities to (1) establish traffic regulations and controls, (2) remove produce markets from central cities, and (3) increase the width of streets. Chicago, for example, relocated the South Water Market, built two-level Michigan Avenue and Wacker Drive, (4) banned left turns in the "Loop," and (5) signalized downtown intersections. Other cities also began to manage their traffic, and the field of Traffic Engineering emerged.

The 1930s and 1940s were characterized by continued growth in automobile traffic congestion on radial highways leading to or within the city center. Figure 3.6 shows congestion patterns in Chicago and its environs in 1942.

Traffic backups were common on many city streets, in some cases extending for miles. Figure 3.7 shows 1940 rush hour congestion on Gratiot Avenue, a major arterial in Detroit.

Fig. 3.2 Traffic congestion in Philadelphia, turn of the century. *Source* National Archives and Records Administration, 30-N-36713

Fig. 3.3 Dearborn street, looking south from Randolph street, about 1910. *Source* Chicago Historical Society

Fig. 3.4 Traffic congestion on a bridge approach, Tokyo, February 16, 1924. *Source* "Wikimedia Commons" http://commons.wikimedia.org/wiki/File:Traffic_congestion.jpg

Traffic congestion in and around the US and Canadian city centers increased in the years following WWII.

A 1950 study of traffic conditions in Chicago's central business district, for example, reported that (on average) traffic delays across the Loop accounted for time losses of 2 min per trip for auto drivers and 2-½ min for bus transit riders.

Fig. 3.5 Typical traffic congestion in downtown Los Angeles, ca. 1920s. *Source* "Art + transportation—pre-crosswalk and stoplight Los Angeles" posted on January 26, 2012 by Larry Ehl (http://www.transportationissuesdaily.com/arttransportation-pre-crosswalk-and-stoplight-los-angeles/)

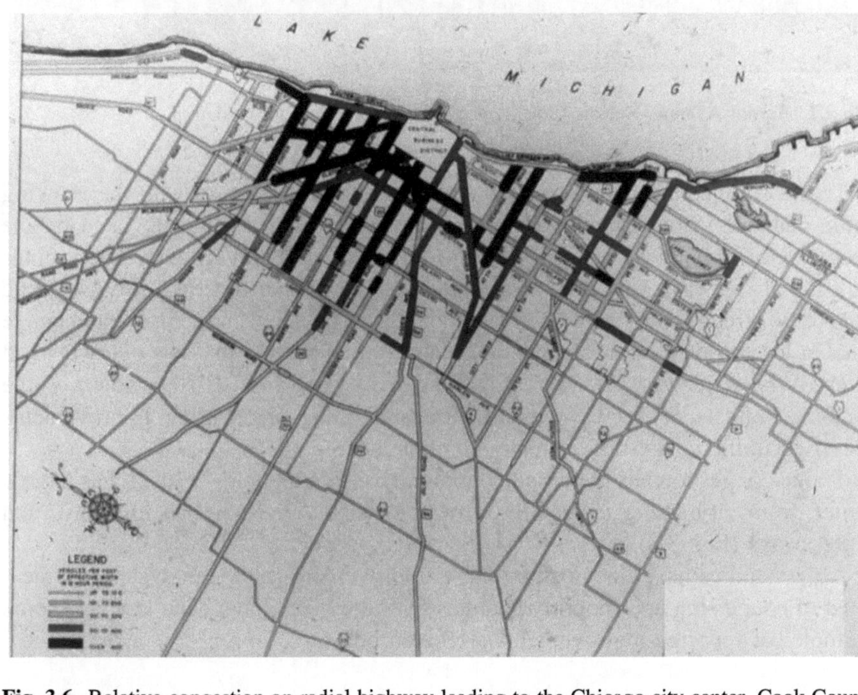

Fig. 3.6 Relative congestion on radial highway leading to the Chicago city center, Cook County 1943. *Source* [2]

Fig. 3.7 Gratiot Avenue, Detroit 1940. *Source* Automotive safety foundation

Traffic engineering control methods alone could not keep up with the growing traffic demands. Congested conditions in cities were eventually alleviated by the freeway construction associated with the Interstate Highway system. Many cities built radial freeways with central area freeway loops that diverted through traffic from city streets. But route convergence, lane balance and close interchange spacing led to operating problems and recurring congestion that has overwhelmed many of these freeways for years. Over the years (1960-mid 1970s), however, the combined effects of freeway construction and traffic engineering improvements alleviated traffic congestion in many communities.

Traffic congestion has increased continuously since the mid-1970s. It has largely shifted from city center to suburbs, from city streets to suburban highways and expressways (Fig. 3.8).

A few studies have documented these improvements although systematic measures of congestion and mobility changes over the years have been lacking. Some examples of reported improvements are given below:

Fig. 3.8 A congested modern freeway. *Source* "Traffic congestion and long commutes cost us dearly in time and fuel", http://www.peachygreen.com/going-green/traffic-congestion-and-long-commutes-cost-us-dearly-in-time-and-fuel

- A 1970 study [4] of mobility in some 17 cities, found that traffic speed increased in 13 cities and decreased in four. Speeds increased more than 15 % in 8 cities and decreased by more than 15 % in two.
- In Los Angeles, despite a tripling of motor vehicle registrations between 1936 and 1967, off-peak travel times between the central business district (CBD) and fourteen outlying locations declined from an average of 33−26 min, largely due to continued freeway development.
- Freeway construction in Los Angeles dramatically reduced traffic on local streets. For example, the daily traffic on S. Figueroa Street declined from 45,000 vehicles in 1955 to 13,500 vehicles in 1958. But it has since been reported to increase again (to over 34,000 in 1993) as a result of increased congestion on the parallel freeway [5].
- Peak hour speeds in Downtown Providence increased from 5−10 mph in 1927, to 10–19 mph in 1978 [6].
- In Boston, motor vehicle traffic into and through the CBD increased by 80 % from 1927 to 1960. Nonetheless average daily speeds on nine major downtown streets increased from 10.5 to 13.3 mph—a 27 % increase over the period.
- A one-way pattern established on north-south avenues in Manhattan produced a travel time saving of 22 % for north-south traffic, a 40 % savings for cross-town traffic, and a 20 % reduction in pedestrian accidents [7].

3.3 Traffic Congestion in the 21st Century

Today congestion is found on circumferential and cross-town roadways as well as radial highways. The growth and dispersion of residences and work places reflects the continued shift of people and jobs to suburban settings; the rise of office buildings and of edge cities [8] along freeways/beltways; and the decline of population and employment in older central cities.

Longer trip lengths and less transit use place greater traffic pressure on many arterial roads and freeways. Consequently the percentage of urban freeways that are congested has grown substantially since 1970. Studies by the Federal Highway Administration show that in 1970 about 30 % of the urban freeways had volume-to-capacity ratio exceeding 0.77; by 1990 more than 50 % of the nation's freeways exceeded this ratio [9]. In Houston, only 10 % of the freeways carried more than 15,000 vehicles per lane in 1980, but just 6 years later (1986) 60 % of the freeways carried more than 15,000 vehicles per lane [10].

Congestion is most acute in urban areas of more than two million people. Today, freeway congestion is found along beltways and many of the radial freeways leading to them. Congestion on freeways can often extend 15–25 miles from the city center during peak periods, and at times (as in Los Angeles) extends throughout the day.

3.3.1 Congestion in Travel Corridors and at Bottlenecks

Area-wide metrics of traffic congestion, while useful to describe trends, do not reflect the severity of congestion along major travel corridors and bottlenecks where traffic moves very slowly for long periods of the day.

The magnitude of metropolitan area traffic congestion is perhaps best measured along freeway travel corridors and at bottlenecks. These measurements are conducted by INRIX annually [11].

3.3.1.1 Corridors

INRIX ([11] p 7, 22) defines a congested corridor as follows:

- Recurring congestion[1] has to occur on multiple road segments totaling at least 3 miles in length.
- At least one segment must be congested 10 h per week, on average, and
- All road segments in the corridor must have at least 4 h a week of congestion, on average.

[1] Where traffic speed is lower than the free-flow speed.

Table 3.1 displays the peak period travel speed for the top ten corridors with the longest peak period delay in 2010. These ten corridors are located in the largest metro areas: Los Angeles, NYC, Chicago, and Washington, DC, and range in length and speed from 11.3 miles at 15.8 mph (NYC: the Cross-Bronx Expressway corridor), to 23.9 miles at 26.6 mph (Washington: I-95 southbound, from I-395 to Russell Road).

There were 341 congested corridors nationwide in 2010, with a total collective length of 2,295 miles. The average congested corridor (6.7 miles in length) operated with a peak period speed of 26.8 mph. But there is a large variance from these average values within metro areas in the same population groups, as well as between groups of different sizes.

The characteristics of congested corridors vary widely among metropolitan areas in extent (number and length), and intensity (average speed)—(INRIX 2010). This variability is mainly related to the size of the urban area.

The largest areas (population of over 5 million) have the largest number of congested corridors (Fig. 3.9) and the longest (Fig. 3.10).

It will be noted that a small number of corridors are congested in metropolitan areas of ½ to 1 million people, and none are found in areas smaller than ½ million.

3.4 Summary and Outlook

Traffic congestion in US cities is a byproduct of their success in attracting people and jobs, and amenities. When growth in economic activities significantly outpaces the growth in transportation infrastructure investments, cities experience congestion to levels that make mobility difficult.

Congestion is a consequence of where we live and work, how and where we travel, and how land is used. It impacts travel cost, the quality of our air, traffic safety, and the fuel consumed by motor vehicles.

Urban growth is likely to continue for the rest of the 21st century. More people are expected to live in metropolitan areas where they will occupy more land and will travel further to dispersed places of work, shopping, and recreation. The population and employment density gradients will continue to flatten—meaning population growth in outlying areas will continue to occupy more land at low density; even as central cities are likely to grow, and the costs of car ownership and driving are likely to increase. Congestion will follow these gradients: increasing in outlying areas and permeating the weak links of the roadway system. Thus freeway and suburban arterial congestion is likely to increase, extending outwards along with land development. Longer trip lengths will place greater traffic pressure on many arterial roads and freeways.

Reducing congestion growth in large urban areas will require the joint implementation of concerted and consistent initiatives that (1) will increase transportation efficiency and capacity through new technologies, (2) will reduce automobile traffic demand through higher land use densities, encourage growth along existing transit

Table 3.1 Average PM peak period speed for the Nation's top 10 congested corridors with longest peak period delay, 2010

Rank	Peak delay (min)	CBSA (Pop rank)	Road(s)	From	To	Length (Miles)	Average PM peak period speed (mph)
1	37	Los Angeles (2)/ Riverside (14)	Riverside Fwy/CA-91 EB	CA-55/costa Mesa Fwy	Mckinley St	20.7	21.8
2	32	Chicago (3)	I-90/I-94 EB (Kennedy/Dan Ryan Expys)	I-294/Tri State Tollway	Ruble St/Exit 52B	15.9	19.5
3	30	New York (1)	I-95 SB (NE Thwy, Bruckner/Cross Bronx Expys)	Conner St/Exit 13	Hudson Ter	11.3	15.8
4	30	Los Angeles (2)	I-5 SB (Santa Ana/Golden St Fwys)	East Ceaser Chavez Ave	Valley View Ave	17.5	22.3
5	29	Washington, DC (8)	I-95 SB	I-395	Russell Rd/Exit 148	23.9	27.6
6	29	New York (1)	Long Island Expy/I-495 EB	Maurice Ave/ Exit 18	Mineola Ave/ Willis Ave/Exit 37	16.0	21.3
7	28	Chicago (3)	Eisenhower Expy/I-290 EB	IL-72/Higgins Rd/Exit 1	Austin Blvd/Exit 23A	21.5	25.3
8	28	Los Angeles (2)	San Diego Fwy/I-405 NB	I-105/Imperial Hwy	Getty Center Dr	13.1	19.2
9	28	Los Angeles (2)	Pomona Fwy/CA-60 EB	Whittier Blvd	Brea Canyon Rd	21.7	26.0
10	28	Los Angeles (2)	Santa Monica Fwy/I-10 EB	CA-1/Lincoln Blvd/Exit 1B	Alameda St	14.9	21.3

Source Reference [11], Table 8, p 24

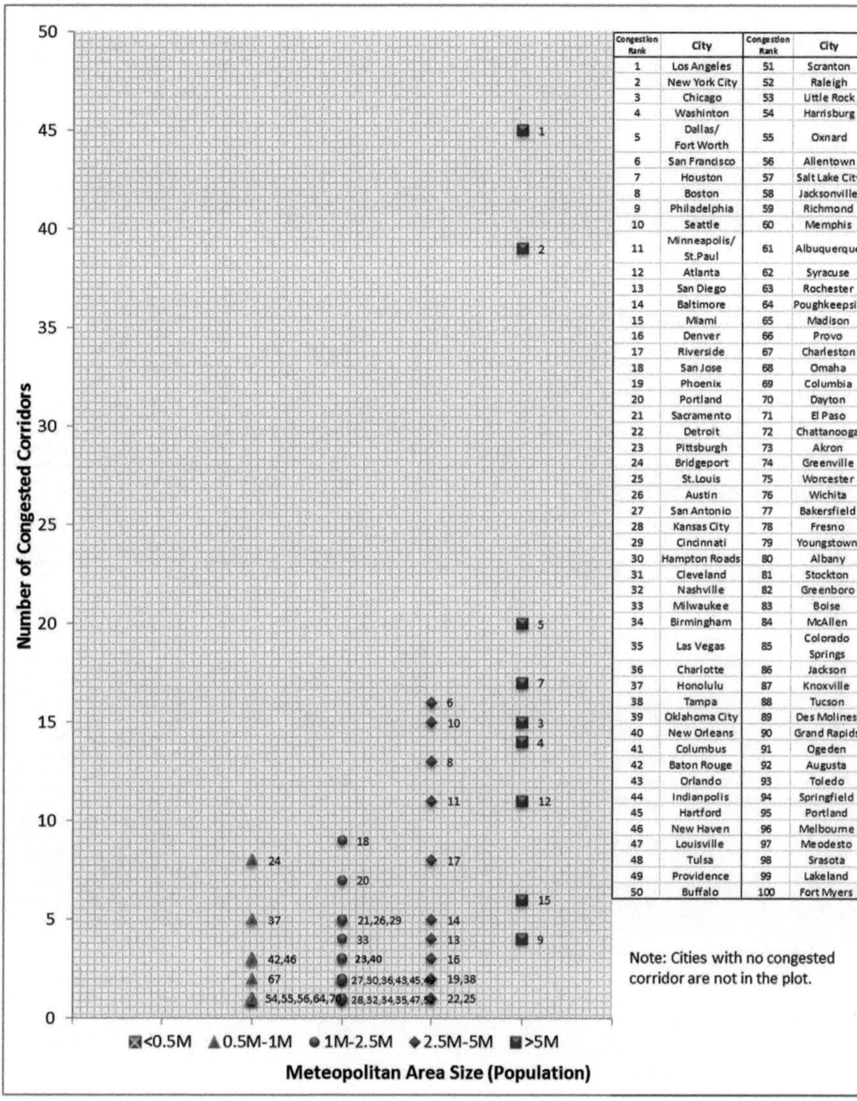

Fig. 3.9 Number of congested corridors in the top 100 most congested metropolitan areas. *Source* Reference [11]

corridors to increase the use of public transportation, walking and biking, and some form of peak period road user charges, and (3) will call for the diverse agencies in a region to work together in a coordinated way in implementing strategies that increase transportation efficiency and reduce automobile dependency. These issues are fully covered in Chaps. 14–24.

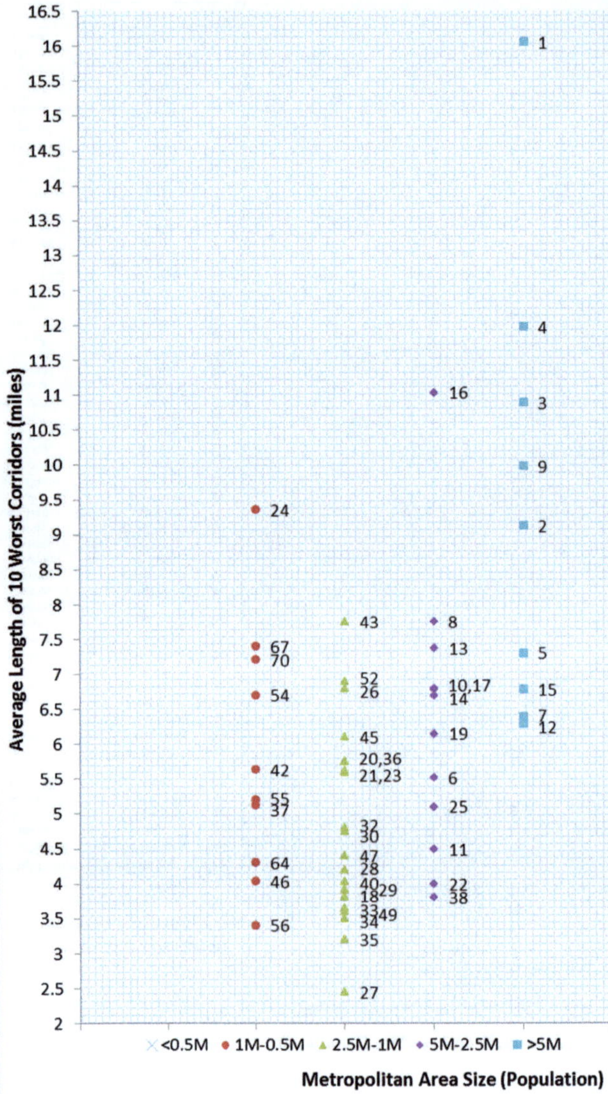

Congestion Rank	City	Congestion Rank	City
1	Los Angeles	51	Scranton
2	New York City	52	Raleigh
3	Chicago	53	Little Rock
4	Washinton	54	Harrisburg
5	Dallas/ Fort Worth	55	Oxnard
6	San Francisco	56	Allentown
7	Houston	57	Salt Lake City
8	Boston	58	Jacksonville
9	Philadelphia	59	Richmond
10	Seattle	60	Memphis
11	Minneapolis/ St.Paul	61	Albuquerque
12	Atlanta	62	Syracuse
13	San Diego	63	Rochester
14	Baltimore	64	Poughkeepsie
15	Miami	65	Madison
16	Denver	66	Provo
17	Riverside	67	Charleston
18	San Jose	68	Omaha
19	Phoenix	69	Columbia
20	Portland	70	Dayton
21	Sacramento	71	El Paso
22	Detroit	72	Chattanooga
23	Pittsburgh	73	Akron
24	Bridgeport	74	Greenville
25	St.Louis	75	Worcester
26	Austin	76	Wichita
27	San Antonio	77	Bakersfield
28	Kansas City	78	Fresno
29	Cincinnati	79	Youngstown
30	Hampton Roads	80	Albany
31	Cleveland	81	Stockton
32	Nashville	82	Greenboro
33	Milwaukee	83	Boise
34	Birmingham	84	McAllen
35	Las Vegas	85	Colorado Springs
36	Charlotte	86	Jackson
37	Honolulu	87	Knoxville
38	Tampa	88	Tucson
39	Oklahoma City	89	Des Molines
40	New Orleans	90	Grand Rapids
41	Columbus	91	Ogeden
42	Baton Rouge	92	Augusta
43	Orlando	93	Toledo
44	Indianpolis	94	Springfield
45	Hartford	95	Portland
46	New Haven	96	Melbourne
47	Louisville	97	Meodesto
48	Tulsa	98	Srasota
49	Providence	99	Lakeland
50	Buffalo	100	Fort Myers

Note: Cities with no congested corridor are not in the plot.

Fig. 3.10 Average length of the 10 most congested corridors in the top 100 most congested metropolitan areas. *Source* Reference [11]

References

1. Levinson HS, Falcocchio JC (2011) Urban development and traffic congestion. In: Proceedings of first transportation and development institute (TDI) congress. American Society of Civil Engineers, Chicago, pp 948–956
2. Levinson HS, Lomax TJ, Turner S (1997) Traffic congestion—past present, future. In: Proceedings traffic congestion and traffic safety in the 21st century, Chicago, 8–11 June 1997
3. Anderson JM (1941) Fares, Please! D. Appleton Century Company, New York
4. Koltnow P (1970) Changes in mobility in American Cities. Highway Users Federation for Safety and Mobility, Washington DC
5. Wilbur Smith and Associates (1966) Transportation and parking for tomorrow's cities. Wilbur Smith and Associates, New Haven
6. City of Los Angeles Traffic Volumes. 1983 and 1993
7. New York City Department of Transportation Data
8. Garreau J (1991) Edge City. s1 Anchor Books, New York
9. Federal Highway Administration, General Accounting Office (1989) Statistics reported in traffic congestion trends, measures, and effects, Washington, DC
10. Lomax T, Bullard D, Hanks J (1988) The impact of declining mobility in major Texas and other US cities. Texas Transportation Institute, College Station
11. INRIX (2011) INRIX National Traffic Scorecard—2010 annual report: back to the road to Gridlock

Part II
Traffic Congestion Characteristics, Causes, and Consequences

Chapter 4
Overview of the Causes of Congestion

4.1 Introduction

Traffic congestion results from the imbalance between the supply of and the demand for transportation facilities:

- The supply is constrained by history and geography, by transportation management and operating practices, and by the level of investment on streets and highways
- The demand results from the concentration of travel in space and in time

Congestion can be classified in two categories: recurring and nonrecurring

- Recurring Congestion is the delay travelers regularly experience/expect during known travel times—such as the morning and evening rush hours
- Nonrecurring Congestion delay is caused by non-predictable (random) events that disrupt traffic flow. These include incidents such as vehicle breakdowns or crashes; road repair and inclement weather; special events that create sudden surges in demand such as the end of a sports event; and natural or man-made disasters. Nonrecurring congestion can either create new congestion (in the off-peak periods), or can increase the delay experienced during periods of recurring congestion.

4.2 Summary of Causes

Nearly a century ago Miller McClintock [1] stated that congestion is due to three general causes: (1) the inability of the streets to hold a sufficient number of vehicles and to process them at an adequate speed, (2) the inclusion of elements in the traffic stream which hamper its free flow, and (3) the improper or inadequate direction and control of traffic.

J.C. Falcocchio and H.S. Levinson, *Road Traffic Congestion: A Concise Guide*,
Springer Tracts on Transportation and Traffic 7,
DOI 10.1007/978-3-319-15165-6_4

Today the causes of traffic congestion are more specifically known and include (1) large concentrations of demand in time and space—including temporal surges in travel demand on roadways of generally constant capacity physical, operational, and design deficiencies that create bottlenecks, (2) traffic demand that exceeds roadway capacity, and (3) physical and operational bottlenecks.

Congestion generally increases with city size. This happens because activity concentrations are larger, and travel distances are longer as cities grow.

Economists view chronic congestion as a pricing-induced problem. They argue that the absence of marginal cost pricing contributes to congestion because average cost pricing makes road use more attractive than it would be if prices would rise with congestion [2, 3].

4.2.1 Concentration of Trips in Space and Time

If all travel demand were evenly distributed among the various sections of the urban area, the traffic congestion problem would be a rare event. Similarly if all travel were evenly distributed to each hour of the day there would be little, if any, congestion.

But travel demand patterns reflect the concentration in time and space of daily activities: where and when people work, shop, recreate, move goods and provide services. It is the peaking of these spatial and temporal travel patterns that contributes to the recurring traffic congestion problem.

4.2.2 Growth in Population, Employment, Car Use and Insufficient Capacity

Growth in population, employment, and car use (vehicle miles of travel—VMT) increase congestion on streets and highways where capacity growth has not kept pace with growth in VMT. The factors that contribute to and shape the growth in population, employment, and vehicle miles of travel (VMT) in urban areas are discussed in detail in Chap. 6.

4.2.3 Bottlenecks

Bottlenecks are perhaps the most common cause of congestion. They result from the convergence of a greater number of lanes in the upstream roadways than are available in the downstream roadways. Bottlenecks delay is typically found in hours of peak flow where the number of lanes converging on a roadway, bridge or a

Fig. 4.1 Holland Tunnel Bottleneck (1940)—27 lanes trying to get into 2 lanes. *Source* Reference [4], p 110

tunnel exceeds the number of lanes these facilities have. An early example of a 1940 bottleneck at the Holland Tunnel in New York City is shown in Fig. 4.1, where traffic from 27 lanes is merging into two Tunnel lanes.

Bottlenecks are also created by roadway incidents that reduce block travel lanes and restrict traffic flow, or they are created by bad weather conditions (e.g., ice on a bridge), a work zone, poorly timed traffic signals, or driver behavior.

The next three chapters discuss the basic causes of congestion in greater detail. Chapter 5 covers the causes of demand concentration; Chap. 6 describes the issues of population and economic growth, growth in car use, and insufficient capacity; and Chap. 7 discusses bottlenecks as the third major cause of congestion.

References

1. McClintock M (1925) Street traffic control s.l. McGraw-Hill Book Company, New York
2. Vickery SW (1963) Pricing in urban and suburban transport. s.l. reprinted in Smerk GM (ed) Reading in urban transportation, Indiana University Press, American economic review: papers and proceedings
3. Button JK, Verhoef TE (eds) (1988) Road pricing, traffic congestion and the environment—issues of and social feasibility. Eder Elgar Publishing Inc, Northampton
4. Geddes NB (1893–1958) Magic motorways. Random House, New York, 1940

Chapter 5
Concentration of Travel Demand in Space and Time

5.1 Introduction

The concentrations of people and their activities (density) in space and in time are a natural consequence of human behavior in urban areas.

If all travel demand were evenly distributed throughout the day, and among the various parts of the urban area, the urban traffic congestion problem would be greatly reduced. But travel demand patterns reflect where and when people live, work, and play. Therefore, they are concentrated in space and time. It is these spatial and temporal concentrations that contribute to the urban traffic congestion problem.

5.2 Concentration of Travel Demand in Space

Major activity and employment centers generate the highest concentration of traffic demand (trips ends per square mile) because of they are places of high development densities (people/employees per square mile).

These centers include long established central business districts and the growing number of large suburban centers. Suburban mega-centers, mainly automobile dependent, generate heavy traffic volumes on major arterials and approach roads which are typically congested. And even where major centers are well served by public transportation, traffic density is high, and congestion is also a problem.

5.2.1 The Central Business District

The central business district (CBD) is usually the single largest urban activity center when measured by employment or floor space. It usually occupies an area of about

J.C. Falcocchio and H.S. Levinson, *Road Traffic Congestion: A Concise Guide*,
Springer Tracts on Transportation and Traffic 7,
DOI 10.1007/978-3-319-15165-6_5

Table 5.1 CBD size and employment in selected cities

Area	Total city population, 2000	Total downtown commuters, 2000	Total transit commuters to downtown, 2000	Downtown land area (square wiles)	Transit share of downtown commuters (%)	Worker density (commuters per square mile)
New York	8,008,278	379,380	290,390	1.08	76.5	351,277.80
Chicago	2,869,016	341,014	210,490	1.13	61.7	301,782.30
Son Francisco	776,733	320,170	156,764	2.55	49.0	125,556.90
Washington, DC	572,059	409,505	154,658	3.99	37.8	102,632.80
Boston	589,141	570,315	137,701	2.32	50.9	116,515.10
Philadelphia	1,517,550	230,358	105,387	2.40	45.7	95,982.50
Seattle	563,374	147,905	54,435	2.99	36.8	49,466.60
Los Angeles	3,694,820	215,340	43,656	3.78	20.3	56,968.30
Portland	529,121	104,810	28,839	2.11	27.5	49,673.00
Houston	1,953,631	155,050	25,874	1.68	16.7	92,291.70
Dallas	1,188,580	91,786	12,493	0.85	13.6	107,983.50
Son Diego	1,223,400	75,850	8,675	2.16	11.4	35,115.70
Sacramento	407,018	64,830	7,959	1.26	12.3	51,452.40
Son Antonio	1,144,646	53,440	3,842	1.15	7.2	46,469.60
Austin	656,562	76,150	2,913	0.95	3.8	80,1 57.90

Note Selected cities with total population 600,000+
Ranked by number of transit commuters to downtown
Source Reference [1]

one to two square miles and has the highest employment densities in the urbanized area. Salient characteristics of selected city centers are shown Table 5.1.

As employment and employment densities increase, there is generally an increase in the public transportation share of commuters. However as employment densities increase, there is also an increase in automobile and pedestrian trips. This tends to overload traffic on the road network making congestion a recurring event during rush hours.

Central business district growth has varied over the years. Economic conditions, efforts to revitalize the center, and availability of good highway and public transit access have been essential in the revitalization of mature centers. As shown in Table 5.2, this growth [2] has generally been modest in most cities (however, any growth will increase congestion unless public and private transport capacity is increased).

5.2.2 *Outlying Mega-Centers*

Over time many of the metro region jobs, shopping and entertainment opportunities have moved closer to their suburban customers at locations near freeway interchanges that are easily accessible by car. The popularity of these suburban

Table 5.2 CBD employment trends

		Year			Ratio
CBD	Area (square miles)	1980	1990	2000	2000/1980
Chicago, IL, CBD	1.15	353,984	325,226	341,014	0.96
Chicago, IL, Expanded CBD	3.55	503,109	503,787	523,382	1.04
Philadelphia, PA	2.4	265,135	287,860	265,838	1.00
San Francisco, CA	2.55	314,100	312,100	341,100	1.09
Oakland, CA	1.72	49,400	54,400	63,100	1.28
San Jose, CA	3.17	44,300	39,000	52,400	1.18

Source Reference [2], p 440, Table 12.2

commercial centers made them grow in size and many became the precursors to the mega-centers of multiuse activities [3].

Examples of reported travel modes are shown in Table 5.3. In contrast to the central business district, outlying activity centers rely largely on automobile access.

More than 90 % of the suburban centers destinations are by automobile (except for Nassau Co., New York). Accordingly, the concentration of commercial and retail activities located along major arterial roads, and at the nodes of major arterial roads and freeways has created excessive concentration of traffic at many locations that far exceeds roadway capacity causing severe congestion for many hours of the day [3].

Because population density of many suburban areas is too low to support effective transit service, it is not generally possible to substantially reduce traffic congestion by providing transit access to many mega-centers. And even where

Table 5.3 Travel modes in suburban centers

Location	Year	Drive alone (%)	Car pool (%)	Total auto user (%)	Transit (%)	Other (%)
Nassau (NY)	1987	79	8	87	13	–
Bellevue (WA)	2000	82	8	90	8	2
Shady grove (MD)	1992	90	6	96	3	1
South coast plaza (CA)	1998	75	20	95	3	2
Parkway center (Galleria) (TX)	1988	90	5	95	1	–
Perimeter center (GA)	1988	94	5	99	1	–
Tysons corner (VA)	1988	89	10	99	1	–
South dale (MN)	1988	92	7	99	1	–
Overall average (rounded)		86	9	95	4	1

Source Reference [3], p 457, Table 12.11

transit service is available, the site design of buildings in these regional centers is usually focused around parking access and not on transit access to the “front doors” of buildings.

Moreover, most suburban developments are mandated by building codes to provide a minimum number of parking spaces per unit floor area of building space. This policy results in a network of parking lots and garages, interspaced between office and commercial buildings, all connected by a network of local and arterial roads. As a result, “suburban gridlock” is common both on freeways and some local streets [3].

City size also affects traffic congestion. Larger urban areas are generally more congested than smaller ones. This condition results from both longer travel distances and larger concentrations of activities (as in the city center).

5.3 Paradox: Reducing per Capita Auto Use Increases Traffic Congestion

Often the debate about urban traffic congestion is framed around the need to reduce widespread use of the personal motor vehicle *and* to increase the use of various public transportation alternatives. One demonstrated way of reducing the dependency on car use is to *increase* the density of land development.

Table 5.4 shows that as densities rise, there are more walking and public transport trips that decrease per capita VMT.

Table 5.4 Average daily travel per person in the united states by population density and mode, 1990 NTPS survey

Density range (persons per square mile)	Daily person trips by mode							Daily person miles	Daily VMT per person
	Auto	Bus	Rail	Taxi	Walk/bike	Other	Total		
0–99	3.35	0.02	0.00	0.00	0.24	0.16	3.77	31.58	21.13
100–249	3.50	0.02	0.00	0.01	0.24	0.13	3.90	29.95	20.73
250–499	3.53	0.02	0.00	0.00	0.29	0.12	3.96	29.33	20.40
500–749	3.52	0.03	0.00	0.00	0.21	0.12	3.88	29.00	20.99
750–999	3.44	0.05	0.01	0.01	0.26	0.13	3.90	26.25	18.35
1,000–1,999	3.48	0.03	0.01	0.00	0.23	0.11	3.86	26.17	18.63
2,000–2,999	3.46	0.06	0.01	0.00	0.28	0.11	3.92	23.45	19.04
3,000–3,999	3.34	0.06	0.02	0.01	0.29	0.09	3.81	24.11	16.89
4,000–4,999	3.51	0.05	0.01	0.00	0.30	0.08	3.95	24.77	17.24
5,000–7,499	3.29	0.09	0.02	0.01	0.36	0.06	3.83	24.56	16.28
7,500–9,999	2.92	0.11	0.05	0.02	0.45	0.07	3.62	20.59	14.15
10,000–49,999	1.90	0.29	0.21	0.03	0.95	0.04	3.42	17.02	8.73
50,000 or more	0.59	0.42	0.61	0.16	1.55	0.07	3.40	12.55	2.31

Source Reference [4], pp 15–21. Table 15.5

Therefore, increasing density can significantly reduce the per capita use of auto travel, and increase the use of alternative modes to the private car.

5.3.1 Population Density, Traffic Density, and Traffic Speed

Although increasing population density *reduces* per capita auto use, Table 5.5 shows that increasing population density also *increases* traffic density (auto trips per square mile)—an indicator of traffic congestion.

5.3.2 Population Density and Traffic Congestion

Figure 5.1 shows how urban density affects average traffic speeds in the NY metropolitan area [5], where average traffic speed is shown to decrease with increasing population density: from 25 mph in the outer areas with population densities of 3,400 persons per square mile to 10 mph in the city's core with a population density of 65,000 persons per square mile.

Hence the paradox: increasing population density reduces per capita auto use (a desirable social and environmental objective) but it also increases traffic congestion (e.g., VMT per square mile increases). However, as will be discussed later, the negative impact of traffic congestion on trip time in high density cities is mitigated by the shorter trip lengths they generate and their beneficial effect of high density on accessibility.

Table 5.5 The impact of population density on traffic density

(1) Population density midpoint Table 5.4 (persons per square mile)	(2) Average daily auto trips per person	(3) Traffic density (auto trips per square mile)
175	3.50	613
375	3.53	1,324
625	3.52	2,200
875	3.44	3,010
1,500	3.48	5,220
2,500	3.46	8,650
3,500	3.34	11,690
4,500	3.51	15,795
6,250	3.29	20,563
8,750	2.92	25,500
30,000	1.90	57,000
65,000	0.59[a]	38,350

Source Calculated from Table 5.4

[a] This drop in auto use may be explained by an increased use of transit, walking and other non-auto modes achievable at population densities approaching those in Midtown Manhattan

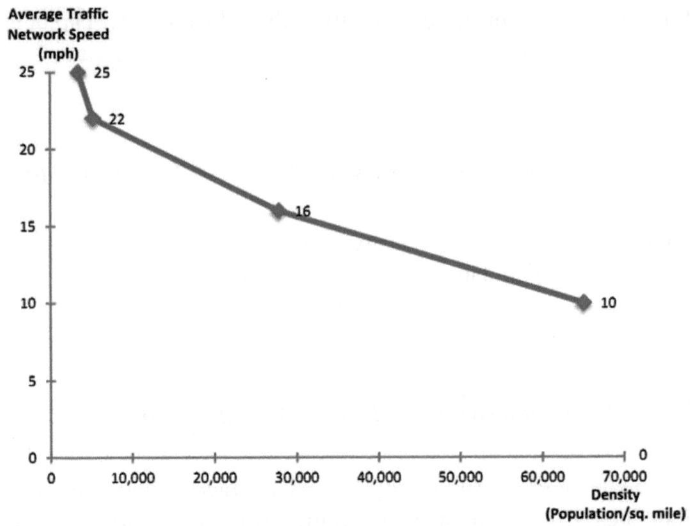

Fig. 5.1 Average traffic speed versus population density in the New York metropolitan area. *Source* Adapted from Reference [5], Table 10

5.4 Concentration of Travel Demand in Time

If all trips made by urban residents were spread equally among the hours of the day traffic congestion would be largely eliminated. But people engage in activities that are scheduled at particular times and not others. Accordingly urban travel patterns are determined by the schedules of human activities (work, school, shopping, vacation, etc.) that vary by time of day, day of week, and month of the year. The concentration of this travel during specific time periods gives rise to traffic and transit congestion.

The time concentration of these trips in time produces peak traffic demands in excess of roadway capacity at various times of the day—typically the morning and evening peak periods. To address this effect the travel tax was introduced by INRIX. It describes congestion as a "surcharge" on free-flow travel time. For example, a driver traveling at the rate of 1.20 min/mi would experience a tax of 20 % of the free-flow travel rate of 1.0 min/mi.

As shown in the example of Fig. 5.2, (Ref. [6], p 11), the travel time tax[1] peaks at 17 % in morning peak hour and at 23 % in the evening peak hour.

The hourly distribution of the travel time tax indicates that the AM peak period begins at 6 a.m. and ends at 10 a.m.; while the PM peak period begins at 3 p.m. and

[1] Travel time tax is term introduced by INRIX. It is meant to describe congestion as a surcharge on free-flow travel time. For example, a driver traveling at a rate of 1.20 min/mi would experience a tax of 20 % if the free-flow travel rate is 1.0 min/mi.

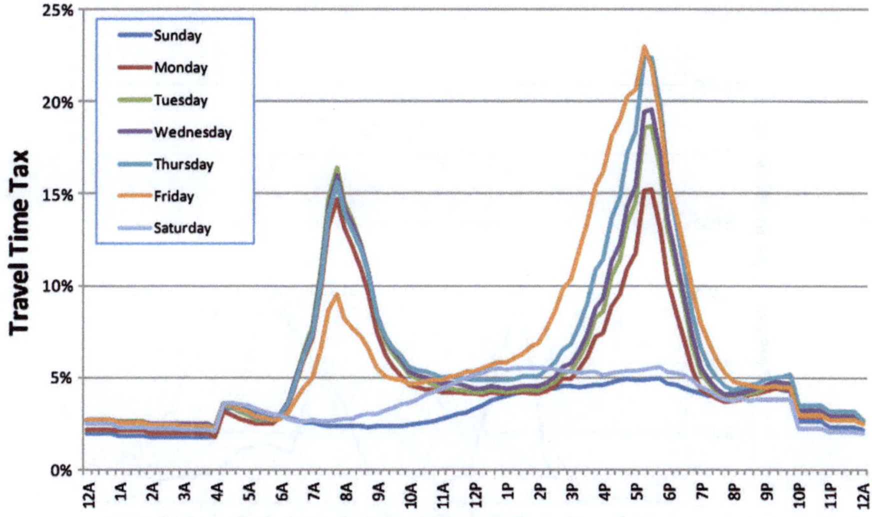

Fig. 5.2 Travel time tax, by hour and day of week. *Source* Reference [6], p 11. Figure 8

ends at 7 p.m. Typically approximately 50 % of weekday person travel occurs in these two peak periods.

It should be noted, however, that peaking patterns vary among the various travel modes. The peak hour accounts for approximately 25 % of commuter rail travel, 14–17 % of rail rapid transit trips, 16 % of bus trips and 7–13 % of highway travel. Approximately 75 % of all transit trips are in the peak direction of travel while 57–64 % of highway travel occurs in the peak direction.

5.4.1 Trip Purpose and Time of Travel

The hourly variations in person trips relate closely to the reasons for travel. On a typical workday in the US, most people leave their home for work at about 7:00 a.m., and leave work at 5:00 p.m. Children start school early in the morning as well, and return home before 4 p.m. Figure 5.3 shows the distribution of trip starting times for various trip purposes: commuting, family—personal business/shopping, school, and social-recreational travel.

5.4.2 Trip Purpose of Peak Period Travelers

Non-work travel is a large component of daily travel—even in the peak commuting hours (Fig. 5.4).

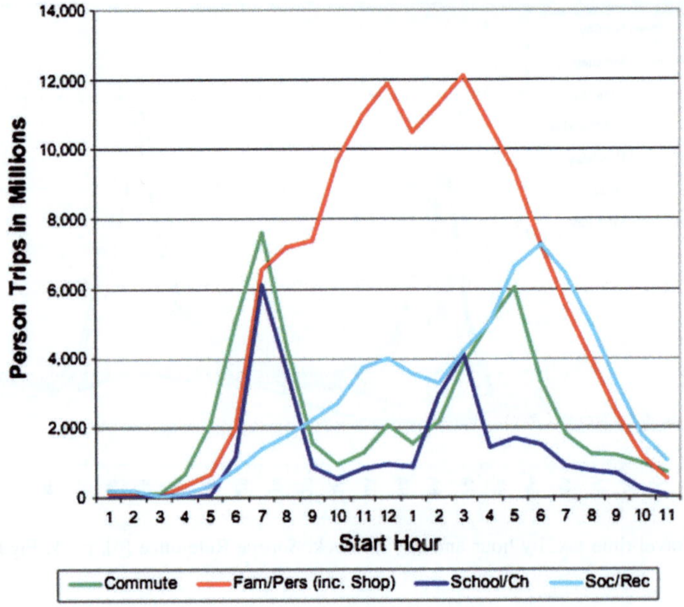

Fig. 5.3 Number of person trips by start hour and trip purposes. *Source* Reference [7], Exhibit 4

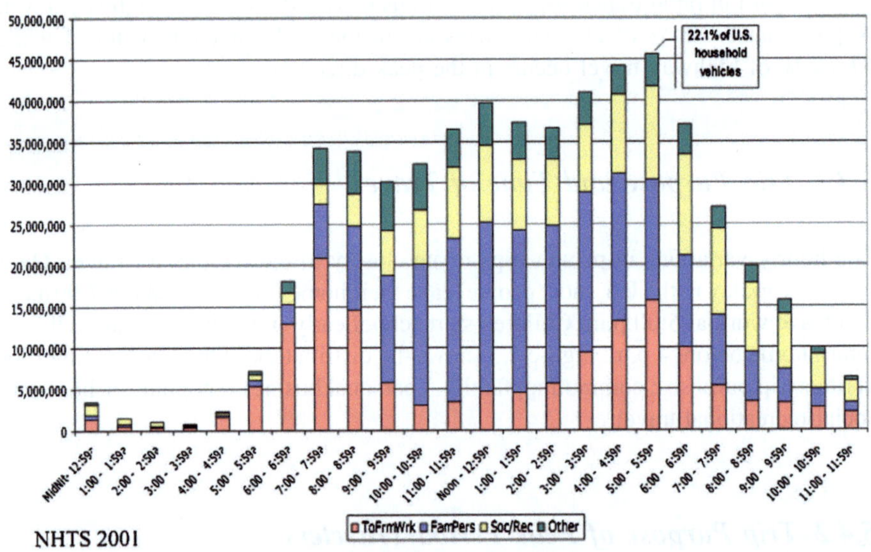

Fig. 5.4 Trip purposes of all vehicles in motion, by time of day. *Source* Reference [8], p 18

The overlap of work and non-work trips during peak and non-peak travel periods is a significant factor in understanding the causes of and remedies for traffic congestion.

Most people might be surprised to learn that more than half of peak period person trips in vehicles are *not related* to work. On an average weekday non-work trips constitute 56 % of trips during the AM peak travel period and 69 % of trips during the PM peak (Fig. 5.5).

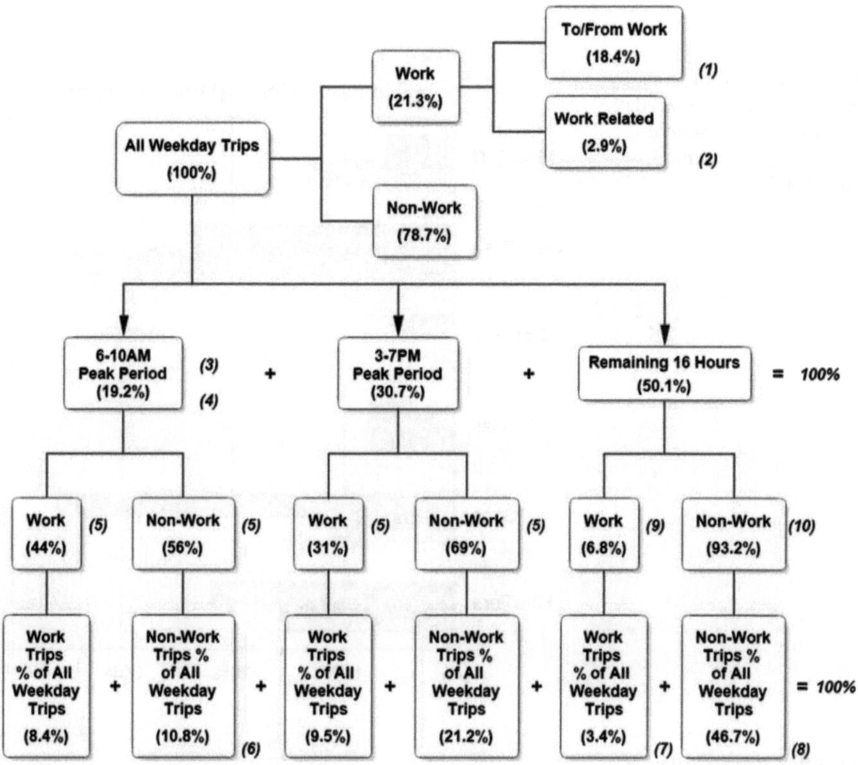

(1) Reference (10), Page 6, Figure 1-5(b)

(2) Reference (10), Page 3, Figure 1-1

(3) Reference (7), page 11, Figure 8

(4) Reference (10), Figure 8 and Table A-12

(5) Reference (8), Exhibit 1 to 4

(6) (19.2%)x(0.44)=8.4%; (19.2%)x(0.56)=10.8%

(7) (21.3%)-[(8.4%)+(9.5%)]=3.4%

(8) (78.7%)-[(10.8%)+(21.2%)]=46.7%

(9) (3.4%)÷(0.501)=6.8%

(10) (46.7%)÷(0.501)=93.2%

Fig. 5.5 Time distribution of weekday trips by purpose

Some non-work trips in the peak periods are part of the daily commute work trip chain involving stops along the way to or from work for other purposes: dropping off or picking up in day care, shopping, going to eat a meal, etc. [9]. For example, Fig. 5.6 shows that approximately 34 % of the stops made in the home-to-work commute were to drop some one off (serve a passenger).

The characteristics of non-work travelers in the AM peak period are shown in Fig. 5.7 (Reference [10], Exhibit 1). While most AM peak period travelers are workers (full or part-time), a large portion of travelers dropping off passengers are

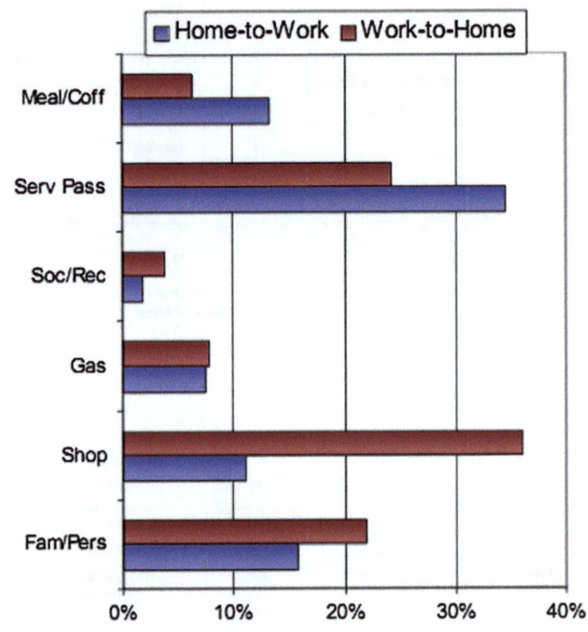

Fig. 5.6 Percent of stops by purpose during the weekday work trip chains. *Source* Reference [9], Exhibit 3, April, 2007

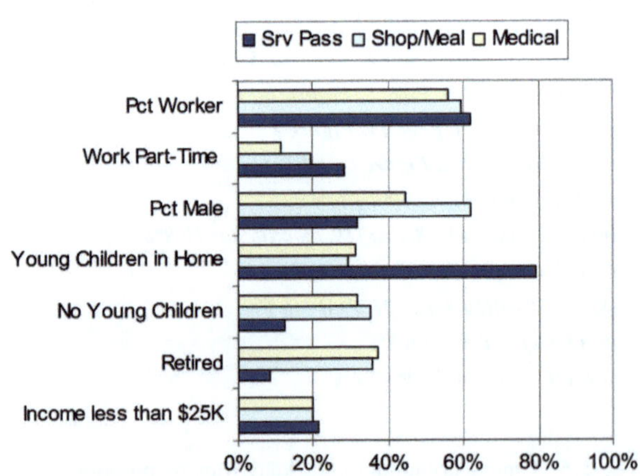

Fig. 5.7 Characteristics of people making AM peak vehicle trips by purpose. *Source* Reference [10], Exhibit 5

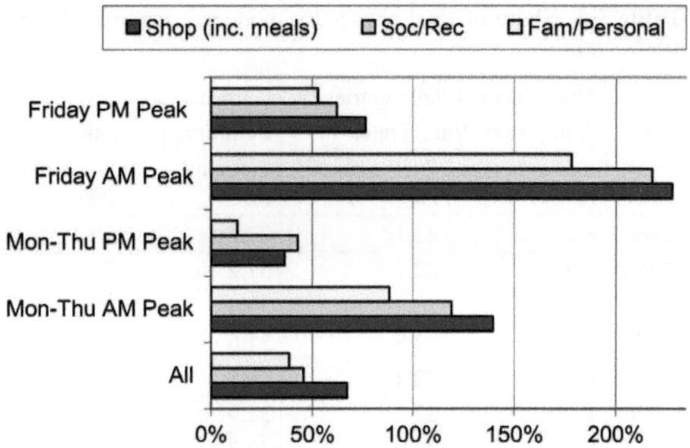

Fig. 5.8 Percent growth in peak period vehicle trips for non-work purposes, 1990–2001. *Source* Reference [9], Exhibit 1, August, 2007

women, while a larger portion of the people shopping (including getting a meal) are men. Nearly 80 % of travelers who drop off a passenger during the AM peak live in households with young children. Retired people are more likely to shop or go to the doctor.

Non-work travel has increased substantially during the workday peak periods. From 1990 to 2001, the percent growth in peak period vehicle trips for non-work travel increased by 100 % in the AM peak and 35 % in the PM peak (Fig. 5.8) [10].

Non-work trips are increasing both in the number of workers making such trips and in the number of stops per worker [11]. Since 1994, 25 % more commuters stop for incidental trips during their commute to or from work, and stopping while commuting to or from work is especially prevalent among workers with the longest commutes [7]. This type of trip chaining is increasing.

Peak period congestion, therefore, is not solely caused by commuter travel but it is also caused by the increasing concentration of non-work trips into the traditional commuter peak period.

5.4.3 Peak Spreading

The growth in non-work travel during the peak period is a major reason why congested conditions extend into the shoulders of the peak hours.

Table 5.6 [12] shows how the shares of total trips entering the Manhattan CBD in the morning peak hour and peak period, have declined over the last half century (1960–2009), as travelers shifted their trips outside the peak periods. Over time, the typical "rush hour" has gradually been transformed into the "rush hours" as travelers leave earlier or later to avoid the worst congestion period.

Table 5.6 Share of total entries by mode occurring in the morning peak hour and peak period: 1960–2008

Year	Share of total daily entries by occupants					
	Auto, Taxi, Van, Truck		Public transportation		Total share	
	8–9 a.m. (%)	7–10 a.m. (%)	8–9 a.m. (%)	7–10 a.m. (%)	8–9 a.m. (%)	7–10 a.m. (%)
1960	8.6	23.1	32.2	59.4	25.3	48.5
1963	8.8	22.7	31.2	58.2	25.1	48.4
1973	8.0	22.2	31.4	60.6	24.3	48.6
1974	8.1	22.8	31.6	59.5	24.8	48.8
1975	8.1	22.4	30.1	59.0	24.3	48.4
1976	8.1	22.2	31.9	59.6	24.5	47.9
1977	8.3	22.9	32.3	60.6	24.8	48.8
1978	8.0	22.5	30.6	58.7	23.5	47.5
1979	8.1	22.4	30.8	59.0	23.8	47.7
1980	8.5	23.1	31.7	60.2	24.6	48.8
1981	8.4	23.4	31.1	60.0	23.9	48.8
1982	8.5	23.4	30.9	59.3	23.7	47.9
1983	8.4	23.5	30.3	59.0	23.2	47.0
1984	8.2	23.1	31.1	59.9	23.3	47.4
1985	7.9	22.2	30.5	59.9	22.6	46.7
1986	7.8	22.1	27.7	56.7	21.0	45.1
1987	7.8	22.0	26.6	56.3	20.2	44.6
1988	7.7	21.6	26.4	56.0	20.1	44.5
1989	7.6	21.6	27.1	56.6	20.8	45.2
1990	7.0	19.9	25.6	54.2	19.4	42.7
1991	6.9	19.8	25.9	54.1	19.6	42.6
1992	6.8	19.5	25.2	53.9	18.9	42.1
1993	6.6	19.2	24.8	52.8	18.7	41.5
1994	6.7	19.5	24.0	52.0	18.3	41.3
1995	6.7	19.3	23.8	51.3	18.1	40.6
1996	7.0	19.9	23.8	50.9	18.2	40.6
1997	7.3	19.9	23.3	50.3	17.6	39.3
1998	7.1	20.0	22.8	48.9	17.2	38.5
2000	7.4	20.2	22.0	49.4	17.0	39.5
2001	6.2	17.9	21.5	48.1	17.1	39.4
2002	6.6	19.3	21.3	47.2	16.8	38.5
2003	7.0	20.2	21.5	47.2	16.7	38.3
2004	6.6	19.5	20.8	45.6	16.2	37.3
2005	6.6	19.0	19.9	45.2	16.1	37.7
2006	6.6	19.1	19.5	43.9	15.8	36.8
2007	6.4	18.8	20.6	45.6	16.8	38.4
2008	6.7	19.2	19.9	43.9	16.7	37.8
2009	6.6	18.9	19.4	43.5	16.0	36.9

Percentages express the peak-period share of private motor vehicle and public transportation to total of 24 h period person entries, respectively

Source Reference [12]

References

1. Commuting to Downtown America—Census 2000 (2005) s.l: Status Report to the Transporation Research Board, subcommittee on Census Data for Transportation Planning. ABJ30(1)
2. Hub Bound Travel Report 2009 (2011) s.l: New York Metropolitan Transportation Council
3. INRIX (2011) INRIX National Traffic Scorecard—2010 Annual Report: Back to the Road to Gridlock. s.l: INRIX, Mar
4. Newman PWG, Kenworthy JR (1988) The transport energy trade-off: Fuel-efficient traffic versus fuel-efficient cities. Transportation Research Part A: General 22(3):163–174
5. Levinson HS (2009) Transportation planning handbook 3rd edn. In: Institute of transportation engineers, vols Chapter 12–activity centers and site impact analysis. Washington, DC
6. McGuckin N, Srinivasan N (2003) A Walk Through Time—Changes in the American Commute. s.l: US Census and the National Household Travel Survey, Cambridge Systematics
7. NHTS Brief (2007) National Household Travel Survey. s.l: US Department of Transportation, April
8. NHTS Brief (2007) National Household Travel Survey. s.l: US Department of Transportation, Aug
9. NHTS Brief (2011) National Household Travel Survey. s.l: US Department of Transportation, Mar
10. Orski KC (1985) Suburban Mobility: The Coming Transportation Crisis. s.l: Transportation Quarterly 39(2)
11. Pikaski AE (2006) Commuting in America III. NCHRP Report 550 and TCRP Report 110
12. TCRP Report 95 (2003) Traveler Response to Transportation System Changes: Chapter 15—Land Use and Site Design. Transportation Research Board, Washington, DC
13. Dunphy RT, Fisher K (1996) Transportation, Congestion, and Density: New Insights. Transportation Research Record 1552. pp 89–96

Chapter 6
Insufficient Capacity, Growth in Population, Employment, and Car Use

6.1 Historical Imbalance of Roadway Supply and Travel Demand

Traffic demand has spread in post-World War II metropolitan America as a result of changing patterns of where people live and work, and how they travel. More people occupying more land have created more activity dispersal and have increased car dependency for mobility and access to activities. During this period, capacity expansion was generally insufficient to efficiently serve growth in population, employment, and car use. This chapter presents some of these trends and their congestion implications.

Congestion increases when the investment in transportation facilities fails to keep up with the growth in travel.

Today's roadway congestion is largely the result of the imbalance over time between the growth of vehicles miles of travel (VMT) and the roadway capacity in lane miles. Examples of this disparity are shown in Figs. 6.1, 6.2 and 6.3 for Interstate, other arterial highways, and local roads, respectively. The trends are shown for the 20 year period between 1980 and 2000 [1]

While the VMT on urban interstate highways grew by 240 %, the lane miles in the system grew by 150 %. The imbalance between VMT growth and roadway capacity growth, however, was not limited to urban interstate highways but was also prevalent throughout the urban roadway system: the VMT for other arterials and local roads grew by an average of 185 % compared to 135 % growth in lane miles.

Where this disparity has been the greatest, so has been its impact on congestion: Fig. 6.4 and Table 6.1 ([2], p. 16) compare the growth in congestion to the ratio of change in demand to change in capacity over a 28 year period (1982–2010) for 101 urban areas ([2], Table 9, p. 52).

These trends show that urban areas where the increase in roadway capacity nearly matched the increases in demand experienced a slower congestion growth than those areas where capacity growth lagged substantially behind the growth in VMT.

J.C. Falcocchio and H.S. Levinson, *Road Traffic Congestion: A Concise Guide*, Springer Tracts on Transportation and Traffic 7,
DOI 10.1007/978-3-319-15165-6_6

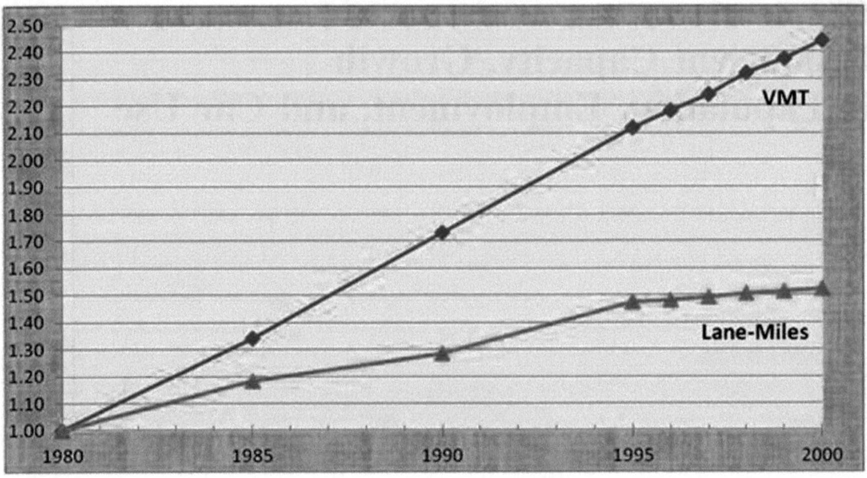

Fig. 6.1 Relationship between vehicle-miles traveled (VMT) and lane-miles for interstate highways in urban areas (1980–2000). *Source* Reference [1], Tables 1-6 and 1-33

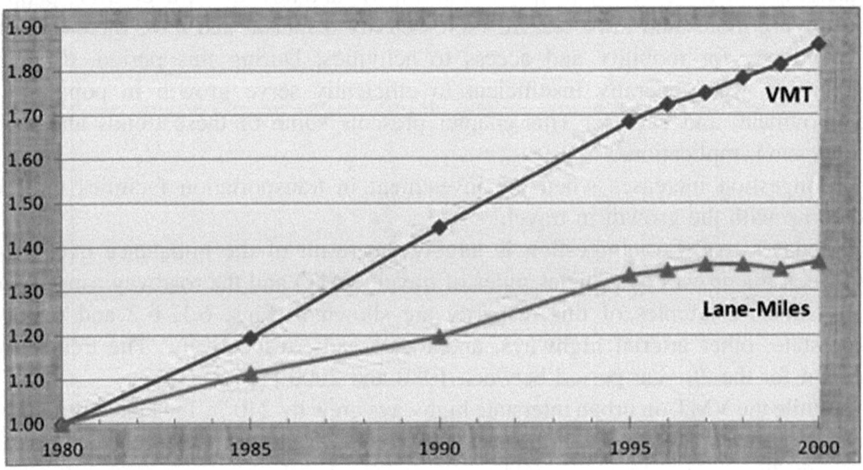

Fig. 6.2 Relationship between vehicle-miles traveled (VMT) and lane-miles for other arterial highways in urban areas (1980–2000). *Source* Reference [1], Tables 1-6 and 1-33

The roadway capacity of many suburban arterial roads built years ago in support of emerging suburban land developments is no longer sufficient to serve the traffic demand in many mature suburbs of today.

The chronic deficiency of roadway capacity relative to increasing traffic demand has grown over time (except during periods of economic slowdown—e.g., 2007–2008). Highway capacity deficiency largely reflects a lack of available space for highway expansion in built up areas where resistance to highway construction

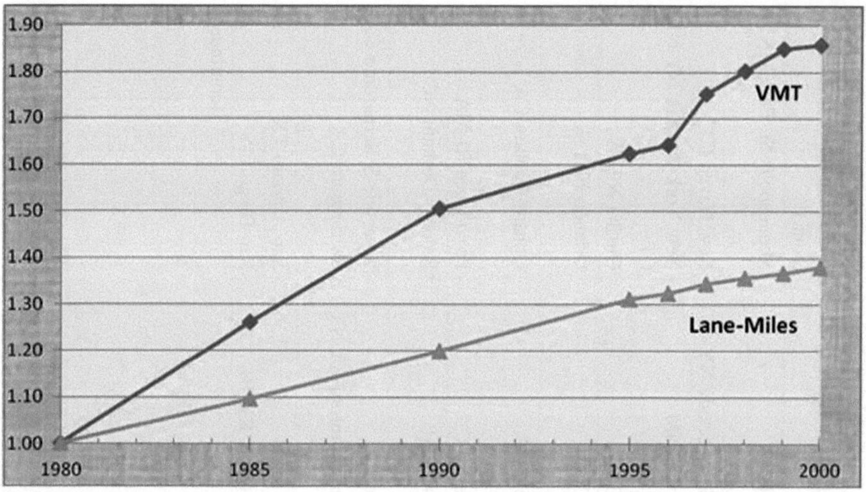

Fig. 6.3 Relationship between vehicle-miles traveled (VMT) and lane-miles for local roadways in urban areas (1980–2000). *Source* Reference [1], Tables 1-6 and 1-33

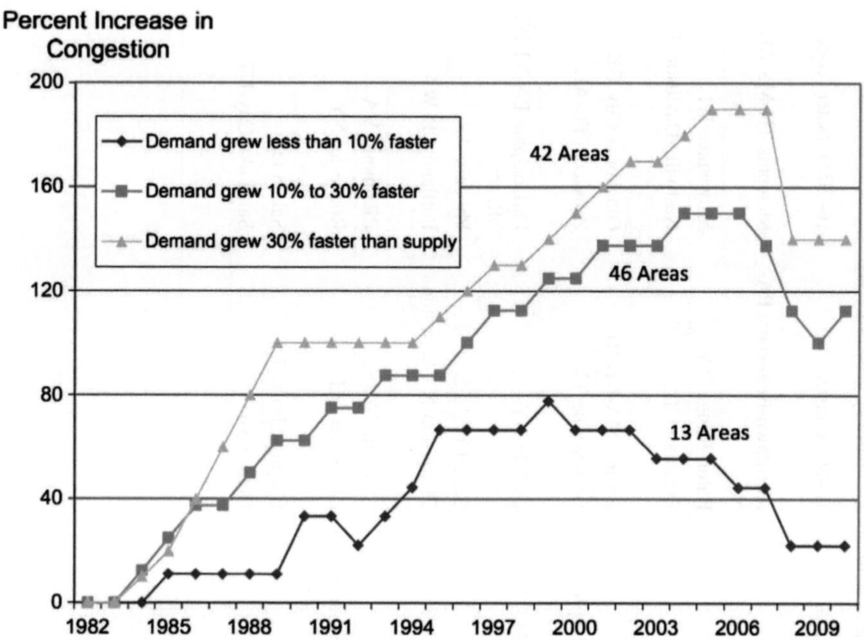

Fig. 6.4 Growth rate in traffic versus growth rate in roadway capacity (1982–2010) and its impact on traffic congestion. *Source* Reference [2], p. B-22, Exhibit B-12

Table 6.1 Urban area demand and roadway growth trends

Less than 10 % faster (13)	10–30 % faster (46)	10–30 % faster (cont.)	More than 30 % faster (40)	More than 30 % faster (cont.)
Anchorage AK	Allentown-Bethlehem PA-NJ	Memphis TN-MS-AR	Akron OH	Minneapolis-St. Paul MN
Boulder CO	Baton Rouge LA	Milwaukee WI	Albany-Schenectady NY	New Haven CT
Dayton OH	Beaumont TX	Nashville-Davidson TN	Albuquerque NM	New York-Newark NY-NJ-CT
Greensboro NC	Boston MA-NH-RI	Oklahoma City OK	Atlanta GA	Omaha NE-IA
Indio-Cath City-P Springs CA	Brownsville TX	Pensacola FL-AL	Austin TX	Orlando FL
Lancaster-Palmdale CA	Buffalo NY	Philadelphia PA-NJ-DE-MD	Bakersfield CA	Oxnard-Ventura CA
Madison WI	Cape Coral FL	Phoenix AZ	Baltimore MD	Providence RI-MA
New Orleans LA	Charleston-N Charleston SC	Portland OR-WA	Birmingham AL	Raleigh-Durham NC
Pittsburgh PA	Charlotte NC-SC	Richmond VA	Boise ID	Riverside-S Bernardino CA
Poughkeepsie-Newburgh NY	Cleveland OH	Rochester NY	Bridgeport-Stamford CT-NY	Sacramento CA
Provo UT	Corpus Christi TX	Salem OR	Chicago IL-IN	San Antonio TX
St. Louis MO-IL	Detroit MI	Salt Lake City UT	Cincinnati OH-KY-IN	San Diego CA

(continued)

Table 6.1 (continued)

Less than 10 % faster (13)	10–30 % faster (46)	10–30 % faster (cont.)	More than 30 % faster (40)	More than 30 % faster (cont.)
Wichita KS	El Paso TX-NM	San Jose CA	Colorado Springs CO	San Francisco-Oakland CA
	Eugene OR	Seattle WA	Columbia SC	San Juan PR
	Fresno CA	Spokane WA	Columbus OH	Sarasota-Bradenton FL
	Grand Rapids MI	Springfield MA-CT	Dallas-Ft Worth-Arlington TX	Stockton CA
	Honolulu HI	Tampa-St. Petersburg FL	Denver-Aurora CO	Washington DC-VA-MD
	Houston TX	Toledo OH-MI	Hartford CT	
	Indianapolis IN	Tucson AZ	Jacksonville FL	
	Jackson MS	Tulsa OK	Laredo TX	
	Kansas City MO-KS	Virginia Beach VA	Las Vegas NV	
	Knoxville TN	Winston-Salem NC	Little Rock AR	
	Louisville KY-IN	Worcester MA	Los Angeles-L Bch-S Ana CA	
	McAllen TX		Miami FL	

Source References [2], p. 51, Table 9

by the impacted communities is strong, environmental regulations limiting the scope of highway improvement projects, and last but not least, a lack of funding sources.

6.2 Causes of VMT Growth

Metro area VMT growth has been driven by population and household growth, by a decreasing population density; a higher labor force participation rate, higher per capita income, and higher car ownership.

6.2.1 City Versus Suburban Population Growth

In 1950 the US population was 151.3 million, with 84.9 million living in metropolitan areas—59 % in central cities and 41 % in suburbs.

The US population grew to 281.4 million in 2000, with 226.0 million living in metropolitan areas—38 % in central cities, and 62 % in suburbs (Table 6.2).

During this period, metropolitan areas grew by 141.1 million people with the suburbs receiving 75 % (105.4 million) of this growth, and central cities the remaining 25 % (35.7 million).

The additional 35.7 million people in central cities live in a higher density environment (with the density of large cities being the higher than that of smaller cities) where they can choose from a variety of mobility options available to them (e.g., car, transit, walking, biking). However the additional 105.4 million people who chose the suburbs, rely mainly on the automobile for daily mobility needs. These location choices and conditions have fundamental consequences on how and where people travel (Table 6.3).

During the 1950–2000 period, metropolitan areas have grown in area as well as in population—with area growth far exceeding population growth (for example, see Table 6.4). Therefore, population densities in the suburbs have remained low, despite their population increase.

Table 6.2 US population trends: 1950–2000 millions

	1950	2000	Change
All US	153.1	281.4	+128.3 (84 %)
Metro areas	84.9 (100 %)	226.0 (100 %)	+141.1 (166 %)
- Central cities	49.7 (59 %)	85.4 (38 %)	+35.7 (25 %)
- Suburbs	35.2 (41 %)	140.6 (62 %)	+105.4 (299 %)

Source Calculated from Reference [3], Table 2-13, p. 27

Table 6.3 Land use and travel indicators—central cities versus suburbs

Indicators	Central cities	Suburbs
Separation of land use activities	Activities located close together	Activities located far apart
Trip length	Shorter	Longer
Car ownership and use	Lower per capita car ownership and use	Higher per capita car ownership and use
Transit availability and use	Higher transit quality and higher use	Transit not available or of poor quality and lower use
Modal alternatives to the car	Many	None to few
Traffic speed	Lower	Higher

Table 6.4 Population growth versus land area growth in selected metropolitan area, 1970–1990

Urban region	Population change (%)	Land area change (%)
Chicago	4	46
Los Angeles	45	300
New York City	8	65
Seattle	38	87

Source References [4], Table 15-50, p. 15-106 and [14]. Copyright © 1996 Island Press. Reproduced by permission of Island Press, Washington, DC

6.2.1.1 Dispersion of Population and Employment

The dispersion of population and employment increases vehicle travel. Tables 6.5 and 6.6, extracted from Transportation Research Board Special Report 298, "Driving and the Built Environment" illustrate those trends [5].

- Table 6.5 shows that while the proportion of people living in the central city has consistently declined from 0.61 in 1940 to 0.38 in 2000, the average metro area density declined from 8,454 to 5,581 people per square mile.
- Table 6.6 gives employment trends inside and outside the Central City for 11 metropolitan areas. This shows that between 1980 and 1990 the central city share declined in all 11 areas while the suburban share increased.

6.2.2 Per Capita VMT Growth

In 2008 the average person contributed an annual average of 9,564 vehicle-miles of travel (VMT) on US roads. This grew from an annual average of about 3,700 vehicle-miles in 1956. By 1998 the VMT per capita reached its peak value of 9,603 vehicle miles, staying within this range until 2005 and beginning to drop slightly to a value of 9,564 in 2008 [6].

Table 6.5 Spatial trends, urban population, 1940–2000

Year	Central city–metro population ratio		Average metro density (persons per square mile)		Density gradient	
	Ratio	Change	Density	Change	Gradient	Change
1940	0.61	–	8,654	–	−0.72	–
1950	0.57	−0.04	8,794	140	−0.64	−0.08
1960	0.50	−0.07	7,567	−1,227	−0.50	−0.14
1970	0.46	−0.04	6,661	−906	−0.42	−0.08
1980	0.42	−0.04	6,111	−550	−0.37	−0.05
1990	0.40	−0.02	5,572	−539	−0.34	−0.03
2000	0.38	−0.02	5,581	9	−0.32	−0.02

Source Reference [5], p. 39, Table 2-1. Copyright © National Academy of Sciences, Washington, DC, 2009. Reproduced with permission of the Transportation Research Board

Table 6.6 Employment trends inside and outside the central city, 1980–1990

	Northeast			Midwest			South	West				
	Buff	NYC	Phil	Chic	Clev	Detr	Hous	Denv	LA	Port	SF	Sea
Percent change												
Total employment	13.2	26.7	28.6	20.3	9.1	19.1	34.9	30.9	48.8	34.8	42.1	48.8
Central city	1.2	22.2	7.7	13.3	−4.0	−6.9	22.4	4.0	32.7	23.8	23.3	21.8
Not central city	21.0	30.5	37.2	25.3	14.7	29.2	61.3	56.2	58.4	43.4	46.8	66.5
Central city share (%)												
1980	39	46	29	41	30	28	68	49	37	44	20	39
1990	35	45	25	39	26	22	62	39	33	41	17	32

Note Buff Buffalo; *NYC* New York City; *Phil* Philadelphia; *Chic* Chicago; *Clev* Cleveland; *Detr* Detroit; *Hous* Houston; *Denv* Denver; *LA* Los Angeles; *Port* Portland; *SF* San Francisco; *Sea* Seattle
Source Reference [5], p. 44, Table 2-2. Copyright © National Academy of Sciences, Washington, DC, 2009. Reproduced with permission of the Transportation Research Board

6.2.3 Factors Contributing to the Rate of VMT Growth

The large change in the annual VMT growth rate from 3,700 in 1956 to 9,564 in 2008, corresponded to a significant increase in the labor force participation rate, an expansion of population growth in suburban areas and an increase in private vehicle ownership at twice the rate of population growth. Table 6.7 summarizes some key trends [7]:

- an increase in the number of workers in the population—from 36.1 % in 1960, to 45.6 % in 2000 due largely to women's increase—whose daily trip rates are 22 % greater than the daily trip rate of non-workers;
- a doubling of per capita motor vehicles ownership from 0.31 in 1960 to 0.63 in 2000 with an equivalent increase in daily vehicle trips per capita.

Table 6.7 Trends in the key factors determining VMT growth in time

	1960	2000
1. Workers ([7], Exhibit 1.6)		
- % of population	36.1	45.6
- % male	67.7	53.2
- % female	32.3	46.7
2. Daily trip rate ([8], Exhibit A-9)		
- Per person	1.9	4.1
- Per worker	2.2	4.5
- Per non-worker	1.7	3.7
3. Ownership of private vehicles ([9], Table 1)		
- Per person	0.31	0.63
- Per 16+ year old	0.53	0.91
- Per driver	0.70	1.06

6.2.3.1 Increasing Trip Length

The average person trip length in low-density areas is considerably higher than that in the higher density areas. As shown in Fig. 6.5 the daily VMT per person increases with decreasing population density from more than 20 miles for the lowest density to about 2 miles for the highest. This is because car ownership is greater in low density areas and vehicle trips to and from these areas are longer.

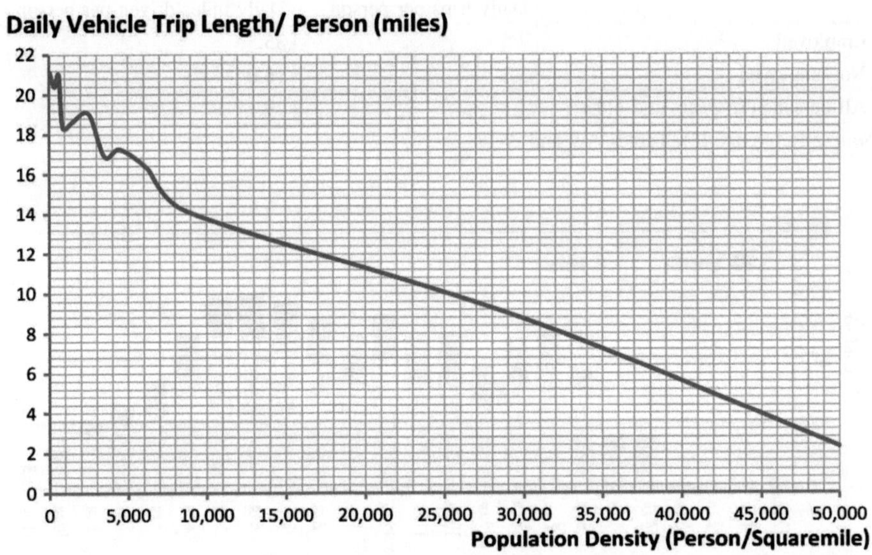

Fig. 6.5 Trip length as a function of population density *Source* Calculated from Reference [4], Table 15.5

Therefore the metro area VMT per capita increases as suburban population growth exceeds center city population growth.

6.2.3.2 Increasing Employment

Employed persons make more trips and travel longer distances than those who are not working (Table 6.8).

While workers constitute about 50 % of the US population, they account for approximately 83 % of the vehicle miles of travel. "It is not (only) their work trips that cause this substantial difference (but also) the other trips and activities engaged in by the working population—on the way to and from work, caring for their families, etc." ([3], p. 6).

This is clearly demonstrated in Fig. 6.6 ([3], Fig. 1-6, p. 7). This figure shows that the workers' travel share by hour of day constitutes the major component of all travel.

However, just as an *increase* in employment increases travel demand, a decrease in employment *reduces* travel demand. As a result of the economic slowdown (2007–2010), the INRIX 2010 National Traffic Scorecard ([10], p. A-1) in fact reported a *reduction* in congestion of 12.7 % for the top 100 metropolitan areas that experienced an aggregate employment loss of 5.8 % from 2006 to 2010.

Table 6.8 Worker and non-worker travel demand, 2001

	Daily trips per person	Daily miles driven per person
Employed	4.5	35.5
Not employed	3.7	16.0
All persons (15 years and older)	4.1	29.1

Source Reference [8], Tables A-9 and A-17

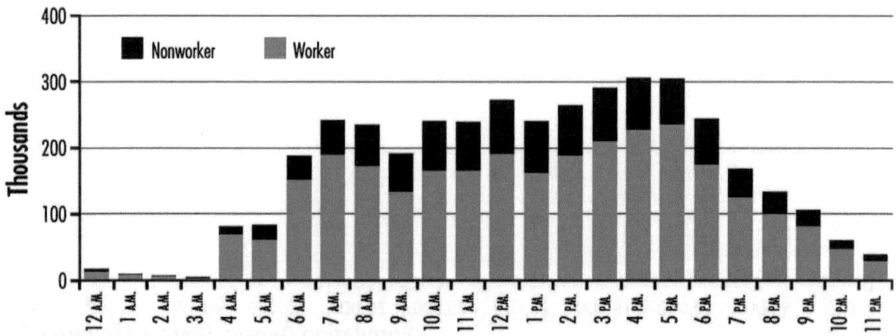

Fig. 6.6 Worker and non-worker shares of travel by time of day. *Source* Reference [3], Transportation Research Board, "Commuting in America III", Fig. 1-6, p. 7

Table 6.9 Daily person trips per person, by vehicle ownership status of household and size of metropolitan area (2001)

Size of metro area	Households with vehicles	Households without vehicles
<250,000	4.45	2.51
250,000–499,999	4.21	2.56
500,000–999,999	4.28	2.69
1–2.9 million	4.24	2.57
3 million+	4.07	2.93

Source Reference [9], Table 33—2001 NHTS summary of travel needs

6.2.3.3 Persons in Households with Autos Generate More Travel

Households have increased their car ownership over time ([3], p. 39, Fig. 2.36). the increases in vehicle ownership has increased travel demand across all metropolitan areas (Table 6.9), with vehicle owning households members traveling significantly more than those living in households without vehicles.

6.2.3.4 Trips by Persons in Higher Income Households Are Longer

Travel demand increases with household income. Table 6.10 shows that the average trip length increases with income—more so for income increments at the lower end of the income scale and reaching a plateau at higher income increments.

As the US personal incomes rise, the value of travel time increases and travelers switch from lower to higher speed modes: from walking—to biking—to bus—to private vehicle. As shown in Table 6.11, the greater mobility provided by the private vehicle has increased the average distance traveled for every age group.

Over this 18-year interval, increasing personal income, smaller household size, and greater ownership of private vehicles, contributed to increasing traveler mobility by an average of 0.82 daily miles per year. This per capita increase, however, was not uniform across all age groups and during the 18-year interval. As shown in Table 6.12, it varied from an average of 1.4 daily miles per year from 1983 to 1990. It was followed by an average increase of 0.76 miles per year from

Table 6.10 Average trip length in private vehicles versus all modes

Household income	Using private vehicles (miles)	Using all modes (miles)
Less than 20,000	6.7	5.6
20,000–39,999	7.4	6.7
40,000–74,999	7.7	7.1
75,000–99,999	7.7	7.1
100,000 and over	7.7	7.1
All households	7.5	6.8

Source Reference [11], Table 11, p. 64, and calculated by the authors from Reference [10]

Table 6.11 Change in average daily person miles of travel per person (1983–2001)

Age	1983	2001	Average yearly change: 2001–1983 (miles/year)
Under 16	16.2 miles	24.5 miles	+8.3/18 = 0.46
16–20	22.2	38.1	+15.9/18 = 0.88
21–35	31.1	45.6	+14.5/18 = 0.81
36–65	29.2	48.8	+19.6/18 = 1.09
Over 65	12.0	27.5	+15.5/18 = 0.86
Total	25.5	40.2	+14.7/18 = 0.82

Source Reference [9], Table 14—2001 NHTS summary of travel trends

Table 6.12 Rate of annual increase in per capita daily person miles of travel (1983–2001)

Age	1983–1990	1990–1995	1995–2001	1983–2001 (miles/year)
Under 16	+0.56 miles/year	+1.00 miles/year	−0.8 miles/year	+0.46
16–20	+1.74	+0.00	+0.28	+0.88
21–35	+0.77	+1.90	−0.07	+0.81
36–65	+0.54	+1.00	+0.62	+1.09
Over 65	+0.31	+1.20	+0.52	+0.86
Total	+1.4 miles/year	+0.76 miles/year	+0.25 miles/year	+0.82

Source Reference [9], Table 14

1990 to 1995; and it slowed down to an average increase of 0.25 miles per year from 1995 to 2001 (see Sect. 6.3, for a discussion of this plateau effect).

6.2.3.5 Growth of Households and Household Vehicles Exceeding Population Growth

As the size of households decreased (e.g., the number of households grew faster than the population—see Table 6.13), the use of private vehicles for daily travel almost doubled from 1969 to 2001 (Table 6.14).

Table 6.13 Daily vehicle trips per person and household size: 1969–2001

Year	Vehicle trips per person	Persons per household
1969	1.21	3.16
1977	1.40	2.83
1983	1.51	2.69
1990	2.22	2.56
1995	2.42	2.63
2001	2.31	2.58
Ratio of 2001 to 1969	1.91	0.83

Source Reference [9], Tables 1 and 2

Table 6.14 Trends in demographics and private vehicle travel, indexed to 1969 (1969 = 1.00)

	1969	1977	1983	1990	1995	2001
Population	1.00	1.08	1.16	1.21	1.32	1.41
Households	1.00	1.21	1.37	1.49	1.58	1.72
Workers	1.00	1.23	1.36	1.56	1.73	1.92
Drivers	1.00	1.24	1.43	1.58	1.71	1.85
Household Vehicles	1.00	1.66	1.98	2.28	2.43	2.79
Household VMT	1.00	1,17	1.29	2.18	2.67	2.93

Source Reference [9]

6.2.4 Trends in VMT and Contributing Factors

From 1969 to 2001, the household VMT more than doubled (2.93/1.41) the growth in population (Table 6.14). The impact on VMT of suburban population growth was magnified by the growth rate in the number of households, workers, drivers, and privately owned vehicles.

6.3 The Plateau Effect of Factors Inducing VMT Growth

The preceding discussion indicated a VMT growth over the last half century that has closely reflected the pattern of growth in its causative factors (population, income, auto ownership, workers, increasing suburbanization of metropolitan areas). It has increased proportionately with the growth of these variables, and it has slowed down when the growth of these variables slowed.

The annual VMT *growth rate* was 5.2 % in the 1969–1990 period, and has decreased to less than 1 % per year since 2004, reaching a negative growth in 2006–2007 (Table 6.15). The last time of a VMT negative growth was in 1980 [7].

Table 6.15 Changes in VMT (1969–2007)

Period	Growth in period (%)	Average annual rate (%)
1969–1990	+110	+5.2
1990–2001	+31	+2.8
2004–2005	+0.8	+0.8
2005–2006	+0.6	+0.6
2006–2007	−0.3	−0.3

Source Reference [7], calculated from Fig. 1a

Table 6.16 Changes in private vehicle ownership (1969–2000)

Year	Vehicles per driver	Average annual (%) change over previous period
1969	0.70	–
1977	0.94	+4.3
1983	0.99	+0.9
1990	1.01	+0.3
2001	1.06	+0.4

Source Reference [9], Table 1

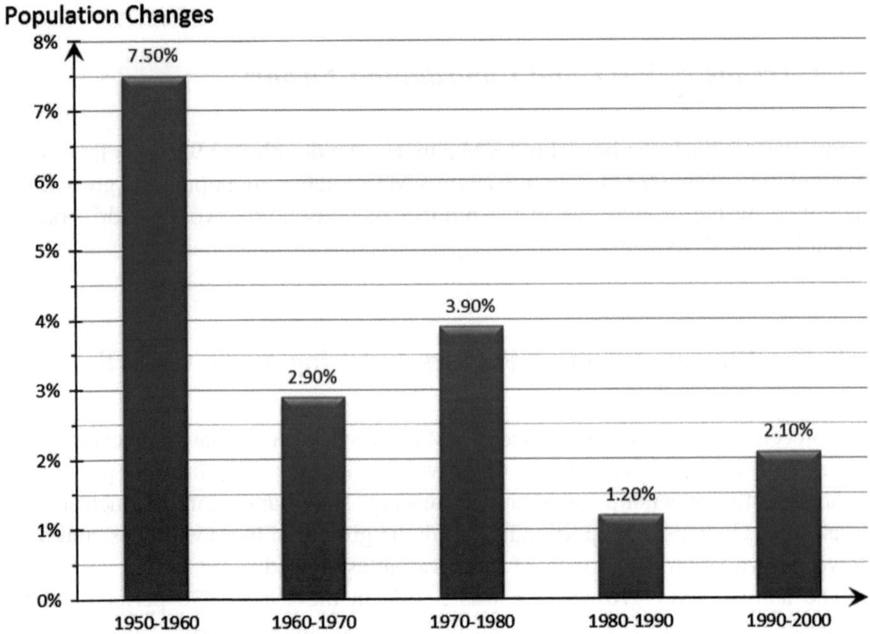

Fig. 6.7 Average annual population changes in suburban areas (1950–2000). *Source* Calculated from Reference [3], Fig. ES-1

Several key factors have influenced a slower annual VMT growth in the 2000–2010 decade. These include: (1) a flattening of the labor force participation rate (Table 6.7), (2) a near saturation of private vehicle ownership (Table 6.16), and (3) a decreasing annual share of population growth in suburban areas (Fig. 6.7)—where automobile dependency is highest. In addition, the extended high levels of unemployment due to the economic recession that began in 2007 (Table 6.17), together with unstable fuel prices (Table 6.18), have reduced the VMT and its congestion impact over the period.

Table 6.17 The impact of employment changes on traffic congestion

Area type	Total employment (thousands)			Travel time tax[a]		
	2006	2010	% Change	2006 (%)	2010 (%)	% Change
Most congested metro area (Los Angeles)	5,695	5,170	−9.2	43.7	35.4	−19.0
Top 100 metros	93.3	87.9	−5.8	11.1	9.7	−12.7
National	136.9	130.7	−4.5	NA	NA	NA

Source Reference [10], INRIX, p. A-1
[a] Per INRIX definition = travel time in excess to free-flow travel time expressed as % of free-flow travel time

Table 6.18 Elasticity of VMT regarding gas prices

	2007	2008	%
$/Gal.	$2.80	$3.25	+16
VMT (Billion)	483	469	−2.9

Source Reference [12], INRIX 2009, p. ES-2

The elasticity of VMT with respect to gas prices can be calculated as

$$E_{VMT}\text{ re : Price} = \frac{-2.9\,\%}{16\,\%} = -0.18 \tag{6.1}$$

This finding is in the range of driving elasticity (the elasticity of VMT with respect to gasoline prices) values of −0.15 to −0.20 values reported by Gillingham [11] for the 2005–2008 period in California.

6.3.1 Rate of VMT Growth per Capita Is Likely to Decrease in the Future

The trends in the factors that generated strong VMT growth in the second half of the 20th century have reached saturation points and have slowed down considerably since the beginning of the new millennium. Including:

- The share of the women in the work force has leveled off to 60 %, and the share of women with driving licenses approaches 85 % [13]
- Increasing household incomes are unlikely to result in corresponding increases in car ownership rates since car ownership rates have reached saturation levels

- Young households' debt burden from student loans makes the purchases of a house and a car more difficult. This is likely to encourage residential locations of young households to places with travel alternatives to the private car
- The slowing pace of suburbanization of the past decades is expected to continue in the future along with a reduction in VMT growth
- The increasing cost of owning and operating a car will reduce car ownership growth and VMT.

6.3.2 *Implications*

Although the per capita automobile growth rate is at near saturation levels, future growth in automobile travel, while slower than in previous decades, will continue place more pressure on urban streets and highways. This condition will increase congestion levels and will increase the need for strategies to make road networks more efficient and will require additional capacity (where possible) to keep pace with this growth.

References

1. Bureau of Transportation Statistics (2002) National Transportation Statistics. US Department of Transportation, Washington, DC
2. Schrank D, Lomax T, Eisele B (2011) TTI's 2011 urban mobility report—powered by INRIX traffic data. s.l.: Texas Transportation Institute, The Texas A&M University System
3. Pikarski AE (2006) Commuting in America III. NCHRP Report 550 and TCRP Report 110. 2006
4. Kuzmyak JR, Douglas GB, Spielberg F et al (2003) TCRP Report 95: traveler response to transportation system changes, chapter 15—land use and site design. Transportation Research Board, Washington, DC
5. Transportation Research Board Special Report 298 (2009) Driving and the built environment: the effects of compact development on motorized travel, energy use, and CO_2 emissions. Washington, DC, s.n.
6. Puentes R, Tomer A (2008) The road…less traveled: an analysis of vehicle miles traveled trends in the US s.l. Brooking
7. McGuckin NA, Srinivasan N (2003) Journey to work in the United States and its major metropolitan areas, 1960–2000. US Department of Transportation, Washington, DC
8. NHTS (2003) Highlights of the 2001 national household travel survey. Bureau of Transportation Statics, US Department of Transportation, Washington, DC
9. NHTS (2004) Summary of travel trends—2001 national household travel survey. US Department of Transportation, Federal Highway Administration, National Technical Information Service, Washington, DC
10. INRIX (2011) National traffic scorecard—2010 annual report. Back to the road to gridlock
11. Gillingham K (2013) Identifying the elasticity of driving: evidence from a gasoline price shock in California. s.l. Dennis J. O'Brian UASEE/IAEE Best Student Paper Award

12. National Traffic Scorecard (2010) 2009 Annual report. s.1. INRIX Inc.
13. McCahill C, Spahr C (2013) VMT inflection point: factors affecting 21st century travel. s.l. State Smart Transportation Initiative
14. Diamond HL, Noonan PE (1996) Land use in America. Island Press, Washington, DC

Chapter 7
Bottlenecks

7.1 Introduction

Travelers and freight movers experience congestion as a result of capacity deficiencies in the roadway system. These congestion problems occur whenever the arrival rate (demand) exceeds the vehicle departure rate (capacity). The demand-capacity imbalance is manifested by a number of conditions: (a) recurring bottlenecks are caused by topographic and physical barriers to movement; the discontinuity of the road network; design and operational deficiencies; (b) nonrecurring bottlenecks are created by incidents, surging demand, inclement weather, work zones/street closures, and driver behavior.

When recurring and nonrecurring bottlenecks happen at the same time their impact on delay is at its worst. The amount of delay caused by recurring events and by nonrecurring events has been estimated by several sources but the estimates provided are not clear as to the amount of congestion produced when recurring and nonrecurring delays overlap.

Table 7.1 shows a national estimate of the causes of recurring and nonrecurring congestion delay for US freeways and expressways.

These estimates represent rough approximations from past and ongoing studies. They show that nonrecurring congestion generates a larger share of delay than recurring congestion (55 vs. 45 %). Incidents cause 45 % of nonrecurring delay and recurring bottlenecks are the most common cause of recurring delay (89 %).

Lockwood [2] breaks down recurring and nonrecurring delay by area type and size of urbanized area. Table 7.2 shows the percentage contribution of recurring and nonrecurring causes to total delay by size and type of area.

Thus the delay in rural areas is predominantly of the nonrecurring type (97 %), while in urban areas nonrecurring delay is a much smaller share of total delay. This share decreases slightly as the size of the area increases (58–72 % in small urban areas versus 55–69 % in areas greater than one million people).

J.C. Falcocchio and H.S. Levinson, *Road Traffic Congestion: A Concise Guide*,
Springer Tracts on Transportation and Traffic 7,
DOI 10.1007/978-3-319-15165-6_7

Table 7.1 Sources of congestion—a national summary

Congestion causes	Recurring (%)	Non-recurring (%)
Bottlenecks	40	–
Poor signal timing	5	–
Traffic incidents	–	25
Work zones	–	10
Bad weather	–	15
Special events/other	–	5
Total	45	55

Source Reference [1]

Table 7.2 Percentage contribution of recurring and nonrecurring causes to total delay, by area type and size

	Cause of delay	Large Urban areas >1 m[a]	Small urban areas 0.1–1.0 m	Rural
Recurring causes	Network demand > capacity	29–37	20–26	0
	Poor signal timing	4–5	7–13	2
Total recurring		33–42	32–33	2
Non-recurring causes	Crashes	35–36	19–26	26
	Breakdowns	6–7	6–10	25
	Work zones	8–19	26–27	39
	Weather	5–6	7–10	7
	Special events/lack of information, other	1	–	0
Total non-recurring		58–67	67	98

[a] Combine estimates for size classes 1–3 m and > 3 m. *Source* Reference [2]. Used by permission

It is important to note that these estimates refer to an area-wide average, and are not intended for estimating conditions at specific highway corridors within the urban area, as these must be estimated from actual experiences. For example, a corridor with older highways and sub-standard designs (e.g., the I-278 corridor in Brooklyn, NY), would experiences higher crash rates, so that the proportion of congestion delay from traffic incidents in the I-278 corridor would be higher than that indicated in the above table.

This chapter describes the various bottleneck factors that contribute to recurring and nonrecurring congestion.

7.2 Recurring Congestion

7.2.1 Physical Bottlenecks

Topographic barriers and physical bottlenecks on streets and highways represent choke points that reduce road capacity and cause peak hour traffic to back up and create congestion of the upstream roadways.

7.2.1.1 Topographic Barriers

A city's physical features can create congestion. Topographic barriers such as hills, mountains, steep grades, and water bodies constrain street patterns and concentrate travel on a limited number of available crossings. Balancing the capacity of the approach roadways with the capacity provided at a limited number of crossings is usually a difficult task seldom achievable. Therefore, in these areas peak hour congestion is a common event on the roadway approaches to bridges, tunnels, and other roadways that traverse such crossings.

Some US examples of topographic barriers in cities illustrate their impact on congestion.

- Manhattan Island in New York City requires motorists to cross the Hudson and East Rivers to reach the business district. AM peak hour inbound traffic backs up forming long queues on the roadways leading to the CBD, often requiring up to 40 min before reaching the crossing. Likewise PM peak hour traffic backs up forming long queues over many blocks on city streets that often require up to 40 min waiting time before reaching the bridge or tunnel crossing.
- San Francisco is located on a peninsula that is separated from Marin County and East Bay communities by San Francisco Bay. Road access from the north and east is limited to the Golden gate and Bay Bridges that constrain traffic demand from the converging freeways leading to the Bay Bridge.
- Los Angeles' San Fernando Valley is separated from the rest of the city by the Santa Monica mountains.
- Seattle is hemmed in by Elliot Bay and Lake Washington.
- Pittsburgh's Golden Triangle is located between the Allegheny and Mongonohela Rivers, and nearby hills to the east and south.
- New Orleans is bounded by the Mississippi River and Lake Ponchatrau.
- The Bronx in New York City, has few continuous east—west streets because of its difficult terrain.

Most cities also have man-made barriers to travel. These include large cemeteries, railroad embankments with infrequent crossings, and large private developments.

7.2.1.2 Design Deficiencies

Traffic bottlenecks are also the outcome of the geometric street layout as well as design deficiencies of critical sections/locations of the street network.

Street Network Geometry

City street patterns are an outgrowth of each city's history, geography and public policy.

1. Pre-automobile Cities—Streets in the central parts of many cities predate the automobile and therefore are not designed to accommodate motor vehicle traffic. Typically they have short, irregular blocks with insufficient capacity in the peak hours for storing vehicles waiting for a green signal. This feature makes these streets prone to spill back traffic creating significant congestion.
2. Many of the post WWII suburban developments have discontinuous street networks and continuous streets that are spaced too far apart with the effect of increasing the number of lanes in each street. This condition concentrate high traffic demand volume where these major streets intersect creating congestion delays.
3. Washington, DC, streets are largely part of the L'Enfant Plan for the National capital. Its combination of multi-direction radial streets superimposed on a rectangle grid creates many complex intersections commonly resulting in traffic congestion. Similar street plans were later adopted in Buffalo, Detroit, and Indianapolis.
4. Converging radial streets are common in many older cities. This pattern often results in peak hour congestion from converging traffic that exceeds the capacity of the intersection.
5. Several cities have a diagonal street system superimposed on a grid. Chicago's for example, has historic plank roads that create six-leg intersections where they cross the grid streets. Historic Broadway in Manhattan cuts across the north-south and east-west grid streets creating complex intersections where it crosses the grid—resulting in reduced intersection capacity.

Facility Design Deficiencies/Constraints

Bottlenecks are created whenever any of the following conditions exist in the road network:

- **Lane Imbalance**: At merge areas, at bridge and tunnel crossings, and where several roadways converge without corresponding increase in travel lanes can create extensive backups and congestion during busy travel periods.

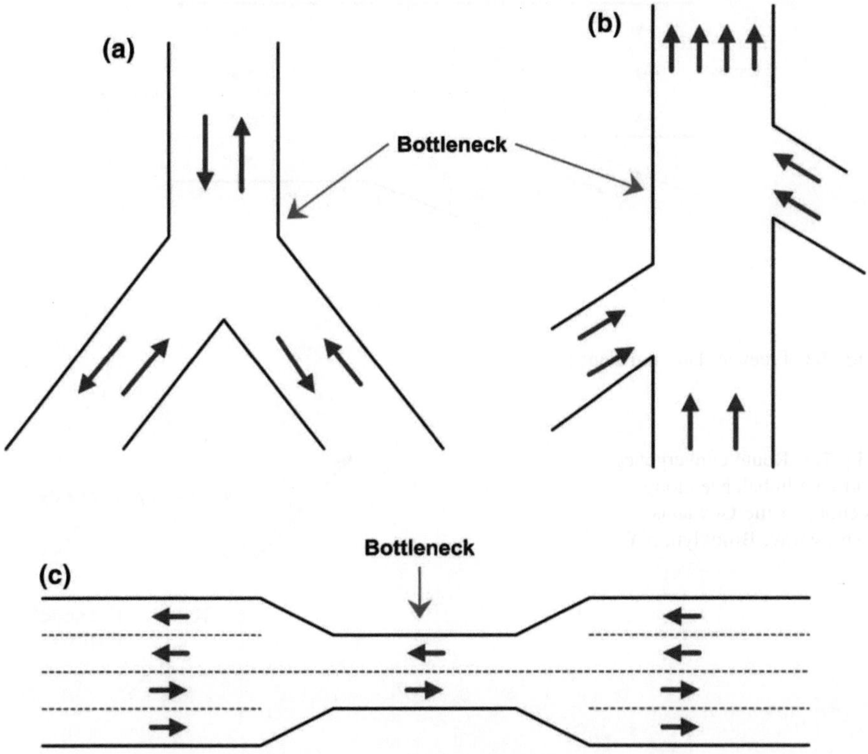

Fig. 7.1 Example of bottlenecks resulting from lane imbalance. **a** Converging roadway two lanes merge into one. **b** Converging roadway six lanes merge into four. **c** Converging roadway four lanes merge into two

Examples of lane imbalance are shown in Fig. 7.1: (a) two travel lanes margining into one; (b) six lanes merging into four; and (c) a 4-lane roadway narrows to 2 lanes for a short distance.

A common source of freeway congestion comes from where the number of entering lanes on two merging freeways exceeds the number of departing lanes (Fig. 7.2). This condition results in recurrent congestion.

A similar problem occurs where a highly traveled entry ramp joins the main freeway lanes without any increase in freeway capacity.

Figure 7.3 shows an example of the lack of lane balance along the Northbound Gowanus Expressway in Brooklyn, NYC. The merge points of the Prospect Parkway and the Belt Parkway with the Gowanus Expressway result in high congestion levels—every weekday morning in the peak hours.

Figure 7.4 shows the pattern of lane convergence along I-95 (southbound) in Connecticut, between Stamford and New Haven. As a result of this condition there is significant amount (intensity, duration, and extent) of daily congestion along this section of the expressway.

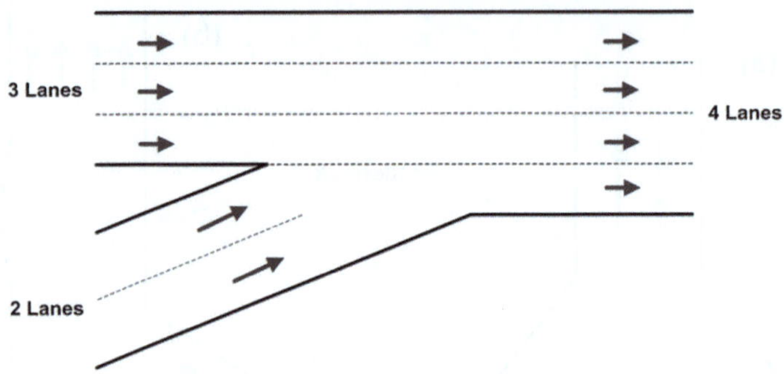

Fig. 7.2 Freeway lane imbalance

Fig. 7.3 Route convergence and lane imbalance along sections of the Gowanus Expressway, Brooklyn, NY

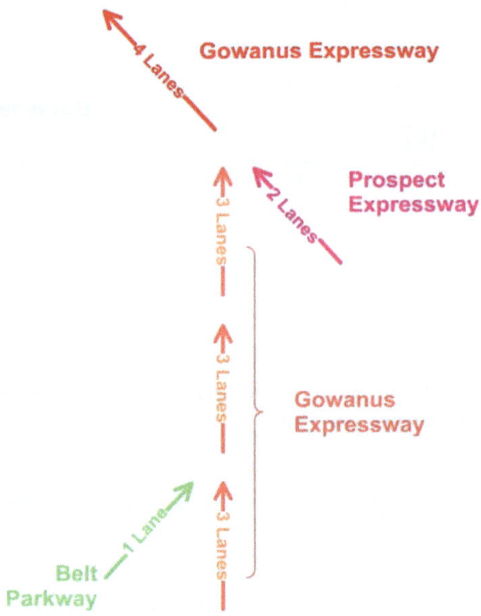

- <u>**Geometric Constraints**</u>: Sharp curves, steep vertical grades, and narrow lanes cause vehicles to slow down (Fig. 7.5). For example, large trucks going on a steep upgrade can create long queues of vehicles in back of the truck.
- <u>**Short Auxiliary Lanes**</u>: Auxiliary lanes on approaches to signalized intersections (Fig. 7.6) are sometimes too short to prevent queued up traffic waiting to turn left or right from blocking the through movement, causing large losses in throughput capacity of the intersection.
- <u>**Inadequate Access Control**</u>: Too many curb cuts and driveways create conflicts between through traffic and vehicles entering/exiting from parking lots/garages or driveways (Fig. 7.7).

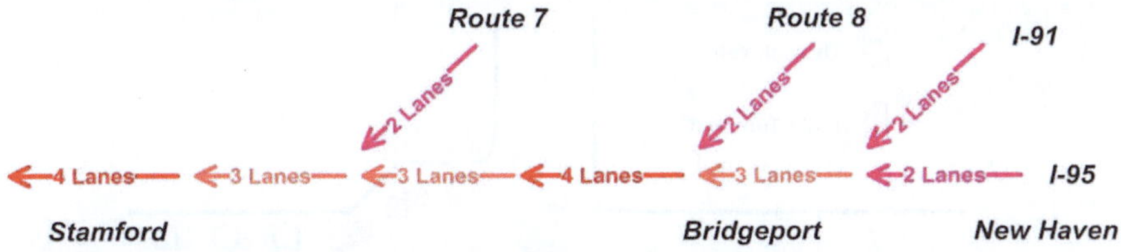

Fig. 7.4 Route convergence and lane imbalances, I-95 southbound (Southern Connecticut)

Fig. 7.5 Examples of sharp *curves* and steep grades

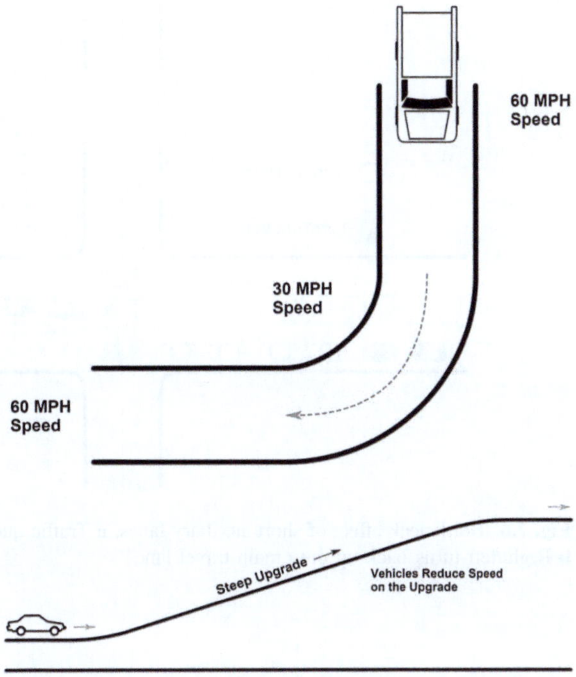

- **Deficient Driveway Geometry**: Commonly found where large entry areas are without lane markings necessary to channel traffic in an orderly way and where the queue space for vehicles entering or leaving the driveway is insufficient to prevent vehicles from spilling back onto the traffic lanes (Fig. 7.8).

Usually there is no provision for protected turning lanes or acceleration lanes along the public road. This condition creates conflicts between vehicles resulting in congested flow and high crash rates. Pedestrian circulation is also problematic due to pedestrian-vehicle conflicts.

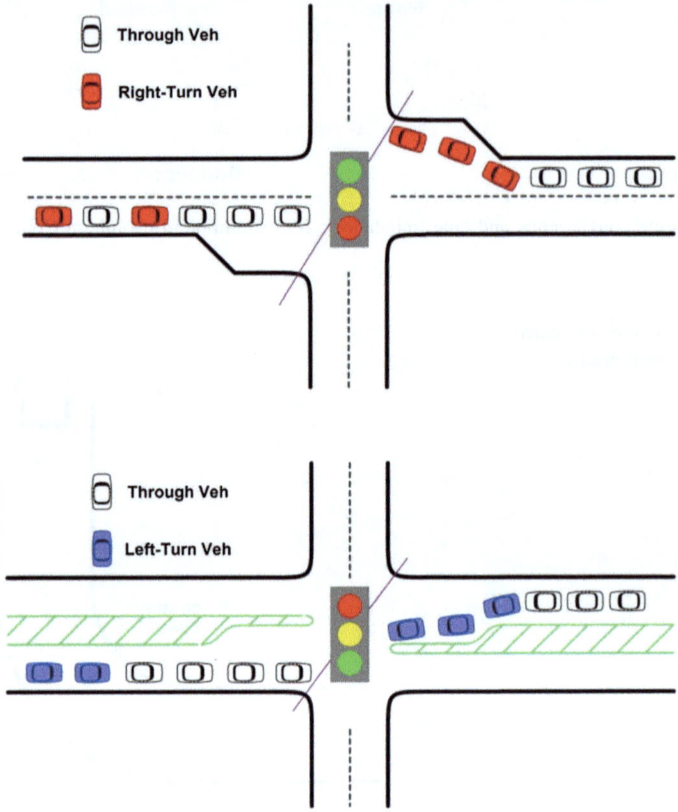

Fig. 7.6 Bottleneck effect of short auxiliary lanes. **a** Traffic queue limits access to auxiliary lane. **b** Right/left-turns back-up onto main travel lane

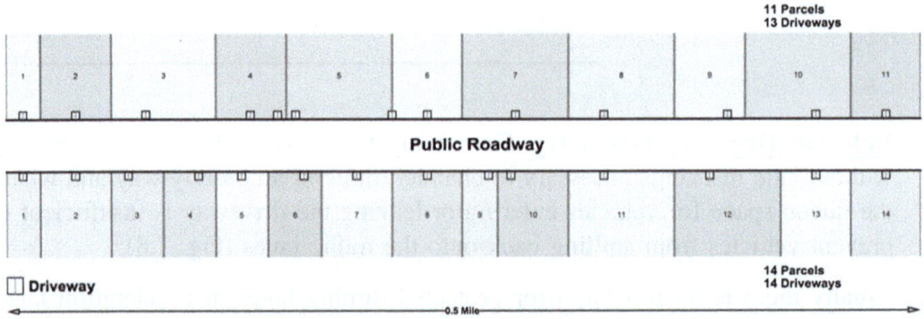

Fig. 7.7 Varying lot sizes with frequent driveways contribute to congestion

- **<u>Multi-leg Intersections</u>**: In many communities complex multi-leg intersections with more than two intersecting streets are the focal point of congestion. For example, an intersection of three streets requires at least three traffic signal phases—up to six phases if special phasing for left turns is required (Fig. 7.9).

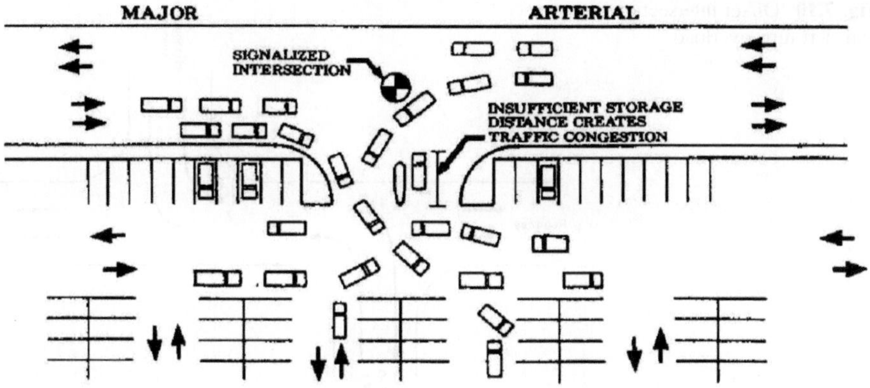

Fig. 7.8 Inadequate driveway geometry creates conflicts and congestion. *Source* Reference [3], p. 86. Fig. 8.32

Fig. 7.9 Six leg intersection

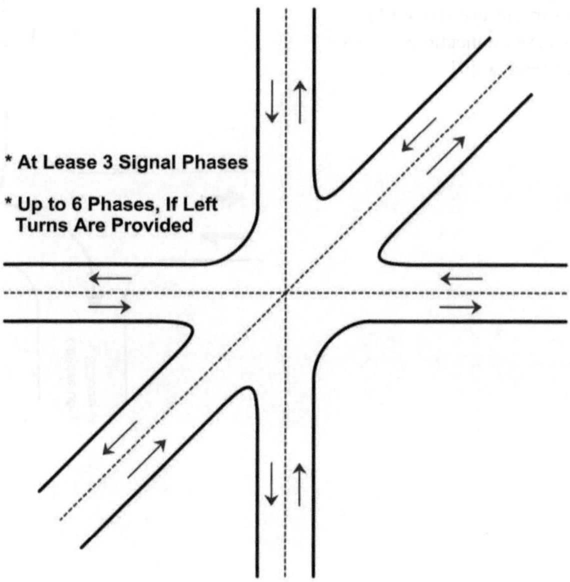

This condition reduces the time available at each approach to serve the approaching volume and creates a capacity deficiency that will result in congested conditions during heavy traffic periods.

- **Offset Intersections**: Offset intersections tend to over-load major roadways, complicating signal timing and sometime creating left-turn storage deficiencies. Examples are shown in (Figs. 7.10 and 7.11).
- **Offset Freeway Alignments**: Offset freeway interchanges result in the double loading of the common freeway segment (Fig. 7.12). Unless carefully designed

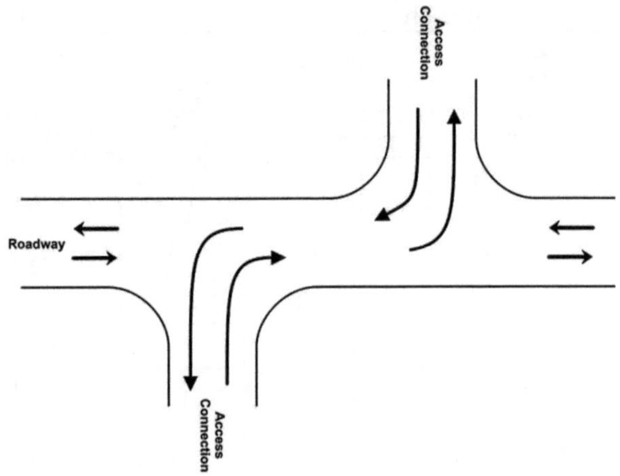

Fig. 7.10 Offset intersection with left-turn overload

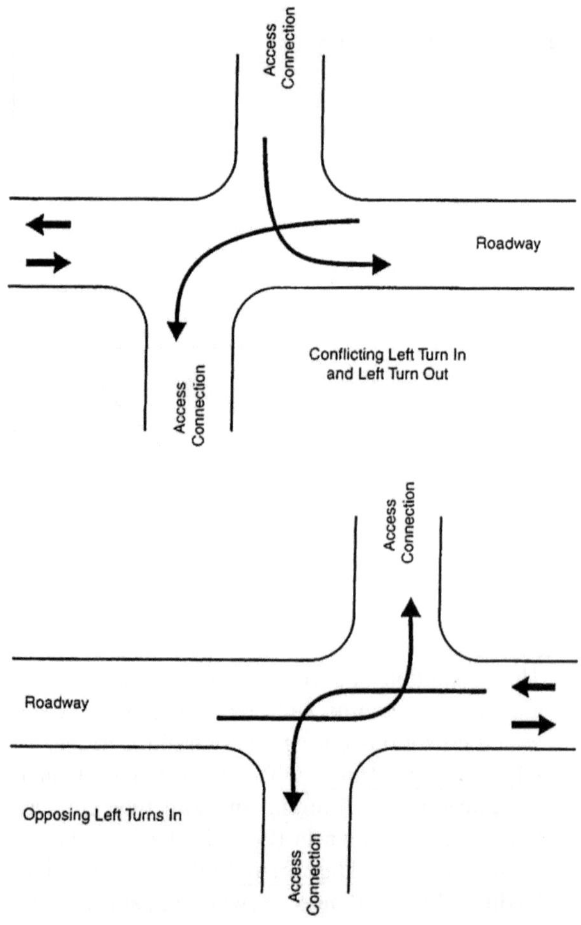

Fig. 7.11 Conflicts arising from closely spaced 'T' access connections. *Source* Reference [4]

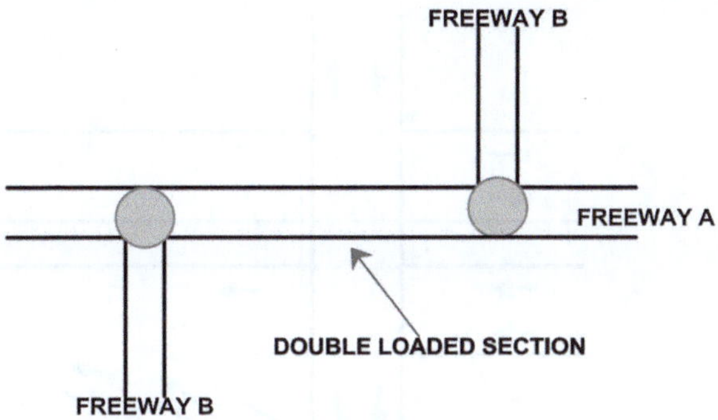

Fig. 7.12 Offset freeway interchange

to prevent cross—weaving maneuvers and lane imbalance, these locations are the source of capacity bottlenecks that cause substantial peak period delays.

- **Complex Weaving Freeway Sections**: Freeway weaving areas, where traffic must merge across several lanes to leave the freeway, can become congestion bottlenecks. This is especially so where the weaving section is too short—requiring traffic to slow down to find acceptable gaps to maneuver between lanes. These weaving conflicts are most problematic where there are left-side entry and right side exit, or right side entry and left side exit ramps, and there are several freeway lanes to cross (Fig. 7.13).
- **Short Freeway On-Ramps**: Short freeway on ramps, especially with inadequate (short) acceleration lanes cause merging traffic to enter gaps by forcing through traffic to slow down. An example is given in Fig. 7.14.
- **Short Freeway Off-Ramps**: Exit ramps with short deceleration lanes require exiting traffic to slow down while still in the through freeway lanes. These

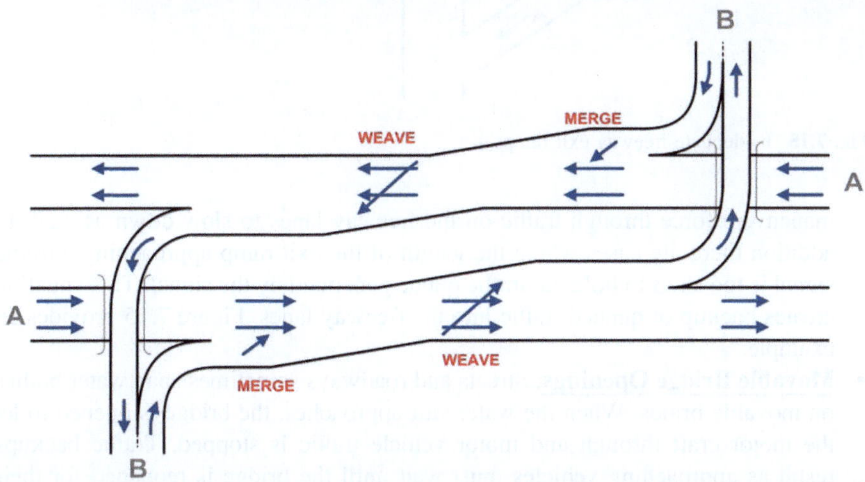

Fig. 7.13 Right side entry ramps and left side exit ramp causing complex weaving maneuvers

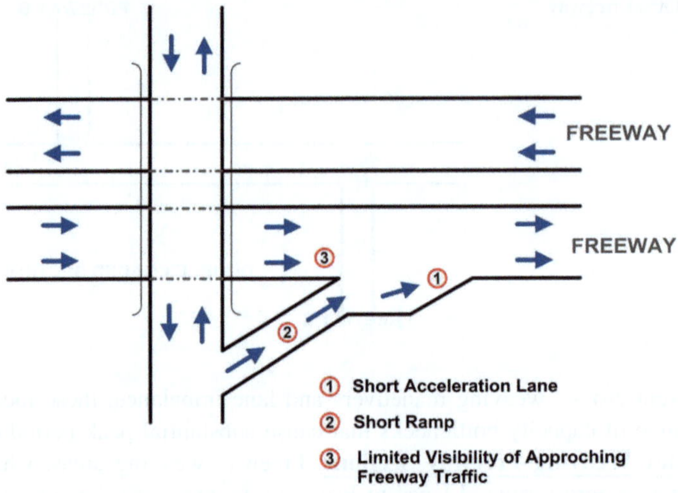

Fig. 7.14 Inadequate freeway on—ramp and acceleration lane

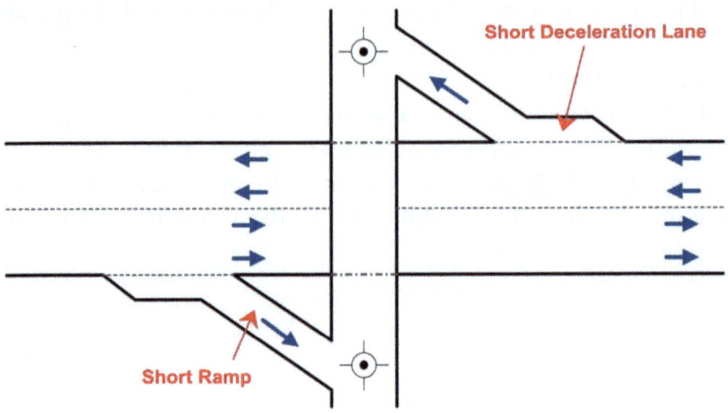

Fig. 7.15 Inadequate freeway exit ramps

maneuvers force through traffic on the freeway lanes to slow down as well. In addition there are cases where the length of the exit ramp approaching a traffic signal is too short to hold the traffic queue generated by the signal. This situation creates backup of queued traffic into the freeway lanes. Figure 7.15 provides an example.

- **Movable Bridge Openings**: Streets and roadways sometimes span water bodies on movable brides. When the watercraft approaches, the bridge is opened to let the motor craft through and motor vehicle traffic is stopped. Traffic backups result as approaching vehicles must wait until the bridge is reopened for their use.

- **Railroad Grade Crossings**: Railroad grade crossings are common in many suburban areas. Whenever trains pass, motor vehicle and pedestrian/bicycle movements are stopped. Also in this situation queues are formed and their length is critical in determining the congestion impacts on the connecting roadways.

7.2.2 Operational Bottlenecks

Midblock and intersection conflicts cause bottlenecks that contribute to congestion especially during heavy traffic periods. These conflicts are the result of loading and unloading of goods from the streets and are the result of cross traffic, as well as turning vehicles and pedestrian conflicts at signals.

7.2.2.1 Curb Parking and Goods-Loading Conflicts

On-street curb parking in business districts frequently results in congestion by reducing the number of lanes that are available for moving traffic. In addition, double parking during peak periods—often by delivery and courier vehicles—have an even more detrimental on movement. They block several traffic lanes, and during heavy periods can cause spillback on approaches to the bottleneck.

This condition is generally found where:

- On-street parking is permitted during busy traffic periods
- There is inadequate enforcement of curb parking regulations
- There are frequent double parkers.

Where a street has two lanes in a given direction, and where parked vehicles occupy one lane there is at least a 50 % loss in capacity.

7.2.2.2 Intersection Conflicts

Intersections of major streets are often the focal points of traffic congestion during peak periods of travel. The many conflicts—between pedestrians, cyclists, and motorized traffic; between through and cross traffic, and between through and turning vehicles are major sources of congestion.

As shown in Figs. 7.16 and 7.17, at a typical four-way intersection there are 32 vehicle–vehicle conflict points and 48 pedestrian-vehicle and bicycle-vehicle conflicts.

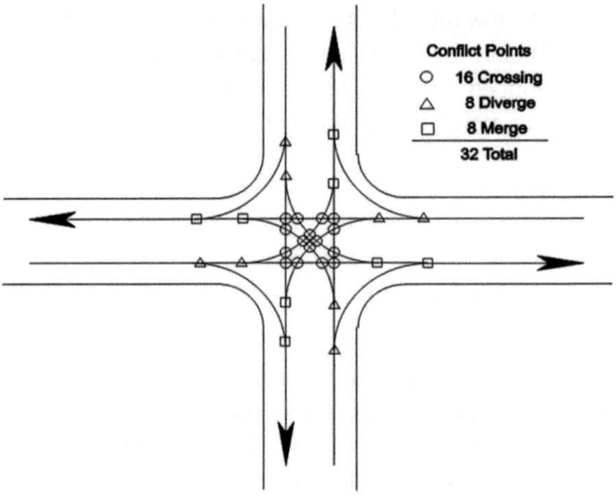

Fig. 7.16 Vehicular conflict points at a typical four-way intersection. *Source* Reference [4]

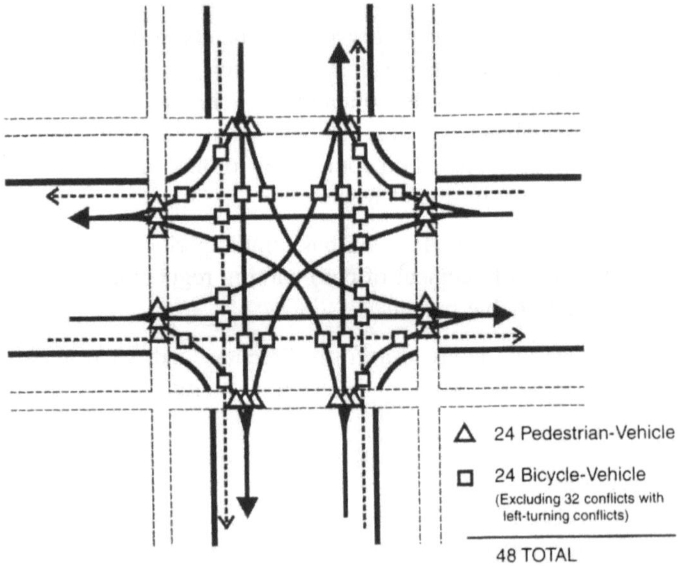

Fig. 7.17 Pedestrian—vehicle and pedestrian- bicycle conflict points at a four-way intersection. *Source* Reference [4]

Conflicting traffic movements through a street intersection are usually separated by a traffic signal that alternatively allocates a proportion (or phase) of the total time available (or cycle) to move traffic and to stop traffic from moving. The stopped time on each approach is a cause of congestion during periods of heavy traffic volume.

The key causes of intersection congestion include:

- an insufficient number of travel lanes on intersection approaches
- the lack of exclusive lanes of adequate length for right and left turns
- heavy traffic volumes and turning movements on the various conflicting approaches
- heavy pedestrian and bicycle movements that conflict with and impede and impeded traffic flow.

7.2.2.3 Traffic Signals

Because traffic signals control conflicting movements they account for much of the traffic delay along streets and roads. Their location, phasing, and timing can substantially increase congestion when:

- the total green time per signal cycle must be shared by conflicting traffic streams
- right-turns conflict with heavy pedestrian volumes
- left-turns operating from a lane shared with through traffic can block through vehicles. When there is one left-turn per cycle, about 40 % of the through vehicles in the shared lane are blocked. When there are three left-turn vehicles per cycle, about 70 % of through traffic is blocked. When protected left turn lanes are provided, there is generally no impedance to through traffic moving in the same direction
- Left turns with exclusive turn lanes must share the green time with the through traffic in the opposing direction.

Traffic signal location, spacing, and timing deficiencies commonly include:

a. Placing signals where they do not fit the progression pattern reduces the width of the through (or green)—band (Fig. 7.18).
b. Although efficient progression can be maintained by increasing the green time on the major street, but this condition would require a reducing the green time on the cross street with a corresponding increase in delay to cross street traffic (4).
c. Providing an excessive number of phases such as a pre-timed exclusive pedestrian phase where there are few pedestrians crossing a highway.
d. Using cycle lengths that are too short to serve peak traffic demands can result in excessive delay.
e. Using cycle lengths that are too long (e.g., over 2 min) make signal coordination difficult to achieve.
f. Operating closely spaced signals that are not coordinated.
g. Operating obsolete traffic signal control systems that limit the ability to establish time-of-day or traffic responsive signal coordination.
h. Locating signals at irregular intervals that limit or preclude coordination.
i. Placing signals too close together, thereby limiting effective coordination and resulting in frequent stops.

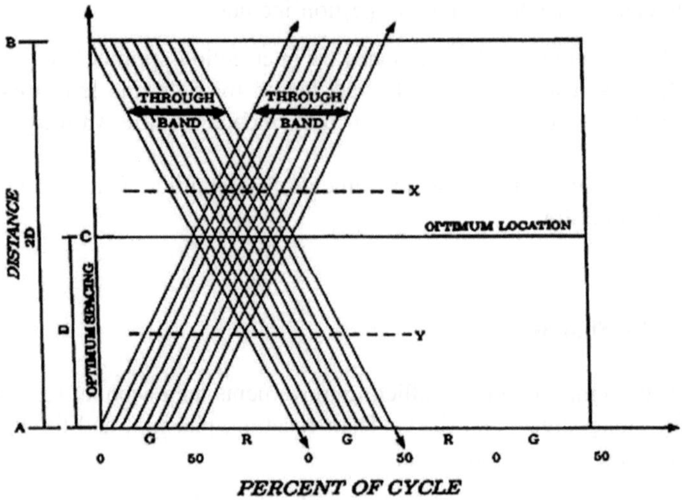

Fig. 7.18 Time-space pattern. *Source* Reference [5]

The key considerations from both capacity and congestion perspectives include (a) the traffic signal cycle, (b) the number of phases, (c) the amount of green time on each phase, and (d) the number of lanes and the traffic volume on each approach. Detailed procedures for estimating intersection capacities, stopped delays at intersections, and "levels of service" are set forth in the 2010 Highway Capacity Manual [6].

Figures 7.19 and 7.20 show the effects of volume-to-capacity ratios and traffic signal spacing on arterial speed. The free-flow speed decreases as the number of signals per mile increases. And the rate of speed drop diminishes after signal density exceeds 5 signals per mile.

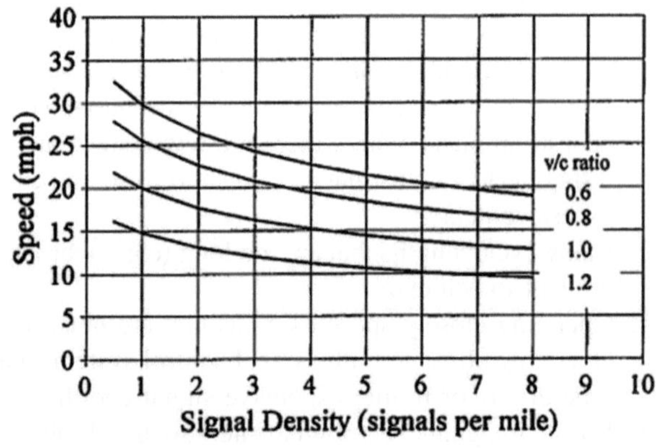

Fig. 7.19 Suggested speed estimation curves as a function of signal spacing and V/C Ratio—Class I arterials. *Source* Reference [7], Fig. 7.10

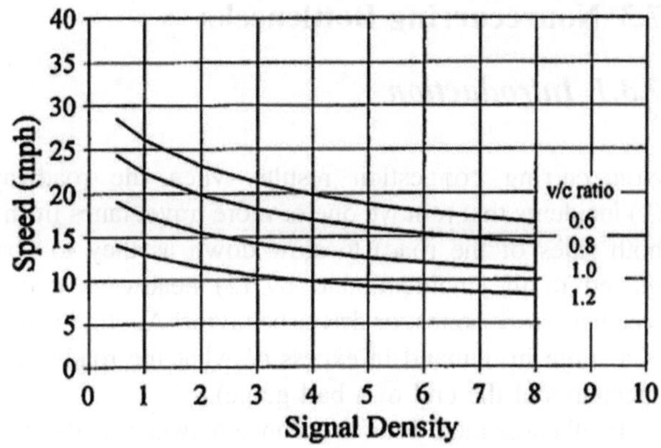

Fig. 7.20 Suggested speed estimation curves as a function of signal spacing and V/C Ratio—Class II and III arterials. *Source* Reference [7], Fig. 7.11

Table 7.3 Percent increase in travel time as a function of traffic signals per mile

Signals per mile	Percent increase in travel time (two signals per mile as base) (%)
3.0	9
4.0	16
5.0	23
6.0	29
7.0	34
8.0	39

Source Reference [8]

The effect of an increase in traffic volume on traffic speed reduction is expressed by a family of curves for various volume-to-capacity (V/C) ratios. For example, for class I arterials, approximately 2 mph drop in speed results for every 0.1 increase in the V/C ratio for roads with fewer than 4 signals per mile. But for roads with 8 signals per mile the drop in speed diminishes to a little over 1 mph for every 0.1 increase in the V/C ratio.

These curves show that traffic signal density has a greater effect on delay than traffic volumes when the volume-to-capacity ratio is less than 0.8. Signal density has its biggest effect of free-flow traffic at 1–3 signals per mile. When traffic demand approaches or exceeds roadway capacity there is a drop in speed at all signal densities.

Using two traffic signals per mile as a base, Table 7.3 provides estimates of the percentage increase in travel time as the signal frequency per mile increases [8].

These relationships suggest that (1) the number of phases should be kept to a minimum, and (2) spacing of signals should permit progression flow to the maximum extent possible.

7.3 Nonrecurring Bottlenecks

7.3.1 Introduction

Nonrecurring congestion results when the roadway capacity is reduced by (1) incidents that remove one or more travel lanes from service, or cause drivers (on both sides of the road) to slow down as they to observe the roadside activities related to the incident; and by (2) headways that are increased by inclement weather, work zones, or driver behavior. Another cause of nonrecurring congestion is a surge in demand in excess of what the roadway can handle (e.g., the exit of spectators at the end of a ball game).

In all cases there is an imbalance between roadway supply and travel demand. In addition to the duration of the above events (e.g., road blockage or demand surge) there is also delay during the recovery time until the normal traffic operation resumes.

7.3.1.1 Traffic Incidents

Traffic incidents reduce roadway capacity and contribute to congestion. The amount of delay depends upon the type/duration of the incident, the number of lanes blocked by the incident, the response times to reach and clear the incident, and the time needed for the roadway (freeway) to resume normal operation.

Traffic incidents reduce roadway capacity [1]. Estimates of the amount of freeway capacity available, as a function of number of lanes blocked by the incident, are provided in Table 7.4, which shows that even when an incident is located at the shoulder of the road it reduces its capacity.

Table 7.4 Freeway capacity available from incident conditions

Number of freeway lanes in each direction	Shoulder disablement	Shoulder accident	Lanes blocked		
			One	Two	Three
2	0.95	0.81	0.35	0	N/A
3	0.99	0.83	0.49	0.17	0
4	0.99	0.85	0.58	0.25	0.13
5	0.99	0.87	0.65	0.40	0.20
6	0.99	0.89	0.71	0.50	0.25
7	0.99	0.91	0.75	0.57	0.36
8	0.99	0.93	0.78	0.63	0.41

Source Reference [1], pp 1–9, Table 1–2

Table 7.5 Estimated speeds from inclement weather

Precipitation condition	Observed speeds (MPH)	Ratio to no precipitation
No precipitation	64	1
Drizzle	51	0.8
Light rain	50	0.8
Light snow	45	0.7
Rain	48	0.8
Sleet	37	0.6
Snow	37	0.6
Thunder showers	53	0.8
Thunder storm	47	0.7
Strong thunder storm	28	0.5

Source Reference [10]. With permission from ASCE.

7.3.1.2 Surge in Demand

Surges in traffic demand include sports events, seasonal shopping, cultural and recreational events, etc. Vehicle traffic demand in excess of roadway capacity creates queues resulting in lower traffic speeds. Delay lasts longer than the duration of the demand surge.

7.3.1.3 Inclement Weather

Bad Weather: rain and snow reduce visibility and causes drivers to reduce speed. The presence of snow and ice on the road can also reduce speeds. Advances in sensor technologies and continued deployment of intelligent transportation system (ITS) architectures provide the means to anticipate, mitigate, and intervene through various traveler advisory and control measures to better manage traffic flow in periods of inclement weather [9]. Light rain could reduce freeway speeds by 20 %; while severe thunder storms could create a speed reduction of about 50 % (Table 7.5).

The effect of pavement conditions from weather events on traffic speed is summarized in Table 7.6.

7.3.1.4 Work Zones/Street Closures

Road Repair: Construction activities on roadways result in physical changes to the roadway including: narrower lanes, lane shifts, reduction in the number of travel lanes. In addition, slower speed limits are also established in construction zones. These changes increase travel time.

Table 7.6 Effect of weather-generated pavement conditions on traffic speed

Condition	Percent speed reduction (%)
Dry	0
Wet	0
Wet and snowing	13
Wet and slushy	22
Slushy in wheel Paths	30
Snowy and sticking	35
Snowing and packed	42

Source Reference [11]

Street Closures: These events result from emergencies or of planned events (marathon, street fairs, visits by heads of state, etc.). Because they reduce roadway capacity, travel speed drops.
Utility Cuts: In many cities utilities are located below the roadway surface, and their repair often involves closing at least one lane to traffic that reduces capacity and travel speed.

7.3.1.5 Driver Behavior

Erratic and improper driver behavior can contribute to a reduction in traffic speed resulting in congestion.

Examples follow:

a. Use of the passing lane by one slow driver reduces the speed of all drivers.
b. Drivers tend to slow down while passing an incident location in the opposite direction (rubbernecking).
c. Loading or Unloading in moving lanes: the use of moving lanes by commercial vehicles for loading and unloading reduces capacity and forces vehicles to slow down as they change lanes. The same goes for bus drivers that don't pull into the bus stop, and for taxi drivers who pick up or discharge passengers from the moving lane.

7.4 Conclusion

As previously shown in Table 7.2, the proportion of total delay attributed to nonrecurring bottlenecks in urban areas far exceeds that from recurring bottlenecks. As much as 2/3 of the total delay in large and small metropolitan areas is attributable to nonrecurring bottlenecks.

Since nonrecurring congestion is often experienced at the same time and locations where recurring congestion occurs, the severity of nonrecurring delay is highly conditioned by the physical and operational bottlenecks inherent in the roadway system. Therefore effective congestion relief measures should be combined to address both nonrecurring and recurring congestion.

References

1. Freeway Management Handbook (1997) Federal Highway Administration, Washington, DC, Report No. FHWA-SA-97-064
2. Lockwood S (2006) The 21st century operation oriented state DOTs, NCHRP Project 20–24. Transportation Research Board American Association of State Highway and Transportation Officials, Washington, DC
3. Koepke FJ, Levinson HS (1992) NCHRP Report 348: access management guidelines for activity centers. Transportation Research Board, Washington DC
4. (2003) Access Management Manual. Transportation Research Board, Washington, DC
5. Koepke FJ, Levinson HS (1992) NCHRP Report 348: Access Management Guidelines for Activity Centers. Transportation Research Board, Washington, DC
6. Highway Capacity Manual (2011) Vol 3—Interrupted flow. Transportation Research Board, Washington, DC
7. (1997) NCHRP (National Cooperative Highway Research Program). Report 398, quantifying congestion, final report, vol 1. Transportation Research Board, Washington, DC
8. Gluck J, Levinson HS, Stover V (1998) NCHRP Report 492, impact of access management techniques. Transportation Research Board, Washington, DC
9. Incorporating Weather Impacts—ITS Report (2009) FHWA-JPO-09-065, EDL #14497
10. Thakuriah P, Tilahun N (2013) Incorporating weather information into real-time speed estimates: comparison of alternative models. s.l. J Trans Eng 139(4):379–389
11. Mahmassani HS, Dong J, Kim J, Chen RB, Park P (2009) Incorporating weather impacts in traffic estimation and prediction systems. s.l. http://ntl.bts.gov/lib/31000/31400/31419/14497.htm. FHWA-JPO-09-065, EDL# 14497

Chapter 8
Measuring Traffic Congestion

8.1 Introduction

Congestion in transportation occurs when the occupancy of spaces (roadways, sidewalks, transit lines and terminals) by vehicles or people reaches unacceptable levels of discomfort and delay. For pedestrians, occupancy is expressed as the number of pedestrians per unit area (or square feet per pedestrian). For vehicles, it is expressed as the number of vehicles per unit length of roadway. As space occupancy increases, the speed of movement decreases.

The traffic definition of congestion has evolved over the years:

- Mc Clintock in his 1925 book "Street Traffic Control," defines congestion in street traffic "as a condition resulting from a retardation of movement below that necessary for contemporary streets users" [1].
- Alan Altshuler [2] indicates that "the term congestion denotes any condition in which demand for a facility exceeds free-flow capacity at maximum design speed".
- Homburger et al. [3] defined congestion as "the level at which transportation system performance is no longer acceptable due to traffic interference. This may vary by type of transportation facility, geographic location, and time of day".
- The Institute of Transportation Engineers "Toolbox" (1996) states that "congestion means there are more people trying to use a given transportation facility during a specific period of time than the facility can handle with what are considered acceptable levels of delay or inconvenience" [4].
- The National Cooperative Highway Research Program Report No. 398 [5, 6] provides a number of definitions of congestion and its correlates of mobility, accessibility, and reliability. These definitions provide a means to measure the effects and consequences of traffic congestion:
  - Congestion—the travel time or delay in excess of that incurred under light or free-flow travel conditions.

J.C. Falcocchio and H.S. Levinson, *Road Traffic Congestion: A Concise Guide*,
Springer Tracts on Transportation and Traffic 7,
DOI 10.1007/978-3-319-15165-6_8

- Recurring Congestion—Occurs every weekday (or weekend day) at the same general location and time.
- Non-Recurring Congestion—A random event (a road incident or inclement weather) that restricts traffic flow.
- Unacceptable Congestion—is the travel time or delay that exceeds established or agreed upon norm. This norm can vary by location in a geographic area, by type of transportation facility, by travel mode, and time of day.
- Mobility—Use of travel time contours (isochrones) to denote the distance covered under congested conditions within a given travel time.
- Accessibility—the achievement of travel objectives within time limits that are regarded as acceptable.
- Travel Time Reliability—The ability to predict the arrival time at the beginning of a trip.

Traffic congestion reflects the difference between the travel time experienced during busy traffic periods and when the road is lightly traveled. It is also expressed as the ratio of actual travel time and uncongested travel time or the ratio of actual versus uncongested travel time rates (e.g., min/mile). The three basic components of traffic congestion include intensity (amount), extent (area or network coverage), and duration (how long it lasts).

8.1.1 Congestion Thresholds

Traffic congestion thresholds can be defined of one of two ways:

1. Using free-flow speed as a congestion threshold.
2. Establishing acceptable minimum speed for various types of facilities and operating environs.

Using free-flow speed as the congestion threshold might be appropriate in rural areas, or in the middle of the night, or on Sunday morning in large urbanized areas. But it might not be realistic to use it as a congestion threshold value to quantify peak periods traffic congestion in large urban areas.

Establishing how much congestion delay travelers are willing to tolerate has been a concern and a challenge to traffic engineers and transportation planners for many years. Key considerations include trip length, city size and facility type.

- Longer trips are impacted more by congestion than shorter trips;
- Congestion is usually greater and lasts longer in larger cities;
- In larger cities congestion is more tolerable than in smaller cities;
- Travelers expect to travel faster on freeways than on city streets.

It is vital, therefore, that standards of tolerable congestion delay are related to the size of the urban area and reflect community input via stakeholders' participation. When this is done the work products of transportation professionals will have a better chance of influencing the decisions of transportation policy makers.

8.2 The Dimensions of Congestion

Congestion can be characterized by four aspects of its occurrence: *intensity, duration, extent, and variability* [6–11].

Intensity reflects the amount of congestion expressed as a rate (e.g. minutes/mile).

Duration refers to the amount of time the road/system is congested.

Extent describes the miles of roads that are congested or the number of travelers affected by the congestion.

Variability measures the variation in the amount, duration, and extent of congestion over time.

These indicators and their metrics are summarized in Table 8.1.

The relationship between intensity, duration, and extent is shown in Fig. 8.1 where the variation in congestion intensity can be measured both in distance (extent) and time (duration). Distance is shown on the horizontal axis, and time is shown on the vertical axis. A series of contour lines, displaying various levels of intensity, are displayed on this distance-time grid. Separate mapping can be used to reflect variability in all three congestion criteria—one for average values, and another for the 95 percentile values.

The above congestion indicators are further described below by various metrics. Along with a description of these metrics, there is also a commentary of their strengths and weaknesses.

8.2.1 Intensity

Intensity measures the amount of congestion delay experienced at an intersection approach, along sections of a given route, several routes, or an entire urban area. Its metrics include: (1) congestion delay rate, (2) vehicle-hours of delay, (3) person-hours of delay, (4) travel time index, and (5) travel time tax.

8.2.1.1 Congestion Delay Rate

This is a measure of the amount of delay experienced on freeways and arterial streets. The equation is as follows:

Table 8.1 Overview of congestion indicators and their metrics

Congestion aspect	System type		
	Single roadway	Corridor	Area wide network
Duration (e.g., amount of time system is congested)	Hours facility operates below acceptable speed	Hours facility operates below acceptable speed	Set of travel time contour maps; "bandwidth" maps showing amount of congested time for system sections
Extent (e.g., number of people affected or geographic distribution)	% or amount of congested VMT or PMT; % or lane-miles of congested road	% of VMT or PMT in congestion; % or miles of congested road	% of trips in congestion: person-miles or person-hours of congestion; % or lane-miles of congested road
Intensity (e.g.. level or total amount of congestion)	Travel rate; delay rate; relative delay rate; minute-miles; lane-mile hours	Average speed or travel rate; delay per PMT; delay ratio	Accessibility; total delay in person-hours; delay per person; delay per PMT
Reliability (e.g., variation in the amount of congestion)	Average travel rate or speed ± standard deviation; delay ± standard deviation	Average travel rate or speed ± standard deviation; delay ± standard deviation	Travel time contour maps with variation lines; average travel/time ± standard deviation; delay ± standard deviation

Note VMT vehicle-miles of travel
PMT person-miles of travel
Source Reference [7], NCHRP 398, Vol. 1, Table S-5, p 7

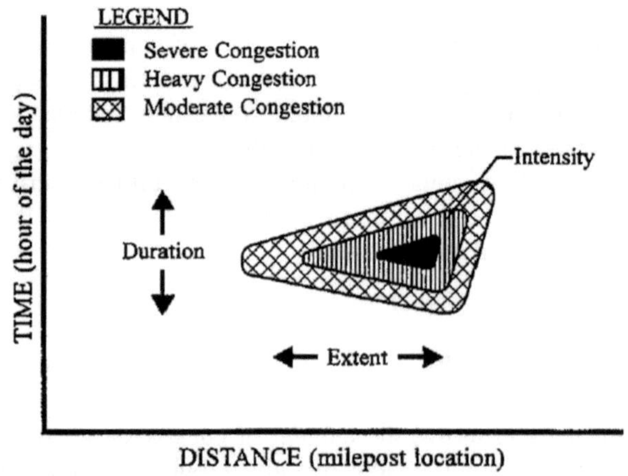

Fig. 8.1 Intensity of congestion in time and space. *Source* Reference [6], NCHRP 398, Vol. 1, Figure 14, p 70

$$\begin{aligned} \textit{Congestion Delay Rate} &= [\textit{Congested travel time rate}\ (\text{min/mi})] \\ &\quad - [\textit{Uncongested travel time rate}\ (\text{min/mi})] \\ &= [1/\textit{congested speed}\ (\text{mph})] - [1/\textit{free}-\textit{flow speed}\ (\text{mph})] \\ &\quad \times (60\ \text{min/h}) \end{aligned} \tag{8.1}$$

Comment:

For large metropolitan areas this definition usually exaggerates the amount of congestion delay calculated for the peak hours of traffic flow, because it assumes that it would be possible to travel at free flow-speeds during the peak hours.

The delay rate also can be used for city streets by comparing the actual off peak and peak travel times. A pioneer study of congestion in the Chicago central business district (conducted during the 1950s) illustrates this approach. The study compared auto and transit travel times in the north–south direction, across the one square mile loop, on a Sunday morning with those during the working day. The results are shown in Table 8.2.

Average auto travel times during weekdays were 50 % higher compared to Sunday morning travel times (6 vs. 4 min). For bus transit they were 44 % higher. The corresponding delay rates were 2.0 min/mile for auto, and 2.5 min/mile for bus transit.

8.2.1.2 Vehicle-Hours of Delay

The average peak hour, daily and annual vehicle hours of delay can be obtained by aggregating the delay incurred in various roadway sections in each direction. Annual delays can be obtained by aggregating the daily delays incurred.

The national reporting of congestion trends by the annual Urban Mobility Report [9] publishes the amounts of hours lost annually by commuters from various cities. This delay is calculated by the following equation:

$$\begin{aligned} &\boldsymbol{Daily\ Vehicle - Hours\ of\ Delay\ (DVHD)} \\ &\quad = [\mathit{daily\ vehicle - minutes\ at\ actual\ speed}] \\ &\quad\quad - [\mathit{daily\ vehicle - minutes\ at\ free - flow\ speed}] \end{aligned} \tag{8.2}$$

Table 8.2 North–South auto and transit speeds and travel times (one-mile distance) in Chicago's central business district, 1950

Auto		
	Speed (mph)	Travel time (min)
Possible (Sunday morning)	12.1	4
Actual working day	7.5	6
Difference	4.6	2
Delay rate	2 min/mile	
Transit		
	Speed (mph)	Travel time (min)
Possible (Sunday morning)	8	5.5
Actual working day	5.4	8
Difference	2.6	2.5
Delay rate	2.5 min/mile	

Source Reference [12], multiple pages

Comment:
In large urban areas, using a free-flow delay rate as a reference to measure delay overstates the time lost because of congestion.

8.2.1.3 Person-Hours of Delay

Person-hours of delay are computed by applying vehicle occupancy factors to the observed vehicle data.

$$\begin{aligned} \textit{Annual Person} - \textit{Hours of Delay} = {} & [(DVHD/weekday)x(250\, weekdays/year) \\ & \times\ x(1.25\, persons/vehicle)] \end{aligned}$$

$$\begin{aligned} \textit{Annual Hours of Delay per Weekday Traveler} = {} & [(\textit{Actual Weekday Travel Time, in minutes}) \\ & - (\textit{FF or PSL Travel Time, in minutes})] \\ & \times [1\ \mathrm{h}/60\ \mathrm{min}] \\ & \times [250\ \textit{weekdays per year}] \\ = {} & \textit{Delay Hours per person, per year} \end{aligned} \tag{8.3}$$

where:
FF = free flow speed
PSL = posted speed limit speed

Comment:
The delay data calculated by Eq. 8.3 are widely used by a variety of sources including the Secretary of the US Department of Transportation in his testimony to the Senate Committee on Housing, Banking, and Urban Affairs, [13]. The data are widely disseminated by the national press each time they are annually updated. A 2011 headline on the Wall Street Journal [14] states "Chicago and Washington, DC drivers idle for an average of 70 h a year in traffic jams, according to the latest Urban Mobility Report from the Texas Transportation Institute."

This headline obviously exaggerates the time lost in congestion—unless one believes that in large urban areas it is possible to travel at free-flow speed during the am and pm peak hours.

8.2.1.4 Travel Time Index (TTI)

The travel time index developed by the Texas Transportation Institute (TTI), compares the travel time rates in the peak period, to travel time rates during free-flow or posted speed limits [10]. The TTI is calculated as shown below:

$$TTI = (Actual\ Travel\ Time\ Rate)/(Travel\ Time\ Rate\ during\ free - flow\ conditions) \quad \text{or} \quad TTI = (Free - flow\ Traffic\ Speed)/(Actual\ Traffic\ Speed) \tag{8.4}$$

For example, a TTI of 1.30 indicates that a trip taken during the peak period will take 30 % longer than if the same trip were made when traffic flows freely. To illustrate: a TTI of 1.30 indicates that a trip that takes 40 min at 3 a.m., will take 52 min if made in the peak period. Typically, a free-flow freeway speed of 1 min/mile has been used as a base.

Comment:

The TTI is a computationally correct and easily understood metric. But it is sensitive to how the base "free flow" speed is applied. As mentioned earlier, in a large city it is not realistic to travel at free flow speed (or at the posted speed limit) in the peak hour. It is not logical, therefore to compare actual peak hour travel times to free-flow peak hour travel times when free-flow in the peak hour is a practical impossibility in a large city.

While the TTI may be an appropriate metric in tracking congestion over time for the same area, it should not be used to compare areas served by road networks with different free-flow speeds. The example below shows what happens when one evaluates the impact of congestion on city and suburban streets using the TTI (see Table 8.3).

Table 8.3 compares free-flow and congested speeds for city and suburban streets. The TTI of 1.33 is the same for city streets and suburban roads—suggesting that congestion impacts city streets and suburban streets equally.

But it does not. The delay rate added to city streets is higher than that added to suburban streets:

- on city streets the delay rate increases by 0.65 min/mile (2.65–2.0), but
- on suburban streets the delay rate increases by 0.5 min/mile (2.0–1.5).

Referring to Table 8.3, city street congestion would reduce the distance that can be traveled in 30 min by 3.7 mile (from 15 to 11.3 mile); while the distance traveled for same 30 min trip on suburban streets is reduced by 7.5 mile (from 22.5 to 15 mile). This example shows that the same proportional change in the travel time

Table 8.3 Measuring traffic congestion on city and suburban streets using the travel time index (TTI)

	City streets	Suburban streets
Free-flow speed	30 mph	40 mph
	(2 min/mile)	(1.5 min/mile)
Congested speed	22.6 mph	30 mph
	(2.65 min/mile)	(2.0 min/mile)
TTI	1.33	1.33

rate (e.g., TTI = 1.33) in city and suburban streets produces a larger impact on the mobility of suburban travelers than on that of city travelers.

Therefore, the TTI should not be used to compare the effects of congestion between areas where the respective road networks have significantly different free-flow speeds.

8.2.1.5 Travel Time Tax—TTT

In its 2010 Annual Report, INRIX introduced the Travel Time Tax, or TTT [11].

The TTT is defined as:

$$TTT = TTI - 1.0 \tag{8.5}$$

The TTT is a surcharge to free-flow travel time. For example, a TTT of 1.**30** represents a tax of **30** % (1.**3**–1.0) added to the free-flow travel time. The TTT is based on the same methodology as the TTI, but it communicates the results in a different way.

8.2.1.6 Comments on the Uses of the Travel Time Index (TTI) and the Travel Time Tax (TTT)

The TTI and TTT can best be used for tracking congestion over time for the same facility type within the same area. They should not be used to compare areas with roadway networks that have different free-flow speeds, or that serve different average trip lengths.

The travel time tax (TTT) represents the actual time loss incurred and it is expressed in minutes per mile. The total time loss (delay) along a roadway can be obtained by weighing the time loss for each section of road by its length and volume. Table 8.4 provides an illustrative example.

The TTT provides a useful metric for measuring and aggregating delay over a given roadway. It is based on real-time measurement of speeds and travel times, and is easy to understand and use. However, it has the same limitations as the Travel

Table 8.4 Illustrative application of the travel time tax (TTT) compared to the TTI

Variables	Arterial street	Freeway
RS (reference free-flow speed)	40 mph (1.5 min/mile)	60 mph (1.00 min/mile)
HS (actual highway speed)	30.76 mph (1.95 min/mile)	46.15 mph (1.30 min/mile)
TTI = RS/HS	1.30	1.30
TTT = (TTI − 1.0)	0.30	0.30
Additional delay rate @ TTI = 1.30 or TTT = 30 %	(1.95 − 1.5) = + 0.45 min/mile	(1.30 − 1.00) = + 0.30 min/mile

Time Index (TTI) in that it is sensitive to the basic assumption of what constitutes congested speed.

8.2.1.7 Illustrative Application of the TTI (I-93 Southbound: Medford to Boston)

Recurrent congestion was common on southbound I-93 before Boston's Central Artery Tunnel and the new cable stay bridge over the Charles River was completed. Figure 8.2 shows the difference between peak (6 a.m.–10 a.m.) and off peak travel times traveling from the Roosevelt Circle off-ramp to Storrow Drive off ramp, for three time periods: Fall 1994, 1998; and Spring 1999. The maximum travel time intensity was experienced at 8:00 a.m., and ranged from 18 min in 1994, to 23 min in the Fall of 1998. The travel time without delays was about 7 min; and the maximum time loss during the am peak period ranged from about 10 min in the Fall 1994, to 17 min in Fall 1998. The travel time index (TTI) ranged from 2.7 to 3.8.

8.2.2 Duration

The duration of congestion depends upon the types of congestion (recurring or non-recurring). It also depends upon city size and type of roadway. Congestion generally is of long duration on major roadways (e.g., freeways) in large urban areas—especially where roadways converge. Duration is less frequent in small urban areas.

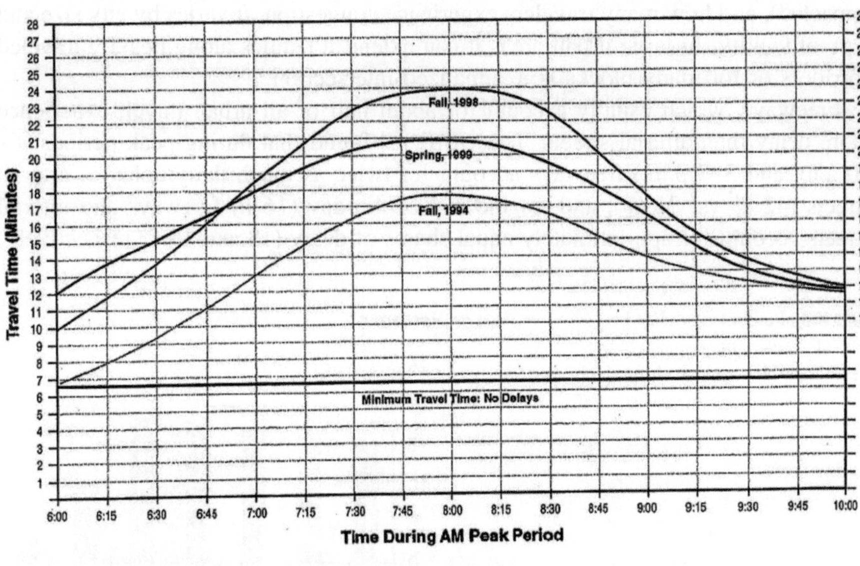

Fig. 8.2 AM peak period travel times for I-93 Southbound: Roosevelt circle off-ramp to storrow drive off-ramp. *Source* [5]

The number of hours each day with congested travel has increased over the years. In contrast to the typical “peak hour” of the 1960s, today’s congestion in large metropolitan areas extends up to 10 h/day, with peak periods lasting 4 h in the morning (6–10 a.m.), and 4 h in the evening (3–7 p.m.). Congested travel corridors, however, tend to experience congestion for longer periods (INRIX 2010), especially on freeways. Thus the delay duration mirrors the day-to-day traffic patterns. Midblock and intersection delays in major business districts frequently last throughout the business day. The delay periods at typical urban signalized intersections usually range from 15 min to more than 1 h, depending on approach volumes, intersection geometry, and traffic signal timing.

8.2.2.1 Daily and Hourly Variations

The average daily duration of delay, for typical days of the week (Fig. 8.3), shows that Friday experiences the largest amount of delay, and Monday the least; while Sunday has the lowest delays. And when delay is distributed by time of day (Fig. 8.4), the worst hour is 5–6 p.m. in the evening, with about 14 % of the daily delay, while the hours with the least amounts of delay are from midnight to 6 a.m.

8.2.3 Extent

Extent measures how far congestion is spread (miles of roadways, or route miles impacted), and how many travelers experience congestion. It varies by city size and type of facility. In large urban areas it can extend for miles along heavily traveled corridors or for many blocks at a signalized intersection.

Freeways, which usually account for about half of all urban travel, experience more delay than arterial streets. The UMR [9] found that during peak periods (6–10 a.m. and 3–7 p.m.) freeways account for twice as much daily delay as arterial streets (42 % vs. 21 %); and for the rest of the day (16 h) freeways and arterial streets account for approximately equal shares of delay (18 and 19 % respectively)

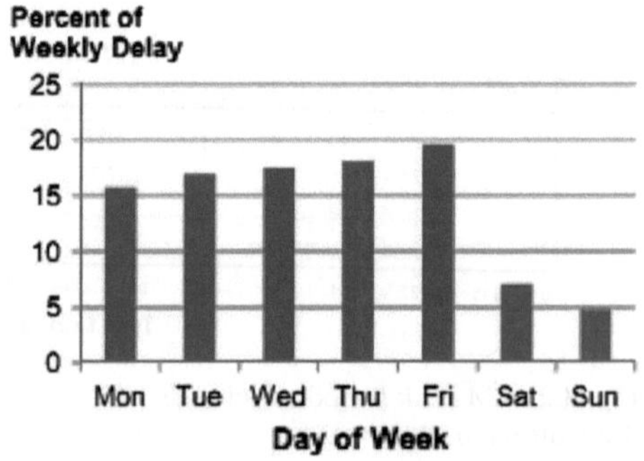

Fig. 8.3 Percent of delay by day of week. *Source* Reference [15], p 7. Exhibit 4

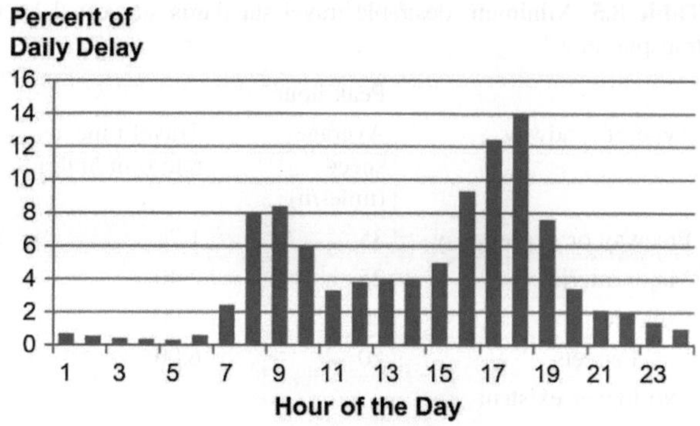

Fig. 8.4 Percent of delay by time of day. *Source* Reference [15], p 7. Exhibit 5

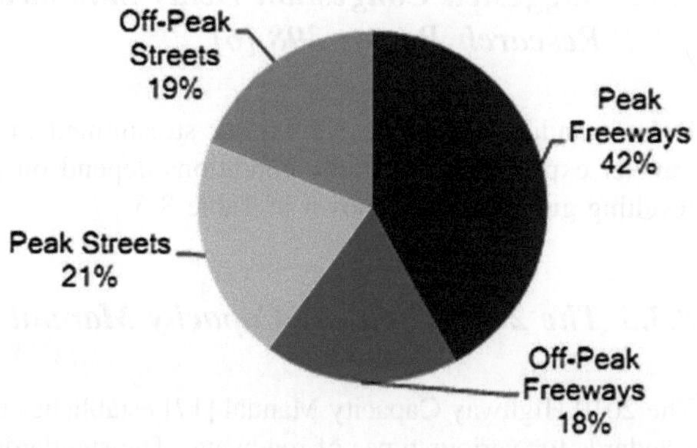

Fig. 8.5 Percent of delay by road type. *Source* Reference [15], p 7. Exhibit 6

(Fig. 8.5). It should be noted that the UMR [9] defines delay as the difference between an actual travel time rate and the travel time rate for free flow conditions.

8.3 Congestion Thresholds Reflecting Travelers Expectations

The thresholds for tolerable congestion levels should reflect traveler and freight movers' inputs, and can be set by policy committees of public agencies for each type of roadway and surrounding environment.

8.3.1 The National Committee on Urban Transportation

In 1958, the National Committee on Urban Transportation [16], first issued peak and off-peak hour standards for "minimum desirable delays" for expressways/freeways, major arterials, collector and local streets. These standards are shown in Table 8.5.

Table 8.5 Minimum desirable travel standards suggested by the national committee on urban transportation[a]

	Peak hour		Off-peak	
Type of roadway	Average speed (miles/h)	Travel time rate (min/mile)	Average speed (miles/h)	Travel time rate (min/mile)
Freeway or expressway	35	1.71	35–50	1.20–1.71
Major arterial	25	2.40	25–35	1.71–2.40
Collector	20	3.00	20–25	2.40–3.00
Local streets	10	6.00	10–20	3.00–6.00

[a] No longer existent

8.3.2 Suggested Congestion Delay Standards from NCHRP Research Report 398 [6]

Delay standards in Table 8.5 further streamlined in this report recognizing that traveler expectations of traffic conditions depend on the size of urban areas. The resulting guidelines are shown in Table 8.6.

8.3.3 The 2010 Highway Capacity Manual Criteria

The 2010 Highway Capacity Manual [17] establishes the "Level of Service" (LOS) standards for various types of roadways. The standards range from "A" (free-flow) to "F" (congested flow). Factors included in establishing levels of service are: (1) speed and travel time, (2) traffic interruptions (delays), (3) freedom to maneuver, (4) safety, and (5) driving comfort and convenience.

How density affects freeway operating speeds is illustrated in Fig. 8.6. As shown, when density exceeds 45 passenger cars per lane per mile, speeds drop rapidly, and the flow rate drops as well. At their critical density (corresponding to the value producing maximum throughput volume), travel times are about 1.15–1.20 times greater than the free-flow travel times. However, at its critical density, traffic flow

Table 8.6 Suggested congestion delay standards for various sizes of urbanized areas

Functional classification	Small urban communities	Mid sized urban areas	Large urban areas
Expressways-freeways	1.0–1.2 min/mile (50–60 mph)	1.3–1.5 min/mile (40–45 mph)	1.7–2.0 min/mile (30–35 mph)
Class I Arterials[a]	1.7–2.0 min/mile (30–35 mph)	2.4–3.0 min/mile (20–25 mph)	3.0–4.0 min/mile 15–20 mph)
Class II/III Arterials[b]	2.4–3.0 min/mile (20–25 mph)	3.0–4.0 min/mile (15–20 mph)	4.0–6.0 min/mile (10–15 mph)

[a] High type arterials with free-flow speeds of 30–35 mph
[b] Other arterials with free-flow speeds of 20–25 mph

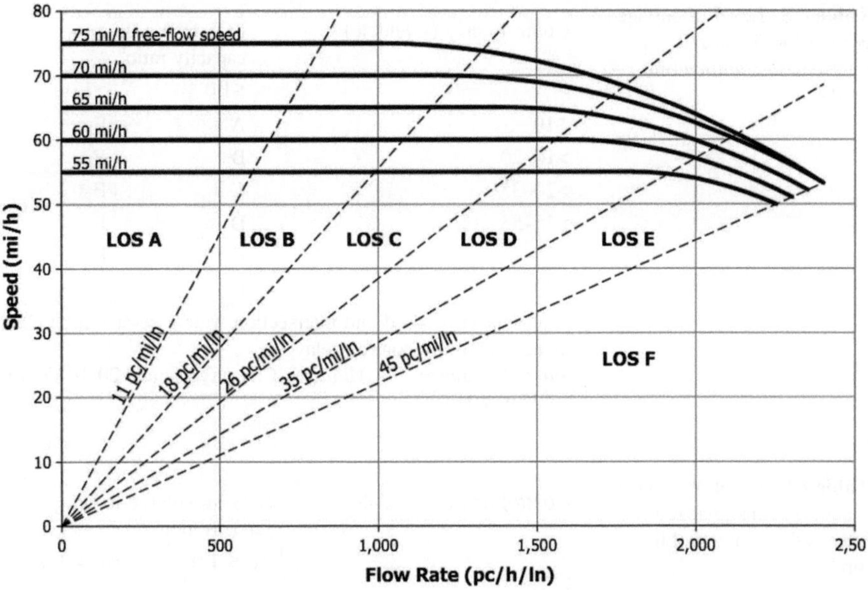

Fig. 8.6 Speed versus flow rate for freeways. *Source* Reference [17], 2010 HCM

becomes unstable and the speed can drop suddenly to reach stop-and-go conditions. The values of Fig. 8.6 represent "ideal" conditions in terms of roadway and interchange design—conditions that seldom exist in many urban areas.

The resulting level of service standards (A–F) for freeways (shown in Fig. 8.1), corresponding to traffic density, are summarized in Table 8.7.

For signalized and unsignalized intersections, the standards are established by the delay time experienced by an average vehicle before crossing the intersection (Tables 8.8 and 8.9).

For intersections that are not signalized, the delay time standards are given in Table 8.9.

Level of service metrics corresponding to speed thresholds for automobiles on city streets are given in Table 8.10. These criteria relate level of service to the observed travel speeds as a percentage of the base free-flow speeds. The thresholds are generally consistent with of the previously established congestion criteria.

Table 8.7 Level of service criteria for freeways

Traffic density (passenger cars/lane/mile)	Level of service
11 or less	A
>11–18	B
>18–26	C
>26–35	D
>35–45	E
>45	F

Table 8.8 Level of service criteria for signalized intersections: automobile mode

Control delay (s/vehicle)	LOS by Volume-to-capacity ratio[a]	
	≤1.0	>1.0
≤10	A	F
>10–20	B	F
>20–35	C	F
>35–55	D	F
>55–80	E	F
>80	F	F

[a] For approach-based and intersection wide assessments, LOS is defined solely by control delay

Source Reference [17], Highway Capacity Manual 2010, Volume 3, Chapter 18

Table 8.9 Level of service criteria for unsignalized intersections: automobile mode

Control delay (s/vehicle)	LOS bv volume-to-capacitv ratio	
	$v/c \leq 1.0$	$v/c > 1.0$
0–10	A	F
>10–15	B	F
>15–25	C	F
>25–35	D	F
>35–50	E	F
>50	F	F

The LOS criteria apply to each lane on a given approach and to each approach on the minor street. LOS is not calculated for major-street approaches or for the intersection as a whole

Source Reference [17], Highway Capacity Manual 2010, Volume 3, Chapter 19

Table 8.10 Level of service criteria corresponding to travel speed on city streets

Travel speed as a percentage of base free-flow speed (%)	LOS By critical volume-to-capacity ratio	
	≤1.0	>1.0
>85	A	F
>67–85	B	F
>50–67	C	F
>40–50	D	F
>30–40	E	F
≤30	F	F

The critical volume-to-capacity ratio is based on consideration of the through movement volume-to-capacity ratio at each boundary intersection in the subject direction of travel. The critical volume-to-capacity ratio is the largest ratio of those considered

Source Reference [17], Highway Capacity Manual 2010, Exhibit 16.4

8.3.4 Congestion Thresholds Established by Transportation Agencies—Three Examples

8.3.4.1 New York Metropolitan Transportation Council (NYMTC)

The New York Metropolitan Transportation Council [8], identifies congested links in the highway network primarily using the Demand Volume-to-Capacity criterion (D/C ratio), calculated by their traffic assignment model.

Based on this criterion, and using the average of a 4-hour weekday morning peak period, NYMTC established three congestion thresholds as shown in Table 8.11.

For demand to capacity (D/C) ratio of 0.8 or less, there is co congestion but where the D/C ratio exceeds 1.0, severe congestion occurs.

8.3.4.2 Washington State Department of Transportation [18]

Congestion thresholds are established as 75 % of posted speed limits. For example:

- For urban Freeways with a speed limit of 60 mph, the congestion threshold speed = 45 mph.
- For arterial streets with a posted speed limit of 40 mph, the congestion threshold speed = 30 mph.

8.3.4.3 Quebec Ministry of Transportation [19]

Congestion thresholds are established as 60 % of the posted speed limit.

8.3.5 Applications

The preceding examples show that urban travelers expect and accept a certain amount of congestion during periods of the day. Congestion becomes unacceptable when it exceeds a threshold value.

While the above examples reflect the need to establish realistic congestion thresholds, local threshold values are not universal: they vary by type of area, time of day, and type of facility.

Table 8.11 Demand-to-capacity ratio thresholds for congestion determination [8]

Demand-to-capacity ratio	Congestion determination
D/C = 0.8	No congestion
D/C = 1.0	Congested
D/C > 1.0	Severely congested

1. Type of Area
   In small towns and rural areas the threshold speed of congestion is (and should be) free-flow speed. Travelers in large cities, however, are accustomed to expect an environment of greater traffic intensity with more vehicle-vehicle and pedestrian-vehicle conflicts. Therefore they are likely to tolerate a lower threshold speed of congestion (Table 8.6).
2. Time of Day
   In large cities lower threshold speeds are tolerated during peak periods than in off-peak periods (Table 8.5).
3. Type of Facility
   One expects faster speeds on a freeway than on a local street. Therefore the congestion threshold speed for a freeway is higher than for a local street (Tables 8.5 and 8.6).
4. Economic Considerations in the Choice of Threshold Congestion Speed
   Some economists and transportation professionals have questioned the use of free-flowing traffic speeds as a basis for calculating peak period traffic conditions in larger urban area. John Meyer and Jose' Gomez-Ibanez in their book "Autos, Transit, and Cities" [20] state: "A highway large enough to allow free flowing traffic during the rush hours in the center of large cities is seldom optimal because the building of such highway capacity is very expensive and the benefits in time savings and operating costs accruing only to a small group of rush hour users will be comparatively small. In short, highway engineers and planners understand that traffic congestion is almost advisable in a well-designed highway system." Countering this position is, perhaps, today's reality that some freeways are congested for many hours each weekday.

8.4 Conclusions

Congestion in cities is a by-product of their success in attracting people to jobs and other amenities, and the inability of cities to improve/expand transportation capacity to keep pace with this growth. The cities' challenge is to keep congestion manageable as their population and economies grow.

To be helpful in congestion management decisions, the definition of congestion should be based on a comparison of "actual travel times" with "expected travel times" for peak hour and off peak conditions. Expected travel times can vary from area to area, by time of day, and by type of routes, and should be established with the input of the area's stakeholders (e.g., travelers and freight movers).

Drivers begin to experience congestion when a time increment of 0.4 min/mile is added to their trip, and it becomes a significant problem when the increment reaches 0.8 min/mile. Congestion mainly occurs in the peak weekday commuter hours—when about 60 % of all congestion is concentrated—but it is also found on summer and holiday weekends when many people travel to or from beaches and other recreational areas.

While using free-flow freeway speed as a threshold for congestion, has the advantage of simplicity in data analysis and in tracking area wide trends, its utility in evaluating traffic congested conditions in large metropolitan areas is questionable because it is nearly impossible to build the capacity necessary to serve peak hour traffic demands at free-flow speeds (e.g., lack of space for new roadways, social and environmental constraints).

A more realistic standard of delay metric should reflect goals that are actionable in terms of social, environmental and financial constraints. These are best reflected in Tables 8.2, 8.3 and 8.4, and through approaches described for the New York Metropolitan area (NYMTC); Washington State DOT; and for the greater Montreal region where congestion thresholds were established to reflect local conditions/ expectations.

Traffic congestion has many impacts. These impacts include (1) longer and less reliable trip times (Chap. 9), (2) decreased mobility (Chap. 10), decreased accessibility (Chap. 11), lower roadway productivity (Chap. 12), and increased costs and environmental effects (Chap. 13). These are discussed in the chapters that follow.

References

1. McClintock M (1925) Street traffic control. McGraw-Hill Book Company Inc, New York
2. Altshuler A, Womack JP, Pucher JR (1979) The urban transportation system—politics and policy innovations. The MIT Press, Cambridge, MA
3. Homburger WS, Hall JW, Loutzenhouser HC, Reilly W (1996) Fundamentals in traffic engineering. Institute of Transportation Studies, University of California Berkeley, California
4. Meyer MD (1993) A toolbox for alleviating traffic congestion and enhancing mobility. Institute of Transportation Engineers, Washington, DC
5. Lisco T, Draft Memorandum to Palewonsky L (1999) RE: AM peak period traffic volumes travel times and speeds on interstate 93. Between Medford and Boston, s.n., 1992–1994 and 1998–1999
6. Lomax T, Turner S, Shunk G, Levonson HS, Pratt RH, Douglas B (1997) NCHRP 398: quantifying congestion, vol 2. Transportation Research Board, National Research Council, Washington, DC. User's guide
7. Lomax T, Turner S, Shunk G, Levinson HS, Pratt RH, Bay P, Douglas B (1997) NCHRP report 398:quantifying congestion, vol 1. Transportation Research Board, National Research Council, Washington, DC. Final report
8. New York Metropolitan Transportation Council (2009) Congestion management process—2009 status report. New York, s.n.
9. Lomax T, Schrank D, Turner S, Geng L, Li Y, Koncz N (2010) 2010 urban mobility report. Texas Transportation Institute, University Transportation Center for Mobility, s.l.
10. Cambridge Systematics, Dowling Associates, System Metrics Group (2008) NCHRP report 618: cost effective performance measures for travel time delay, variation, and reliability. Transportation Research Board, Washington, DC
11. INRIX (2011) National traffic scorecard—2010 Annual Report. pp 163–174
12. City of Chicago study and recommendations for improving traffic movement in the central business district (1950) The Mayor's Traffic Survey Committee for the Central business District, s.l.

13. La Hood R (2009) Secretary of transportation. Senate Committee on Banking, Housing and Urban Affairs, Washington, DC
14. White JB (2011) American idle: on the road. Wall Street J : s.n.
15. Schrank D, Lomax T, Eisele B (2011) TTI's 2011 urban mobility report—powered by INRIX traffic data. University Transportation Center for Mobility, Texas Transportation Institute, s.l.
16. National Committee on Urban Transportation (1958) Better transportation for your city—a guide to the factual development of urban transportation plans administration services. National Committee Urban Transportation, Chicago
17. HCM (2010) Highway capacity manual. Transportation Research Board, Washington DC
18. Washington State Department of Transportation. The 2011 Congestion Report, 23 Aug 2011
19. Transport Canada (2006)
20. Gómez-Ibáñez JA, Meyer JR(1981) Autos, transit, and cities: a twentieth century fund report. Harvard University Press, Cambridge

Chapter 9
The Impacts of Congestion on Trip Time

9.1 Introduction

Congestion metrics that focus only on *network* performance (e.g., see Chap. 8) are necessary but not sufficient in addressing the impact of congestion on *travelers*. Reliance just on network congestion metrics provides only a partial understanding of the congestion problem. This is because network performance metrics are not connected with trip length. A delay rate of 1 min per mile has a different impact on a 5 mile trip (common in compact cities) than on a 10 mile trip (common in suburban areas). Therefore, one should not ignore the effect of trip length in analyzing the impact of congestion on travelers.

9.2 Travelers with Different Trip Lengths

How trip time is affected by congestion depends on the length of the trip. The length of the trip, in turn, depends on the size of the urban area, its population density, and its development patterns.

The effect of area size on trip length is shown in Fig. 9.1 [1]. This relationship is from data of the 60s and 70s, which does not account for growth in car ownership and population in the expanding suburbs of metropolitan areas in the last 50 years, thus the trend shown would underestimate average trip lengths for current conditions (see Chap. 6).

The effect of population density on trip length is shown in Fig. 9.2.

When population densities are less than 5,000 people per square mile, the average trip length exceeds 16 miles. As densities increase there is a corresponding decrease in trip lengths—a decline to about 10 miles per average trip at densities of 25,000 persons per square mile, and to less than 5 miles when densities exceed 40,000 people per square mile.

J.C. Falcocchio and H.S. Levinson, *Road Traffic Congestion: A Concise Guide*,
Springer Tracts on Transportation and Traffic 7,
DOI 10.1007/978-3-319-15165-6_9

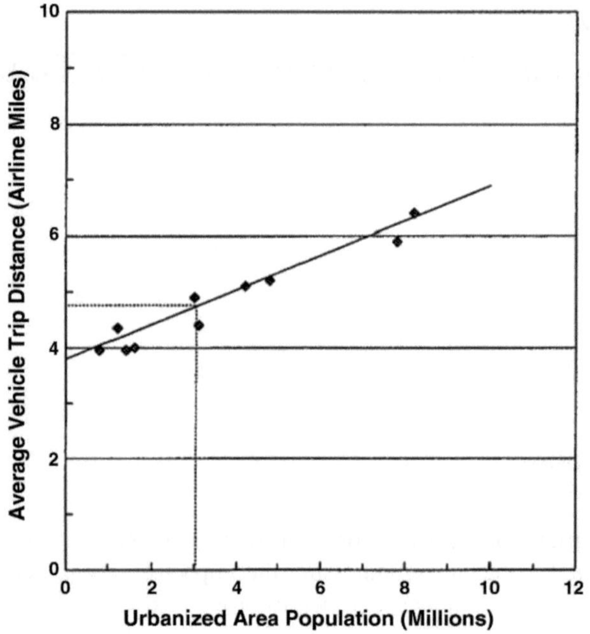

Fig. 9.1 Average trip length as a function of urbanized area population. *Source* Reference [1], p 128. Figure 40

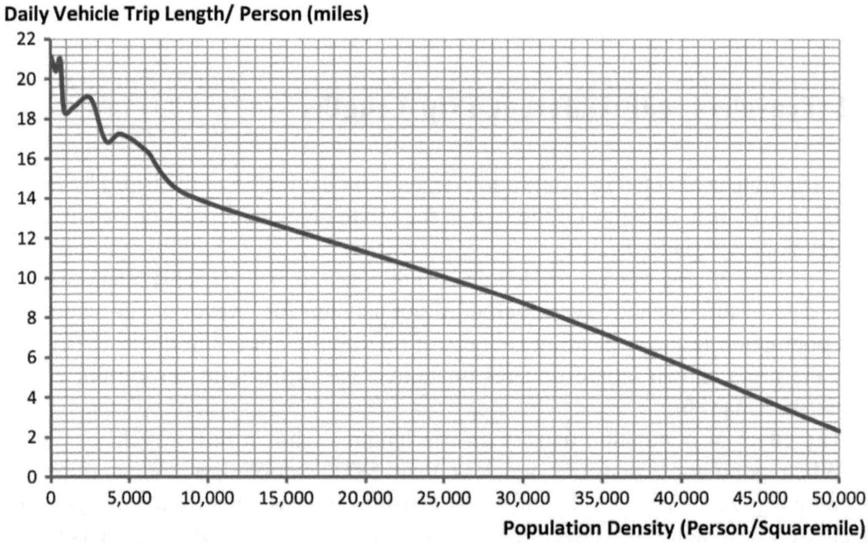

Fig. 9.2 Average trip length as a function of population density. *Source* Reference [2], Table 15.5, p 15–21

Therefore, land use patterns and area size play a key role in analyzing the traffic congestion problem. Travelers who live in low-density suburban areas, where trip lengths are longer, will be penalized more by congestion than those travelers who live in more compact urban areas where trip lengths are shorter. A delay rate of 1 min per mile would add 15 min to a 15 mile trip, but only 5 min to a 5 mile trip. Therefore trip length should be included when analyzing the effect of congestion on travelers.

Because average trip length varies with population density (Fig. 9.2), land use density should be considered as a potential mitigating factor in reducing the effects on congestion travel. But to consider changing land use patterns as a mitigating factor makes it necessary to change the perspective of congestion—from one focused primarily on traffic speed to one that also includes trip time.

9.3 Travel Time Reliability

Travel time reliability is often viewed as more important than average travel time. The ability to predict travel time is highly valued by travelers and the business community. Travel time variability affects trip starting time, the choice of routes, and travel modes. This condition requires travelers and freight carriers to add a time buffer in planning and scheduling their trips. A route that takes a longer average trip time but experiences a smaller variability in travel time may be preferred to one with a shorter average travel time but with a larger travel time variability.

Travel time reliability is a key performance indicator of traffic congestion. It can be defined as the degree of certainty that a trip will be completed within a specific time. While other delay metrics quantify congestion using average values (for a specific time period) reliability metrics focus on the ranges and distributions of travel times that are not likely to be exceeded—usually the 80 and 95 percentiles.

9.3.1 Sources of Travel Time Variability

Sources of travel time variability include the following:

- Travel Demand Volume—travel time varies with traffic demand on the roadway: more traffic increases travel time and less traffic reduces it
- Traffic incidents such as crashes and debris on roadways block travel lanes, and increase travel times
- Work zones reduce vehicle speeds
- Environmental conditions such as inclement weather that reduce vehicle speeds and increase vehicle spacing

- Special events increase traffic demand beyond capacity—especially when they occur during peak hour traffic flow
- Driver Behavior such as the presence of slow vehicles in the fast lane that block ability to change lanes, or failure to maintain speed in an upgrade, reduces speeds

9.4 Reliability Metrics

The recommended metrics shown in Table 9.1 are useful in evaluating both traffic congestion and travel time reliability.

9.4.1 The Buffer Time Index

The Buffer Time Index (*BTI*) is a measure of trip time reliability that expresses the amount of extra time to be added to the average trip time in the peak hour if one aims to arrive on time, 95 % of the time (e.g., being late for work 1 day out of every 20 work days). The BTI can be calculated using the Eq. 9.1 ([4], p 16):

$$\begin{aligned} \text{BTI}(\%) = {} & [(95th\ Percentile\ Travel\ Time \\ & - Avereage\ Travel\ Time) \\ & /(Avereage\ Travel\ Time)] \times 100\,\% \end{aligned} \tag{9.1}$$

Tabel 9.1 Recommended reliability metrics

Reliability performance metric	Definition	Units
Buffer index	Difference between 95th percentile TTI and average travel time, normalized by average travel time Difference between 95th percentile TTI and median travel time (MTT), normalized by MTT	%
Failure and on-time measures	Percentage of trips with travel times <1.1 MTT and <1.25 MTT Percentage of trips with space mean speed less than 50, 45, and 30 mph	%
Planning time index	95th percentile TTI	None
80th percentile TTI	Self-explanatory	None
Skew statistic	(90th percentile TTI – median)/(median – 10th percentile TTI)	None
Misery index (modified)	Average of highest 5 % of travel times divided by freeflow travel time	None

Source Reference [3], p 6, Table ES.4

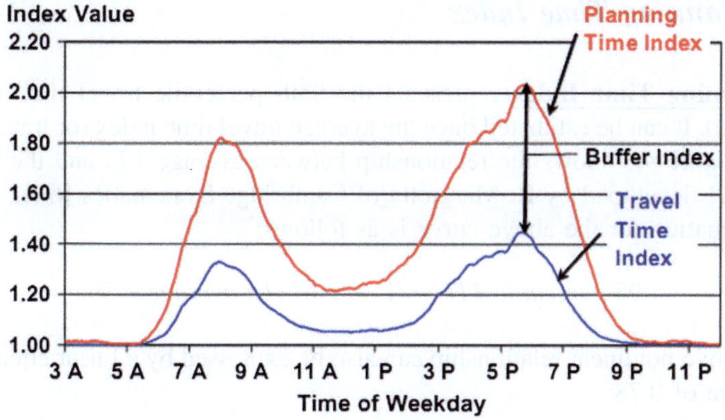

Fig. 9.3 The extra "buffer" time needed when planning important trips. *Source* Reference [5], p B-53. Exhibit B-33

For example, if the average travel time in the peak hour takes 30 min, a BTI of 1.20 means that a traveler should allow for an extra 20 % travel time, or 6 min, in order to arrive at the destination on time for 95 % of the trips.

The BTI tends to be larger during peak periods than in off-peak periods: Fig. 9.3 ([5], TTI Mobility Report, 2011, Exhibit B-33),

An example of peak hour travel time distribution is provided in Fig. 9.4 for a section of I-75, northbound, in Atlanta.

This example shows that peak hour travel times range from 6–32 min, with an average of 12.2 min. If a commuter wanted to arrive to work on time 95 % of the time, he or she would have to allow 18.54 min to the trip (e.g., the 95th percentile).

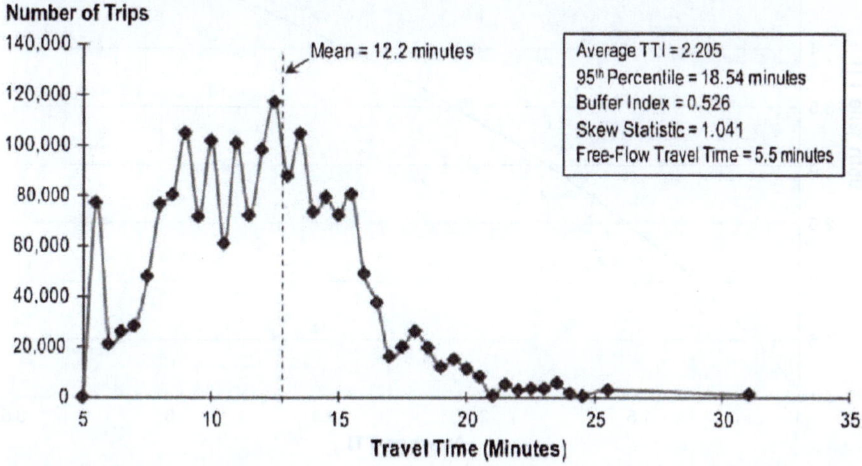

Fig. 9.4 Distribution of peak hour travel time, atlanta, I-75 NB, I-285 to SR 120 (2007). *Source* Reference [3], p 55. Figure 4.4

9.4.2 Planning Time Index

The Planning Time Index represents the 95th percentile travel time index (or travel time). It can be estimated once the average travel time index (or travel time) is known. Figure 9.5 shows the relationship between average TTI and the 95th percentile TTI developed by R. Margiotta of Cambridge Systematics [6].

The equation for the above curve is as follows:

$$95\ percentile\ TTI = 1 + 3.67\,(Ln\ average\ TTI) \tag{9.2}$$

The above nonlinear relationship can also be expressed by a linear equation with an r square of 0.78:

$$Planning\ Time\ Index\ (PTI) = 1.7 \times (Average\ TTI) - 0.39 \tag{9.3}$$

Example

If the average TTI is 1.2, the PTI = 1.7 (1.2) – (0.39) = 1.65

This means that if the free-flow travel time is 20 min, it would take an average of 24 min to travel in the peak hour. But if it is important not to arrive late (95 % of the time) at the destination, then one should schedule the peak hour trip to last 33 min [(20) x (1.65)].

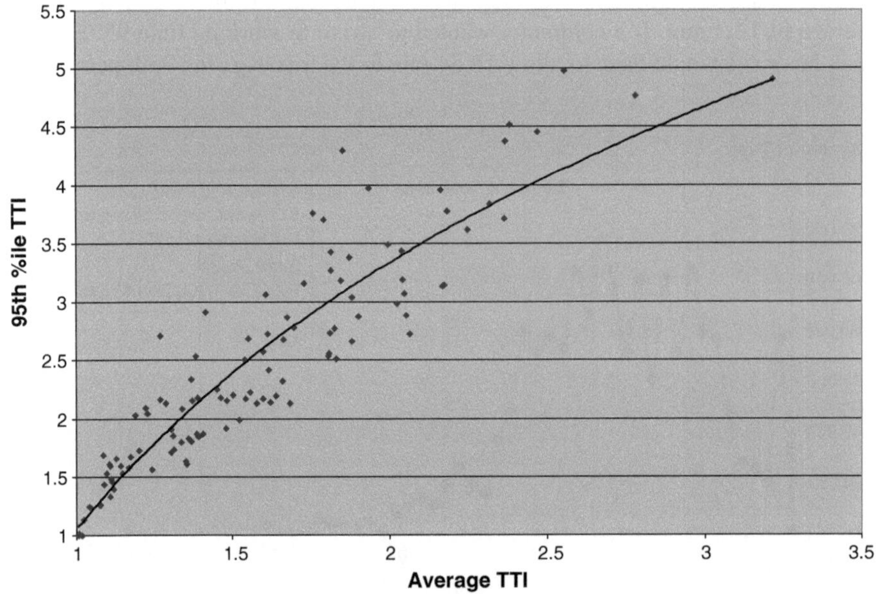

Fig. 9.5 Planning time index (95th percentile TTI) as a function of average TTI. *Source* Reference [6], slide 23

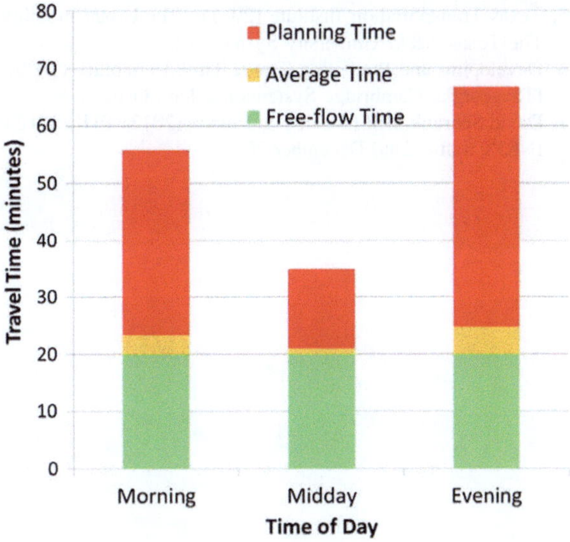

Fig. 9.6 Extra time to make important trips. *Source* Reference [7], David Schrank, Bill Eisele, Tim Lomax, TTI's 2012 urban mobility report—powered by Inrix traffic data, December 2012

9.4.3 Implications

Travelers and freight movers typically adjust their trip times to the prevailing (expected) congestion at the time of their trip, and they tend to adjust their daily schedules to the prevailing traffic congestion as long as they can predict their arrival times to a destination. The ability to know how the average travel time for a trip is likely to vary over time is essential for planning a reliable arrival time. This can be done by adding an appropriate time buffer in trip planning.

Figure 9.6 shows an example of how the components of total trip time can vary by time of day from an average trip [7]. In this case motorists making important trips during peak travel periods have to plan for a travel time about three times what it would take to make the trip in light traffic conditions.

References

1. Martin WA, McGuckin NA (1998) NCHRP report 365: travel estimation techniques for urban planning. Transportation Research Board, Washington, DC
2. Kuzmyak JR et al (2003) Transit cooperative research program report 95, chapter 15, land use and site design, traveler response to transportation system changes. Transportation Research Board, Washington, DC
3. Cambridge Systematics, Inc. with Texas A&M Transportation Institute, University of Washington, Dowling Associates, Street Smarts, Herb Levinson, Hesham Rakha (2013) SHRP 2 Report S2-L03-RR-1, analytical procedures for determining the impacts of reliability mitigation strategies. Washington DC
4. Cambridge Systematics, Inc., Dowling Associates, Inc., System Metrics Group, Inc., Texas Transportation Institute (2008) NCHRP Report 618: Cost-effective performance measures for travel time delay,variation, and reliability. Transportation Research Board, Washington, DC

5. Texas Transportation Institute (2011) 2011 Urban mobility report—powered by INRIX data. The Texas A&M University System, S.l.
6. Developing and Predicting Travel Time Reliability (2009) Margiotta, richard. presentation to ITS georgia. Cambridge Systematics, Inc., October 5
7. David Schrank, Bill Eisele, Tim Lomax (2012) TTI's 2012 Urban mobility report—powered by INRIX traffic data, December 2012

Chapter 10
The Impact of Traffic Congestion on Mobility

10.1 Defining Mobility

Mobility is the ability of people and goods to travel easily, safely, quickly and reliably. Trip mobility varies with the speed of travel, and it may be defined as the number of trips taken and their distance (trip-miles) within the traveler's daily travel time and cost budgets. Therefore, lower speeds resulting from traffic congestion reduce mobility.

This chapter has two basic objectives: (1) to define and quantify traveler mobility, and (2) to determine how mobility is impacted by traffic congestion.

10.2 Factors Influencing Mobility

Mobility depends on three key factors: (1) the type of transportation modes available to the traveler (e.g., walk, bicycle, bus, rail, private vehicle); (2) traveler's requirements and needs that influence mode selection (e.g., travel time budget, physical ability in using a given mode, affordability, trips purpose, and the license to drive); and (3) the operational characteristics of the travel modes (e.g., speed, reliability, cost, safety, comfort/convenience, and the absence of physical barriers to using the mode). Figure 10.1 shows how these factors interact in determining a traveler's mobility.

10.2.1 Traveler Requirements/Needs

Traveler requirements/needs include:

Travel Time Budget: the time one allocates daily to travel. Daily travel time budgets have remained relatively stable over time and across metropolitan areas

J.C. Falcocchio and H.S. Levinson, *Road Traffic Congestion: A Concise Guide*, Springer Tracts on Transportation and Traffic 7,
DOI 10.1007/978-3-319-15165-6_10

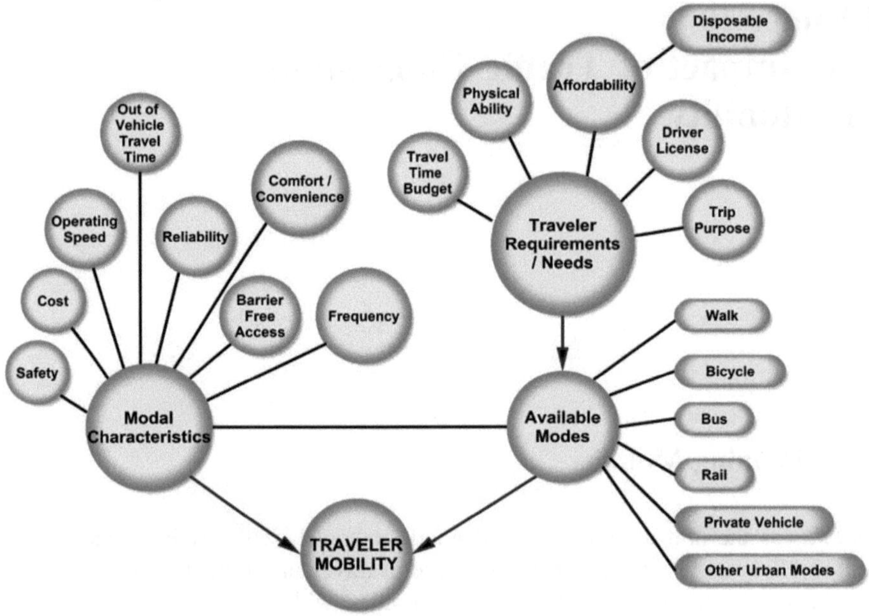

Fig. 10.1 Factors impacting traveler mobility

[1–6]. Time is a finite resource, and it is allocated to various activities in finite amounts in accordance with the individual's physiological, social, educational, and economic needs.

- Zahavi [1] in his pioneering study of travel by car, found that average daily auto travel time is stable in all urban areas, with a slight tendency to increase with the size of the urban area
- Reno et al. [2], found that the average time spent by persons who drove private vehicles on their travel day was within a very narrow range (69–74 min) and it was independent of the size of urban area (see Fig. 10.1)
- In the New York Metropolitan area [3], the average time spent in travel was reported as 77 min/day
- Hanson and Giuliano [4] report the results of a study by Hupkes [6] who found that the time spent for travel by all modes of transportation is nearly constant over time—from decade to decade. It was reported that "although shifts occur over time in the modes used (walking, biking, busing, driving), a remarkable stability in the overall number of trips and in total hours devoted to travel was documented"
- Within these aggregated values, however, not all travelers have the same travel time budgets. Garrison and Levinson [5] reported that for the Washington, DC metropolitan area, adults (age 18–65) have higher travel time budgets, and employed adults spend more time traveling than those who are not employed. And women spend less time traveling than men

- There is also variability in the time allocated for different travel activities (trip purposes). For example, more time is spent for work trips than non-work travel. In the New York Metro area work trip took an average of 33 min and non-work trips 20 min [3].

Trip Purpose: some trips (e.g., work) tend to use a larger share of the travel time budgets than other trips (e.g., convenience shopping).

Physical Ability: to enter, ride, and exit the vehicle/mode.

Affordability: the choice of mode is often based on its trip cost. Affordability is related to the disposable income of the traveler.

Driver's License: only those of legal age are allowed to obtain a driver's license. Those who cannot drive can only use private vehicles as passengers.

10.2.2 Availability of Travel Modes

Many modes of travel are usually available to people living in metropolitan areas. They include walking, bicycles, private vehicles, taxis, schedule-based public transit (including: buses, street cars, light rail, bus and rail rapid transit, commuter rail and ferry). The modes chosen are among those that will satisfy travelers' requirements/ need. However, in specific neighborhoods all modes might not be available. Transit coverage is limited in many suburban areas. Rapid transit lines are sometimes too far from where people live, and some households do not own automobiles.

10.2.3 Modal Characteristics

Key modal characteristics that influence the choice of private and public transport include frequency, operating speed, reliability, out-of-vehicle travel time, door-to-door travel time, trip cost, perceived safety, barrier-free access, and comfort/ convenience.

Frequency: how often the service is available during a particular time of day. It is an indicator of waiting time and travel mode convenience.

Operating Speed: faster modes provide greater trip mobility than slower ones.

Reliability: modes/routes with stable speeds are preferred to those which experience variability in speed.

Out-of-Vehicle Travel Time: this is the time spent walking to/from vehicles and waiting for vehicles, while on the trip. Travelers value *excess travel time* from 2 to 2.5 times the value of line-haul travel time. Thus an excess trip time of 10 min is equivalent to a line-haul travel time of 20–25 min.

Trip Cost:

- out-of-pocket trip costs: fare, toll charges, parking charges, fuel, etc.
- vehicle ownership costs: purchase/lease, maintenance, insurance, etc.,

Safety: Travelers prefer modes that they perceive to be safe, and avoid those that they perceive to be unsafe.

The perceived safety of using the mode reflects:

- probability of personal injury,
- probability of fatality,
- risk to personal security.

Barrier-Free Access: vehicle entry/exit that allows access to the physically disabled.

Comfort/Convenience—Includes the following variables:

- Walking distance
- Number of transfers
- Frequency of service
- Waiting time for vehicle ("it's not just how long you wait; it's how you spend the time waiting")
- Physical comfort: cleanliness, temperature and humidity, cleanliness, ride quality, space per passenger, weather protection
- Psychological comfort: sense of being in control, availability of real-time information
- Availability of vehicle parking
- Availability/dependability of travel mode
- Ability to carry packages, tools, etc. when needed.

10.2.3.1 Door-to-Door Travel Time and Speed

The door-to-door trip speed of each travel mode is a key measure of modal mobility, and it depends on the door-to-door trip distance and trip time. It is expressed as follows:

$$\begin{aligned} Door\text{–to–Door Travel Speed} &= (Door\text{–to–Door Trip Distance}) \\ &\quad /(Door\text{–to–Door Travel Time}) \end{aligned} \tag{10.1}$$

Door-to-door trip time is typically the sum of (1) vehicle riding/driving time, and (2) "out-of-vehicle" or excess time. Excess time represents the time spent outside the vehicle. It includes the walk time from the trip origin to the vehicle location, waiting time for the vehicle, transfer time from one vehicle to another for trips involving multiple vehicles, and walk time from one leaves the vehicle to the trip destination.

This distinction in travel time components is important because travelers value the out-of-vehicle time 2–2.5 times the time spent riding the vehicle. The exception is across the platform transfer at rapid transit stations.

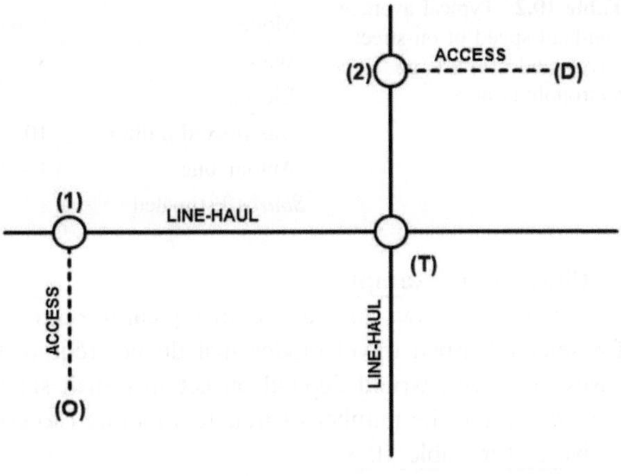

Fig. 10.2 Door-to-door trip components

Figure 10.2 illustrates a trip distance from origin [O] to destination [D] using two line-haul transit lines, with access at [1], requiring a transfer at [T], and an egress at [2] for access to a destination at [D].

The travel time components for this trip are shown in Fig. 10.2 and summarized in Table 10.1.

In-vehicle Travel Time

This time is calculated by dividing the distance traveled in a vehicle by the average vehicle speed:

The average private vehicle speed is a function land use density, road design speed and traffic volume. For public transportation vehicles, the average speed is also affected by service patterns, bus stop/station frequency, dwell times and rights-of-way.

Typical average modal speeds in a large urban area are shown in Table 10.2. Operating speeds of rapid transit and commuter-rail lines range between about 20–30 miles/h.

Table 10.1 Time components of the door-to-door trip in Fig. 10.2

Trip components	Out-of-vehicle travel time	In-vehicle travel time
Access to or from line-haul vehicle: • [O]–[1] • [2]–[D]	• Walk time • Wait time	Time in auto, bus, or bicycle to [1] or from [2], if distance to/from line-haul is beyond walk distance
Line-haul vehicle • [1]–[T] • [T]–[2]	For multivehicle trips: • Transfer walk time • Transfer waiting time	Time riding the main line vehicle (transit or auto)
Total	Out-of-vehicle time	In-vehicle time

Source Fig. 10.2

Table 10.2 Typical average line-haul speed of on-street travel modes in a large metropolitan area

Mode	Urban (mph)	Suburban (mph)
Walk	3	3
Bicycle	8	12
Bus (mixed traffic)	10–12	12–15
Automobile	15–20	25–30

Source Estimated

Illustrative Example

Table 10.3 shows an example of typical average speeds and excess travel times for selected urban travel modes that do not require transfers. Values for specific cases, however, would depend on actual transit schedules or service areas, route structures, and the number of transfers required to complete the trip.

Notes for Table 10.3:

(1) Actual average mode speed is typically determined by the volume of traffic in the road network, the roadway's design speed and capacity, and the operational characteristics of the mode (e.g., local or express bus; on local or arterial road, freeway, etc.).
(2) Access to and from the bus mode assumes an average walk distance of 0.25 miles for each trip segment.
(3) Waiting time is assumed ½ the vehicle headway.
(4) In addition to the above components of excess travel time, travelers also consider the travel time reliability of the individual travel modes. When using a mode that is subject to random but significant delays, excess travel time would be larger than what is shown in the table.

The travel time of an urban trip made by bicycle or by a private motor vehicle typically does not include significant "excess time components." But the excess time components of urban trips by transit can be typically a significant share of door-to-door travel time.

Table 10.3 Typical average speed and excess travel time for line-haul modes

Line-haul mode		Average speed of line-haul mode (mph)	Excess travel time (min)			
			Vehicle access	Waiting time	Destination access	Total
Walk		3	–	–	–	0
Bike • Urban • Suburban		8 12	–	–	–	0 0
Bus local/surface	Urban	12	5	3	5	13
	Suburban	15	5	8	5	18
Auto	Urban	20	1	2	3	6
	Suburban	30	1	1	1	3

Source Estimated

10.3 Measuring Mobility

Traveler mobility is defined as the distance traveled with a chosen mode within an acceptable travel time. The time acceptable or budgeted for a trip is related to trip purpose (e.g., the time acceptable for a work trip is likely to be greater than that for a shopping trip), and the size of the metro area: for example, in a small urban area an acceptable trip to work is 20 min long, but in a metro area of over 5 million people an acceptable trip time to work is 45 min.

The distance traveled is calculated as follows:

$$\begin{aligned} \textit{Traveler Mobility} &= \textit{Distance Traveled} \\ &= \{[\textit{speed of line by haul mode}] \\ &\quad \times [(\textit{time budgeted for the trip} - (\textit{excess travel time})]\} \\ &\quad + \textit{Distance from the origin to line} - \textit{haul vehicle} \\ &\quad + \textit{Distance from line} - \textit{haul vehicle to trip destination} \end{aligned} \tag{10.2}$$

Using Eq. 10.2, traveler mobility provided by each of the four travel modes shown in Table 10.2, is plotted in Fig. 10.3 for trip times up to 60 min.

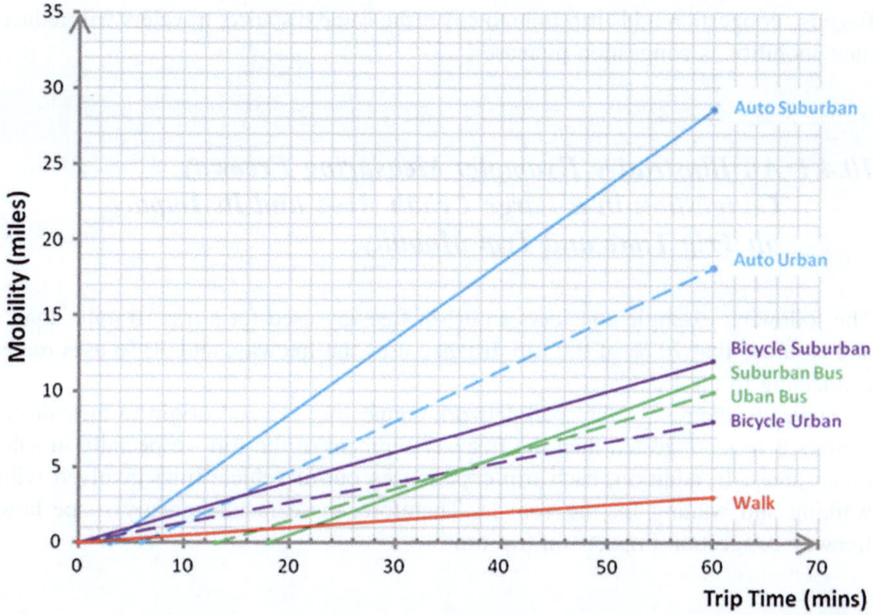

Fig. 10.3 Trip mobility of selected modes and trip time

Thus if one is willing to allocate 20 min to a shopping trip, the trip mobility of to a shopping destination would be: 1.0 mile if by walking, 1.5 miles if by bus, 2.5 miles if by bicycle, and 4.8 miles if by auto.

This example shows that the mode's operating speed is not always an indicator of mobility. From the viewpoint of travelers what is important is distance traveled within an acceptable door-to-door travel time. For example, the mobility advantage of the bicycle over the bus for trips less than 30 min points to the need for implementing bicycle networks in urban areas.

The above comparisons do not include rail rapid transit lines operating on grade-separated or private guide—ways, whose line haul speeds are usually at least double those of surface transit. Long established US rapid transit lines in New York City, Chicago, Philadelphia, and Boston have typical line-haul speeds of 20–25 miles/h; newer regional lines have speeds in excess of 30 miles/h. Commuter rail have line haul average speeds that can reach 40 miles/h. Rail modes were excluded in Table 10.3, however, to simplify the discussion.

Actual modal speeds will vary both within and among urban areas. Communities interested in measuring modal mobility could develop curves similar to Fig. 10.3 to reflect local conditions (e.g., area size, travel barriers, availability of freeways, or public transportation operating on exclusive right-of-way).

10.4 Congestion Impacts on Mobility

Because congestion adds trip time to cover the same distance, travelers will reduce their mobility as congestion increases.

10.4.1 An Illustrative Example: Measuring Freeway Congestion in a Large Urban Area and Its Impact on Trip Time and Trip Mobility

The following example considers an urban freeway used for a trip 10 miles long., and assumes that 70 % of the trip distance uses the freeway, and 30 % uses roads connecting to the freeway.

Typically the rest of the road network connecting drivers' origin to their destinations is less congested than the freeway segment of the trip—especially in suburban areas. In addition, each trip experiences out-of-vehicle time involved with walking and waiting for the vehicle. Therefore, these factors greatly shape how freeway congestion impacts on trip time.

Table 10.4 Trip components for a trip distance of 10 mile

Trip segment	Distance (miles)	Speed (mph)	Time (min)	% Trip time	% Trip distance
Other roads	2.9	**20**	8.7	32.6[a]	29
				27.6[b]	
Uncongested freeway	7.0	35	12.0	44.9[a]	70
Congested freeway	7.0	25	16.8	53.3[b]	70
Out of vehicle travel time (walking, waiting)	0.1	3.0	6.0	22.5[a]	1
				19.1[b]	
Total—(uncongested freeway)	10.0	22.5	26.7	100[a]	100
Total—(congested freeway)	10.0	19.0	31.5	100[b]	100

[a] No congestion on freeway
[b] Freeway congested

Other assumptions are:

- Acceptable peak period average freeway speed = 35 mph (see Table 8.6, Chap. 8)
- Congested freeway speed = 25 mph.
- Average speed of the roads connecting to the freeway = 20 mph.
- Out-of-vehicle time for the trip = 6 min (includes walking to and from the vehicle and waiting for the vehicle).

The questions are:

I. What is the impact of peak period freeway congestion on trip time?
II. What is the impact of peak period freeway congestion on mobility?

Table 10.4 lays out the assumed values of the trip components under congested and uncongested trip conditions for the 10 mile trip.

10.4.1.1 Findings

Key findings are as follows:

Impact of Freeway Congestion on Trip Time

In this example, an increase in freeway congestion of 40 %, (4.8/12) increases trip time by only 18 % (4.8/26.7). This means that one cannot estimate the trip time lost to congestion by only measuring the freeway delay rate because the time spent on the congested freeway segment comprises only a fraction of the total trip time [7].

Considering that over 65 % of urban trips is shorter than 5 miles [8], and that trips shorter than 5 miles are unlikely to be freeway users [9], measuring the impact of traffic congestion on trip time using data only from freeways and other principal arterials, is likely to overestimate average trip congestion delay.

Therefore, to accurately evaluate the impact of traffic congestion on trip time, all segments of the trip (as shown in Table 10.4) must be considered in the analysis.

Impact of Freeway Congestion on Trip Mobility

As shown in Fig. 10.4, a traveler's response to increased freeway congestion would be either to reduce trip mobility by over 15 % (from 10 to 8.46 miles)—for a constant trip time—or to increase trip time by 18 % (from 26.7 to 31.5 min) in order to maintain the same level of trip mobility (10 miles).

Data on traveler responses to an increase in travel speed shows that travelers generally have used the travel time reductions to increase the trip distance instead of reducing their trip times [10]. So it may be inferred that the converse is true: when congestion increases travel time, trips distance (mobility) would decrease.

Comments

The example uses 35 mph freeway speed to represent acceptable peak hour speed conditions—a realistic assumption for many freeways in large metropolitan areas. This value (35 mph) is considerably less than the freeway free-flow speed (e.g.,

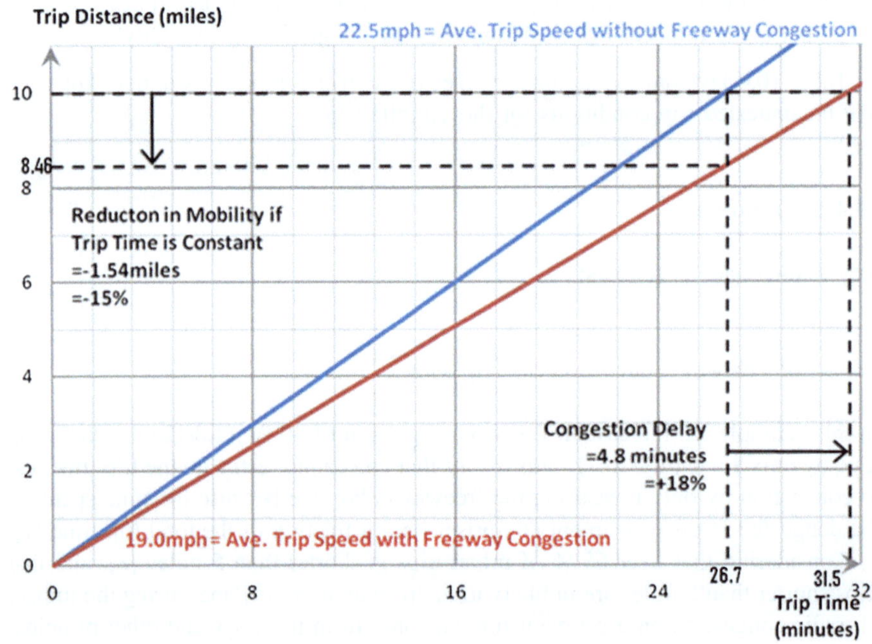

Fig. 10.4 Congestion impacts on mobility and trip time for a trip 10 miles long

60 mph) used as the basis in measuring the Travel Time Index. In practice each urban area should set its own congestion threshold criterion (Chap. 8).

10.4.2 Not All Travelers are Impacted by Traffic Congestion

Some travelers see their mobility increase even as roadway traffic congestion increases. They include those who move from the city to the suburbs and switch from transit and walking to driving, and those who get their driving license as they become of age.

10.4.3 Impacts of Traffic Congestion on the Mobility of Transit Riders, Pedestrians, and Bicycle Users

10.4.3.1 Transit Riders

Because the out-of-vehicle travel time for transit riders is a large component of door-to-door travel time (see Table 10.3), network traffic congestion which affects primarily in-vehicle-travel time, tends to have a smaller impact on the mobility of transit users than it has on that of automobile users.

10.4.3.2 Traffic Congestion Impact on the Mobility of Pedestrians and Bicycle Users

Pedestrian and bicycle trips tend to be shorter than vehicle trips. Therefore traffic congestion tends to have a smaller impact on pedestrian/bicycle trip times than it has on motorized vehicle trips.

10.5 Trends in Traffic Congestion and Traveler Mobility

The Texas Transportation Institute (TTI) annual metropolitan traffic congestion reports [11] indicate that traffic congestion has been steadily increasing since 1980—except for the years of economic slowdown (see Chap. 6). The news media and elected/appointed officials rely on these annual reports to inform the public about the increasing cost of traffic congestion [12].

But according to the National Household Travel Survey [13] and reported by Pisarski and Alan [14] increasing freeway/expressway traffic congestion does not seem to have reduced average traveler mobility (Table 10.5).

Table 10.5 Trip mobility and trip time trends—1980 to 2000

Commuter trips	1980	1900	% change (from 1980)	2000	% change (from 1990)
Trip length[a]	8.54 miles	10.65 miles	+24.0	12.41	+13.7
Trip time[b]	21.7 min	22.4 min	+3.2	25.5 min	+13.8
Average speed	23.6 mph	28.5 mph	+20.7	28.5 mph	0
All trips					
Trip length	8.68 miles	9.47 miles	+9.1	10.03 miles	+5.9
Trip time	21.7 min	23.4 min	+7.8	25.5 min	+9.0
Average speed	24.0 mph	24.3 mph	+1.3	23.6 min	−2.9

[a] *Source* [13 p 51]
[b] *Source* [13 p 101]

10.5.1 All Trips

- 1980–1990: Trip mobility increased more (+9.1 %) than trip time (+7.8 %). Reflecting higher trip speed and an small increase travel time budge.
- 1990–2000: Trip mobility continued to increase (but at a slower rate—+5.9 %) however, trip time increased more (+9.0 %)—indicating that the travel time budget increased more than trip mobility.

10.5.2 Commuter Trips

For commuter trips the findings are even more dramatic:

- 1980–1990: Trip mobility increased eight times more than trip time (24 % vs. 3.2 %)—reflecting a relatively stable trip time budget, and big gains in mobility due to higher travel speed (+20.7 %).
- 1990–2000: Trip mobility and trip time kept increasing but at a slower pace (+13.75 % and +13.8 %, respectively).

Possible explanations for the discrepancy between network speed trends and trip speed trends are as follows:

(1) Network speed is not synonymous to door-to-door trip speed. Network speed reflects the performance of the system under observation. It does not measure the performance of all trip components that affect door-to-door trip speed.
(2) Freeways and other principal arterials are the most congested roads in large metropolitan areas. However, because they serve only a fraction of the metro area's traffic, their speed performance cannot be the sole indicator traffic congestion of the entire roadway network that includes collectors and local streets.
(3) Typically, traffic speed on city streets is lower than traffic speed on suburban roads. Greater population and job growth in suburban areas has increased the growth of vehicle trips in low density areas where traffic speeds are higher,

and has decreased the growth in the number of trips within the city where traffic speeds are lower. In addition, many of the trips that in the city were made by walking and transit (slower travel modes), in suburban areas they are substituted by the auto—a higher speed mode.

References

1. Y Zahavi (1974) Traveltime budgets and mobility in urban areas. s.l. Technical report, Federal Highway Administration
2. Reno A, Kuzmyak R, Douglas B (2002) Characteristics of urban travel demand. TRCP report 73
3. Schaller Consulting (2000) Travel in the New York-New Jersey metropolitan area. s.l. prepared for New York metropolitan transportation council and North Jersey transportation planning authority
4. Hanson S, Giuliano G (2004) The geography of urban transportation, 3rd edn. s.l. The Guilford Press, New York
5. Garrison WL, Levinson DM (2006) The transportation experience—policy, planning, and deployment. s.l. Oxford University Press, Oxford
6. Hupkes G (1982) The law of constant travel time and trip rates. s.l. Futures 14:38–46
7. Taylor B (2002) Rethinking traffic congestion. Access Mag1:21
8. New York Metropolitan Transportation Council (2000) Council contact—household interview survey
9. Levinson HS (1979) Characteristics of urban transportation demand—appendix. Department of Transportation, Mass Transportation Administration, Washington, DC
10. Reno AT (1988) Personal mobility in the United States: a look ahead—year 2020, Special Report 220. s.l. Transportation Research Board, National Research Council
11. Shrank D, Lomax T, Turner S (2011) 2010 Urban mobility report—powered by INRIX data. s. l. Texas Transportation Institute
12. La Hood R (2011) Statement of the honorable ray la hood, secretary of transportation. Senate Committee on Banking, Housing, and Urban Affairs, Washington, DC
13. NHTS (2004) Summary of travel trends-2001 national household travel survey. s.l. US Department of Transportation, Federal Highway Administration. National Technical Information Service
14. Alan PE (2006) Commuting in America III: the third national report on commuting patterns and trends. NCHRP report 550 and TCRP report 110

Chapter 11
The Impact of Traffic Congestion on Accessibility

11.1 Introduction: Defining Accessibility

Accessibility is a widely used term. In transportation, it can refer to a traveler's physical or economic ability in using a given travel mode; it can refer to describe a traveler's access to one or more destination opportunities available within a specific distance, travel time, or travel cost from the traveler's origin; or it can be used by the marketing department of a retail store chain to describe/quantify the number potential customers within a 20 min travel time to a store.

The number of destination opportunities accessible from a given location is determined by (1) a traveler's mobility (the door-to-door distance one can cover within a travel time and cost budgets), (2) the connectivity of the street network that determines the directness of travel between an origin and a desired destination, and (3) by the number of desired opportunities located within this distance.

Therefore, the same zone or area can have different accessibility measures: one for those who walk; another for those who ride transit, and yet another for car users.

As shown in Fig. 11.1, a traveler's *trip distance* is determined by her/his trip time budget, the modal door-to-door trip speed, and the degree of network connectivity; while the number of destination opportunities is determined by the land use density and mix of the desired activities located within this distance.

For example, a drug store located 10 min away by a slow mode (e.g., walking) is just as accessible as one located 10 min away by a faster mode (e.g., auto). The reason why the walking mode provides the same accessibility than a faster mode is because the two drug stores are located in areas of different land use density: the city drug store with a walk access of 10 min is located ½ mile away while the suburban drug store with a drive access of 10 min is located 5 miles away.

Therefore, focusing on mobility alone to improve accessibility ignores the role played by land use policies in the urban area. This chapter describes the impact of traffic congestion on accessibility via its impact on mobility and the patterns of activities in the urban area.

J.C. Falcocchio and H.S. Levinson, *Road Traffic Congestion: A Concise Guide*, Springer Tracts on Transportation and Traffic 7,
DOI 10.1007/978-3-319-15165-6_11

Fig. 11.1 Elements of accessibility

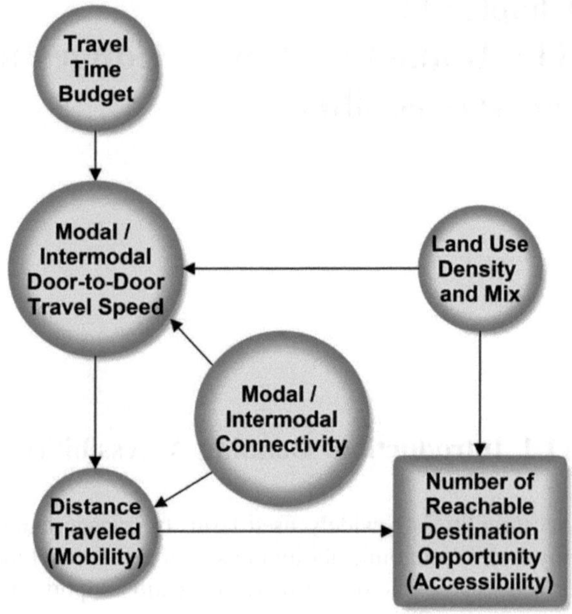

Hanson and Giuliano in their book *The Geography of Urban Transportation,* [1] note that "urban planners and scholars have long argued that accessibility should be a central part of any measure of the quality of life. In contrast, the goal of transportation planners (and traffic engineers) has been to increase people mobility, sometimes equating increased mobility with increased accessibility." This evokes the need to recognize that transportation and land use are interconnected elements of the urban system that impact on accessibility.

This interconnection is a function of the location of activities (close together or far apart) and the directness of travel to reach them. Road networks that are designed to serve travel by private motor vehicles are based on a hierarchical classification system of road types that guide a typical trip from local roads to collector roads and from collector roads to higher capacity/higher speed arterial roads—including freeways. This type of hierarchy favors auto mobility for long trips but it penalizes mobility by alternative modes for short trips because it reduces the number of direct path choices needed to encourage walking and biking. For example, as shown in Fig. 11.2 [2] traveling between points A and B involves a longer distance (3.6 miles) with a hierarchical network, but a much shorter distance (1.3 miles) with a strongly interconnected network.

The distance (1.3 miles) between A and B could easily be traveled by biking on local streets but the same trip made on a hierarchical network with numerous disconnected dead end streets connecting to high speed arterials in a circuitous manner, is considerably longer (3.6 miles). This condition is less conducive to

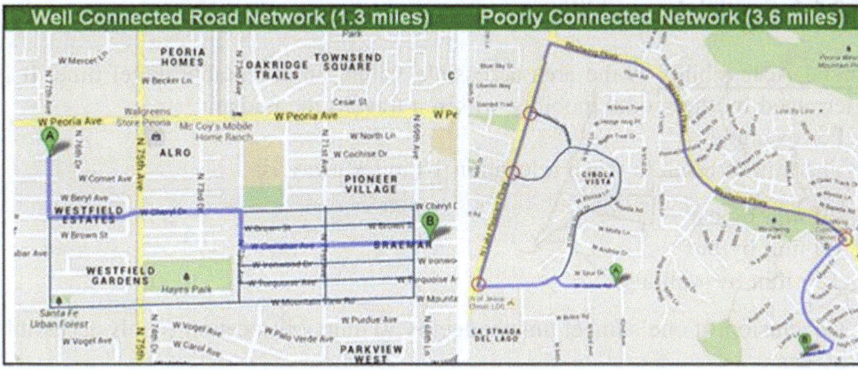

Fig. 11.2 Hierarchical versus well-connected road networks. *Source* Reference [2]

walking or biking than it is to driving. Thus the importance of providing network connectivity that allows walking and biking for short trips becomes a key factor in determining the accessibility potentially provided by these modes.

11.2 Measuring Accessibility

Accessibility of activity sites can be measured for a variety of conditions. For example: (1) from one's home location to a number and size of activity sites (employment locations, retail and service outlets, or recreational opportunities), and their size, that can be reached at a given time from the person's home; (2) from a store location to potential customers; or (3) from a particular location (e.g., the Central Business District) to residential zone.

The number of activity sites that can be reached is determined by how much time one is willing to spend on the trip, and the speed of the mode used. The mode chosen for the trip also depends on the travel options available and their level of service. In addition to speed, other factors considered in mode choice are trip cost, frequency, reliability, comfort, convenience, security, etc., and availability of space for vehicle parking (if needed).

The travel time needed to arrive at a chosen location is the product of the door-to-door speed of the travel mode and the travel distance.

11.2.1 Examples

Several examples illustrate how accessibility can be determined are provided below.

11.2.1.1 Spatial Accessibility

Spatial Accessibility is the area accessible within an acceptable travel time. It is determined by the modal mobility within a travel time budget:

For example:

The time it takes to reach a location 10 miles away is:

- 30 min by car
- 60 min by bus
- 180 min by walking.

Conclusion: If one's travel time budget is 30 min, this location is only accessible to by car.

Two examples illustrate the application of this accessibility concept:

Example 1

This example determines the Destination Opportunity Area accessible in a 30 min trip, by the given mode:

(a) For walking, biking, and motor vehicle, the area is approximated by a circle with a radius equal to the distance traveled in 30 min trip time. However, often the 30 min travel time contour is elliptical to reflect different speeds for each direction of travel.
(b) For a bus trip, the accessible area is approximated by: (the distance traveled by bus) × (0.5 miles service area—i.e., 0.25 miles based on each side of the bus line).
(c) Distance traveled in 30 min = [speed of travel mode] × [30 − (excess travel time)]/60.
(d) Typical modal speeds and modal excess travel time are found in Chap.10, Table 10.3.
(e) Excess travel times (from Table 11.3) are subtracted from total travel times to calculate distance traveled.

1. **Destination Opportunity Area Accessible by Walking** (assuming a fully interconnected street grid)

   (a) Distance traveled at 3 mph for a 30 min trip = 1.5 miles (in all directions)
   (b) Walking destination opportunity area = 7 square miles (1.5 × 1.5 × 3.14).

2. **Destination Opportunity Area Accessible by Bicycle** (assuming a fully interconnected street grid)

   Urban area

   (a) Distance traveled at 8 mph for a 30 min trip = 4.0 miles
   (b) Destination opportunity area reachable by bicycle = 50 square miles (4 × 4 × 3.14).

Suburban area

(a) Distance traveled at 12 mph for a 30 min trip = 6 miles
(b) Destination opportunity area reachable by bicycle = 113 square miles (6 × 6 × 3.14).

3. **Destination Opportunity area accessible by Bus** (assuming two intersecting bus routes)

Urban Area

(a) Distance traveled at 12 mph, for 17 min (30 − 13 min) = 3.4 miles
(b) Destination opportunity area by bus = [(3.4 × 0.5 miles) × 2] × 2 routes = 7 square miles.

Suburban Area

(a) Distance Traveled at 15 mph, for 12 min (30 − 18 min.) = 3.0 miles
(b) Destination opportunity area by bus = [(3.0 × 0.5 miles) × 2] × 2 routes = 6.0 square miles.

4. **Destination Opportunity Area Accessible by Private Motor Vehicle**

Urban Area

(a) Distance traveled at 20 mph for 24 min (30 min. door to door travel time-6 min. excess travel time) = 8 miles
(b) Destination opportunity area reachable by motor vehicle = 201 square miles.

Suburban Area

(a) Distance traveled at 30 mph for 27 min (30 − 3 min.) = 13.5 miles
(b) Destination opportunity area accessible by motor vehicle = 572 square miles.

These results in rank-order are shown in Table 11.1.

Example 2

Another measure of spatial accessibility is a mapping of the area accessible by each mode as a function of travel time. Travel time contours can then be developed for each mode and the area accessible within a desirable travel time (i.e., 30 min) can be used to compare the accessibility provided by each mode. An illustrative example is shown in Fig. 11.3.

Alternatively, door-to-door travel times for a given trip distance from a major focal point, such as the city's central business district (CBD), can be estimated for each travel mode. As shown in Fig. 11.4 modal door-to-door travel times are

Table 11.1 Accessible area within 30 min trip time, by travel mode

Mode	Area accessible in 30 min trip time (square miles)
1. Car	
Urban	201
Suburban	572
2. Bicycle	
Urban	50
Suburban	113
3. Bus	
Urban	7
Suburban	6
4. Walk	
Urban	7
Suburban	7

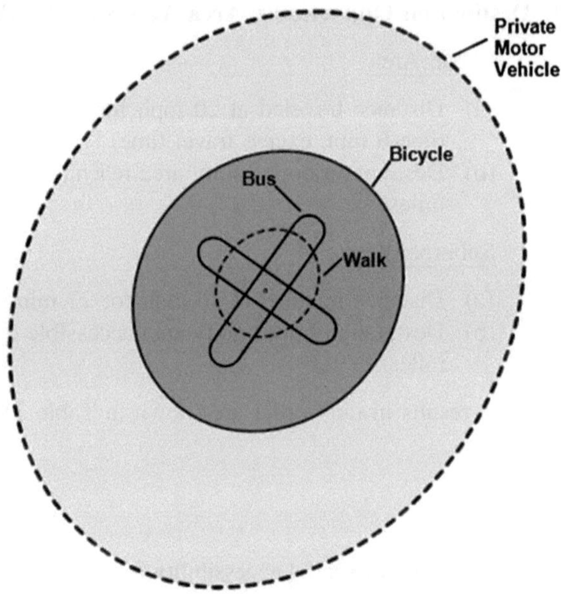

Fig. 11.3 Trip distance isochrones by various modes for a 30-min trip

generally determined by the type of routes and by the amount of excess time (walking, waiting, and connecting) required by a particular mode. Thus arterials/ freeways and rapid transit routes entail the shorter travel time, while arterials and surface transit require the longest door-to-door travel times.

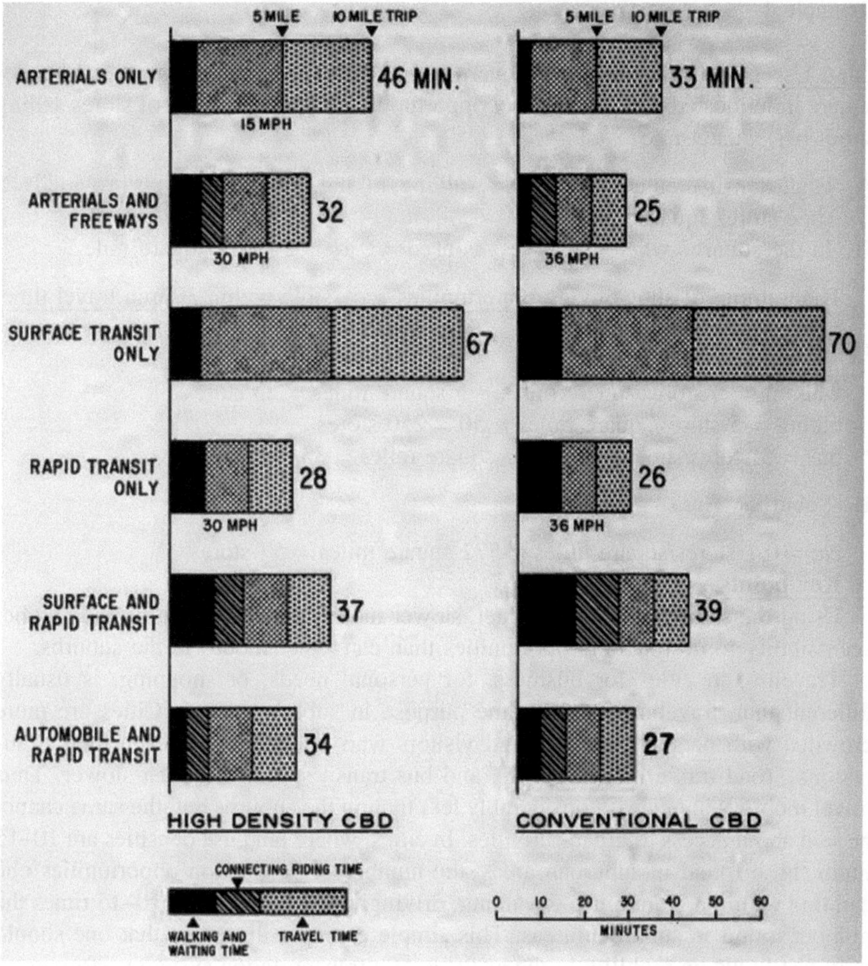

Fig. 11.4 Comparative travel times from the central business district if traveling by transit or automobile for trips lengths of 5 and 10 miles. *Source* Reference [3], p 150, Fig. 61

11.2.1.2 Access to Destination Opportunities Within Accessible Area

The following examples show the interdependence between mobility and density of development. While increasing density could decrease mobility, it also tends to increase accessibility.

Example 1

Land Use Density Assumptions: average number of destination opportunities per square mile located within the travel opportunity area (e.g., number of stores selling shoes per square mile)

A. High-density center city—where walking, biking, and bus transit is provided = 5 shoe stores/square mile
B. In the suburbs where car access is used = 0.1 shoe stores/square mile

The number of shoe buying opportunities accessible within 30 min travel time:

A. Center City

walking = 5 stores/square miles × 5 square miles = 25 stores
biking = 5 stores/square miles × 50 = 250 stores
bus = 5 stores/square miles × 7 square miles = 35 stores

B. Suburban Area

car = 0.1 stores/square miles × 572 square miles = 57 stores
Key points:
Using the same travel time budget, slower modes in the city can provide higher accessibility to destination opportunities than can faster modes in the suburbs.

Traveling in cities for business, for personal needs, or shopping, is usually different than traveling for the same purpose in suburban areas. Cities are more crowded with people (residents and visitors who often walk to reach their destinations), road traffic moves slowly, and bus transit speeds are even slower. Thus travel mobility in cities is considerably less than in the suburbs but, the same cannot be said about access to urban activities. In cities, where land use densities are 10–15 times those found in suburban areas, the number of destination opportunities one can find within a 20 min trip—walking, driving, or by transit—are 10–15 times the number found in suburban areas. This simple example illustrates that one should only just focus on mobility in the analysis of a transportation system performance, without considering access to activities.

Example 2

From empirical data of urban travel behavior, it has been observed that travelers tend to keep their trips as short as possible (see example in Fig. 11.5).

This means that destination opportunities closer to the traveler are likely to be selected more frequently than those located far away.

The number of activity sites and the size of their activity (employment locations, retail and service outlets, and recreational opportunities) that can be reached at a

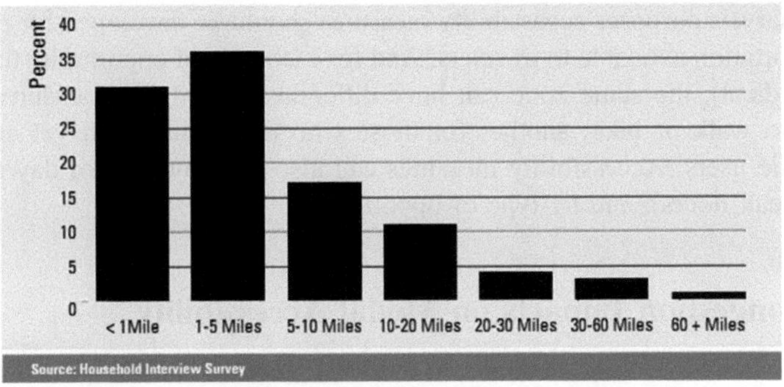

Fig. 11.5 Distribution of person trip lengths for the NY metropolitan area (all modes, all purposes, 1997/1998). *Source* Reference [4]

given distance and "discounting" that number by the intervening travel time, cost or distance [1] can be estimated by Eqs. 11.1 and 11.2:

$$A_i = \sum_{j=1}^{z} \sum_{,k=1}^{n} (DO_j)_k \Big/ (tt_{i-j})^{b} m \quad (11.1)$$

$$A_{ik} = \sum_{j=1}^{z} (DO_j)_k / ((tt_{i-j})^b m \quad (11.2)$$

where:

1. $\mathbf{A}_i$ = an accessibility index of a reference zone (i) to destination opportunities (j) aggregated across all of types of destination opportunities ($k - n$) (e.g., work, shopping, etc.) in the area. This zonal accessibility index is a metric used to compare and rank zones by their aggregate accessibility using a given mode (m). Thus the accessibility value of each zone is mode-dependent.
2. $\mathbf{A}_{ik}$ = the accessibility index of reference zone (i) to destination opportunities (j) of a given type (k). The value of A_{ik} is used to compare the modal accessibility of one zone to that of another zone or to compare the zonal accessibility provided by different transportation modes.
3. i = the reference zone
4. $\boldsymbol{j}$ = the zone of destination opportunities
5. $\boldsymbol{k}$ = the type of destination opportunities (work, shopping, medical, etc.)
6. $(\mathbf{DO_j})_k$ = the number of destination opportunities of a given type k at zone j,
7. $(tt_{i-j})^b$m = travel time measures separating zone i and zone j using mode m. The exponent of travel time differs with trip purpose [5]. Thus for work travel $b = 2.0$; while for shopping travel b could = 3.0, to reflect less willingness to tolerate longer travel times for shopping trips than for work trips. As was discussed earlier, $(tt_{i-j})_m$ is determined by the door-to-door travel speed of the mode used.

For analysis purposes accessibility measures should be developed for each mode of transportation available to travelers. And for each type of opportunity (e.g., work, shop, medical), the same zone can have different accessibility measures: one for those who walk or bike; another for those who ride transit, and yet another for automobile users. Accessibility measures can also vary by time of day (i.e., peak and off peak hours), and by type of opportunity.

11.3 Congestion Impacts on Modal Accessibility

Evaluating the impact of road traffic congestion on accessibility requires measuring the effects of road congestion on traveler mobility within the impacted area. For travelers, increasing congestion reduces the number of destination choices, and for freight carriers with a fixed fleet size, it reduces their market areas.

11.3.1 Examples

Several examples illustrate the likely congestion effects on accessibility.

11.3.1.1 Example 1

Assumptions:

1. trip time budget = 30 min
2. base-line average peak speed;
   - 30 mph private vehicles
   - 15 mph buses
3. ten-year traffic growth is projected to reduce peak traffic speed:
   - 20 mph private vehicles
   - 10 mph buses

<u>Base Line Condition</u>

1. If Trip is by Auto:
   - Out-of-vehicle time = 3 min
   - Distance traveled in 27 min (30 − 3) at 30 mph = 13.5 miles
   - Accessible area = 572 square miles
2. If Trip is by Bus:
   - Out-of-vehicle time = 13 min
   - Distance traveled in 17 min (30 − 13) at 15 mph = 4.25 miles
   - Accessible area = (4.25) × 2 × 0.5 = 4.25 square miles

<u>Congested Conditions</u>

1. If Trip is by Auto:

   - Out-of-vehicle time = 3 min
   - Distance traveled in 27 min at 20 mph = 9.0 miles
   - Accessible area = 254 square miles

2. If Trip is by Bus:

   - Out-of-vehicle time = 13 min
   - Distance traveled in 17 min at 10 mph = 2.83 miles
   - Accessible area = (2.83) × 2 × 0.5 = 2.83 square miles

These impacts are summarized in Table 11.2. This table shows that while congested roads reduce the mobility of bus riders and auto drivers by the same percentage, congested roads reduce private vehicle drivers' accessibility by a greater amount than they do for bus riders.

11.3.1.2 Example 2

Another way of identifying the accessibility impacts of traffic congestion is to compare peak and off-peak trip distances contours (by mode) from major focal points such as the city center, major outlying residential areas, or commercial centers.

Figure 11.6 shows the area boundaries within 30 min travel time contours by automobile during the peak hour, when there is congestion and during the off-peak, when there is no congestion. The shaded area between the two contours represents the reduction in accessible area due to traffic congestion.

Table 11.2 Traffic congestion impacts on mobility and area accessibility for the case example of a 30-min trip

Mobility				Area Accessibility		
Travel mode	Base-line speed (miles)	Congested speed (miles)	% Change	Base-line speed (square miles)	Congested speed (square miles)	% Change
Private vehicle	13.5	9.0	−33	572	254	−56
Bus	4.25	2.83	−33	4.25	2.83	−33

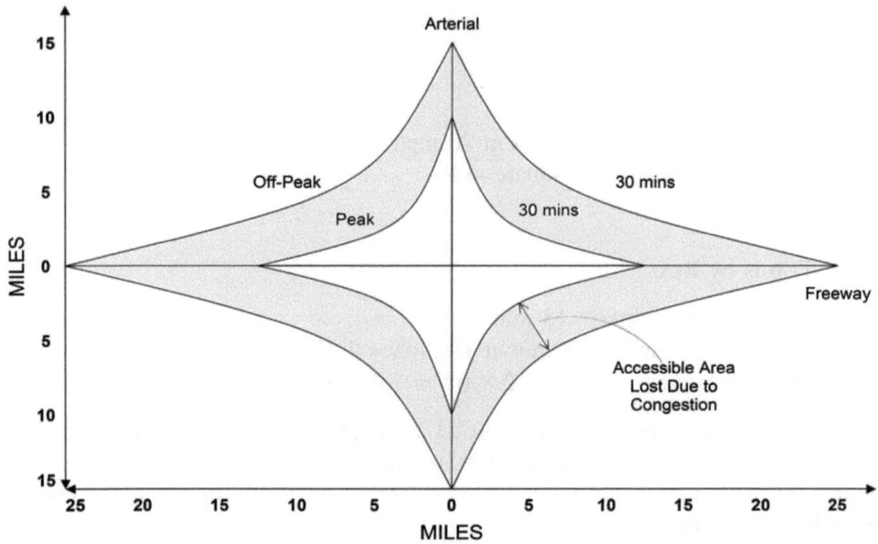

Fig. 11.6 Accessible areas for a 30 min automobile trip during congested and uncongested traffic

11.4 Barriers to Modal Accessibility

The previous examples assume that each travel mode was free of barriers to its use. In reality, however, the choice of a travel mode is sometimes affected by the presence of barriers to its use. Some of the most commonly found barriers to mode access are listed below for the walking and cycling mode; public transportation modes; and auto drivers.

11.4.1 Accessibility Barriers to Walking and Bicycle Use

- Steep grades and steps
- Risk exposure to vehicular traffic
- Long stretches of freeway that divide neighborhoods
- Crossing wide streets with heavy traffic
- Discontinuous networks/cul de sacs
- No sidewalks
- Streets not conducive to bicycle lanes
- Lack of traffic control signals at crossings of busy streets

11.4.2 Accessibility Barriers to the Physically Disabled

- Inaccessible vehicles
- Inaccessible transit stations
- Inconvenient access to transit routes
- Excessive walking distance to or from a transit stop
- Lack of real-time information about schedules and transfers

11.4.3 Accessibility Barriers to Auto Users

- No parking available at destination
- Auto-free zones
- High cost of car use/ownership

11.5 Conclusions

This chapter has described the impacts of traffic congestion on two types of accessibility: activity accessibility and spatial accessibility.

Spatial Accessibility is clearly dependent on door-to-door travel speed—the faster one travels the greater the area covered. Therefore increasing traffic congestion reduces spatial accessibility for all vehicle users. However, traffic congestion tends to have a greater impact on private vehicle users than for public transit users.

Activity Accessibility is a measure resulting from a combination of spatial accessibility and the number of desired destination opportunities within the accessible area. In this case the impact of traffic congestion on activity access cannot be determined by focusing on travel speed alone—without considering the density (and number) of activities located within the accessible area.

References

1. Hanson S, Giuliano G (2004) The geography of urban transportation, 3rd edn. The Guildorf Press, New York
2. Litman T (2014) Victoria Transportation Institute, personal correspondence, s.l. Accessed 25 Nov 2014
3. Wilbur Smith and Associates (1961) Future highways and urban growth. Accessed Feb 1961
4. New York Metropolitan Transportation Council (2000) Council contact—household interview survey. New York Metropolitan Transportation Council, New York
5. Martin WA, McGuckin NA (1998) Travel estimation techniques for urban planning. National Cooperative Highway Research Program—Report 365

Chapter 12
The Impacts of Congestion on Roadway Traffic Productivity

12.1 Introduction

Road productivity can be defined as the throughput traffic volume (vehicles or persons per hour) of a roadway at a given point, or as the person-miles or vehicle-miles per hour that can be moved on a roadway segment, or on an area-wide road network.

This chapter identifies the various traffic elements (throughput volume, capacity, speed, and density) that determine road productivity. Through illustrative examples it shows how productivity of freeways and arterial streets is impacted by congestion.

12.2 Fundamentals of Traffic Flow

Traffic flow theory provides an analytical means of evaluating capacity, congestion, and productivity. It is useful in analyzing the effects of changes in traffic demand, flow, and speeds over time. For example, it can assist in answering questions such as: what are the congestion and productivity impacts of a new land development along a heavily traveled roadway?

12.2.1 Basic Relationships

Traffic speed is a basic indicator of congestion. Speed depends upon (1) type of roadway (freeway, expressway, arterial, or collector streets); (2) the roadway geometry and controls (from a roadway that is fully accessible from adjacent land use activities, to a roadway with full access control); (3) the nature, extent, and duration of traffic conflicts/incidents that interfere with traffic flow; and (4) the

J.C. Falcocchio and H.S. Levinson, *Road Traffic Congestion: A Concise Guide*,
Springer Tracts on Transportation and Traffic 7,
DOI 10.1007/978-3-319-15165-6_12

traffic demand volume along the roadway at a given time; and (5) the capacity of the road to serve the demand.

Traffic speed varies with traffic volume: as traffic volume increases, speed drops. The speed at which volume reaches its maximum value is the *critical speed*—so called because when speed continues to drop below this critical value the throughput volume of the roadway begins to drop as well. Therefore the critical speed becomes a useful metric in achieving the goal of maximizing/protecting roadway productivity, and it is a useful traffic congestion management tool.

12.3 Freeway and Expressway Productivity

12.3.1 Introduction

Figure 12.1 shows the relationship between average speed and average rate of throughput volume for five freeways with different free-flow speeds: 75, 70, 65, 60, and 55 mph as set forth in the 2010 Highway Capacity Manual, Chap. 11 (2).

These speed-flow curves indicate that traffic speeds on freeways do not begin to drop perceptibly from free flow speed until their volume reaches about 80 % of their maximum throughput volumes or when their respective volume-to-capacity ratio (V/C) = 0.8. When freeways reach their maximum throughput volumes their free flow

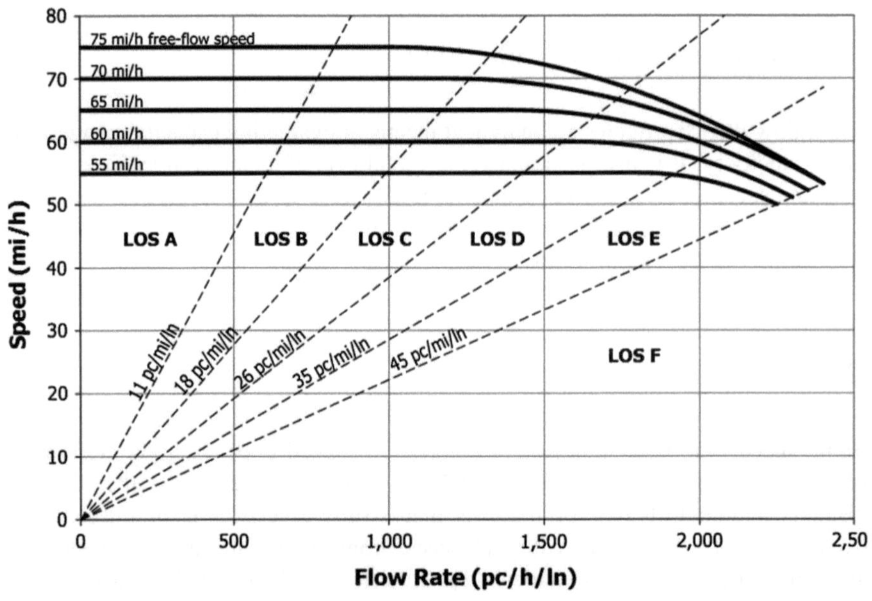

Fig. 12.1 Typical speed-flow curves for freeways when traffic density is below 45 passenger cars per lane-mile. *Source* Reference [1], highway capacity manual 2010, Chap. 11

Table 12.1 Maximum throughput volume (V_m), critical speed (S_c) and critical density (D_c) for freeways with different free-flow design speed (S_f)

Design speed	Maximum throughput volume	Critical speed	Critical density
S_f (mph)	V_m (pcphpl)	S_c (mph)	$D_c = V_m/S_c$ (pcplm)
75 and 70	2,390	53	45
65	2,340	52	45
60	2,290	51	45
55	2,250	50	45

Source Approximate values estimated from Fig. 1

traffic speed drops to from 75 or 70 to 53 mph, from 65 to 52 mph, from 60 to 51 mph, and from 55 to 50 mph, respectively, depending on the initial free-flow speed.

At these maximum throughput volumes (V/C = 1.0) and at their critical speeds (S_c), their critical density (D_c) of 45 passenger cars per lane mile is reached. these relationships are shown in Table 12.1.

When critical density is reached, however, traffic flow becomes unstable, vehicle speeds are apt to drop unpredictably below their critical speed (S_c), and productivity also declines. This speed-flow pattern is shown in Fig. 12.2 for an urban freeway [2]:

Density contours (vehicles per lane per mile) along a freeway have been useful in identifying the location and extent of high densities (i.e., congested flow). The contours can also identify bottleneck points both upstream and downstream of a given location. Density contours such as shown in Fig. 12.3, have been developed

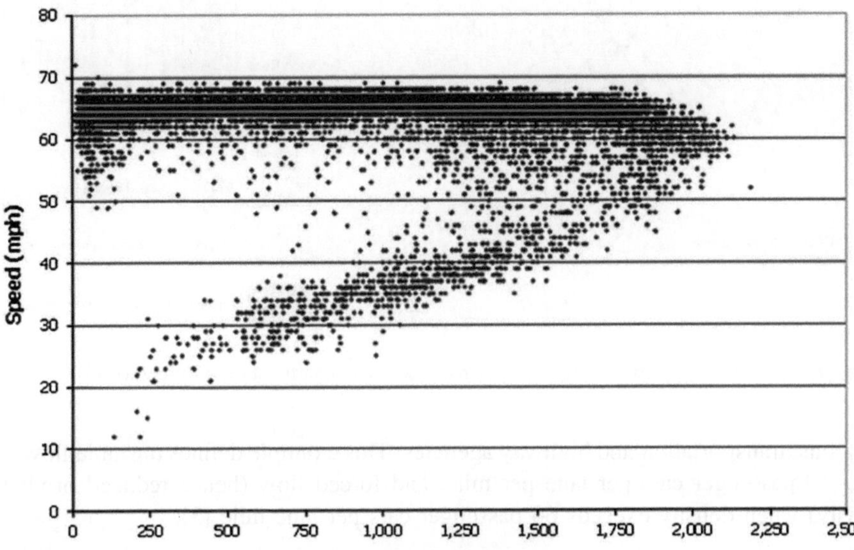

Fig. 12.2 Speed flow data for an urban freeway. *Source* Reference [2], Fig. 5

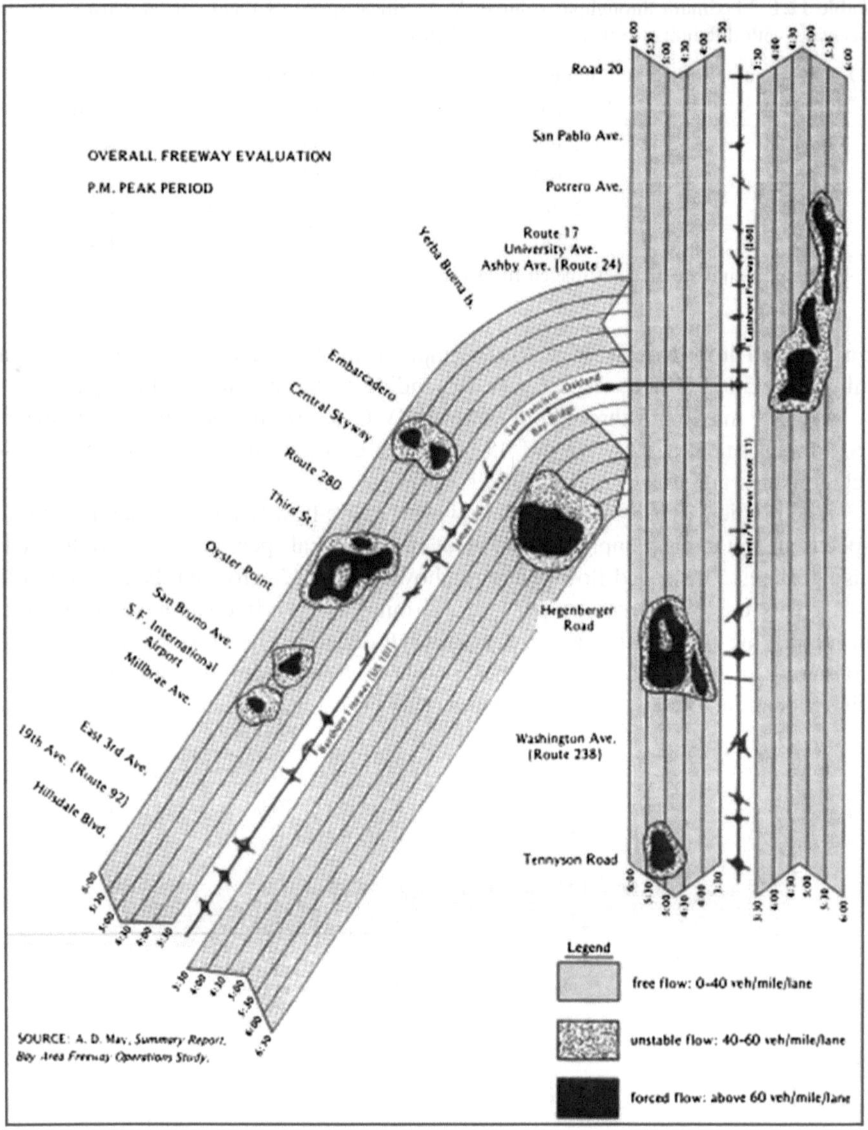

Fig. 12.3 Freeway system evaluation by density contours charts. *Source* Reference [3]

by state transportation and highway agencies. This example defines unstable flow as 40–60 passenger cars per lane per mile, and forced flow (hence reduced productivity) when density exceeds 60 passenger cars per lane-mile [3].

12.3.2 Skycomp Analysis

An example showing what happens to freeway speed when traffic density increases beyond critical density (D_c), was captured through aerial surveys of freeways and expressways in the New York City area by Skycomp (5). These patterns are shown in Fig. 12.4.

From Skycomp's observations, correlations between traffic speed and traffic density were developed to illustrate (1) the impact of increasing traffic density beyond its critical value of 45 pcplpm, on traffic speed; and (2) the impact of traffic speed on productivity (throughput volume) at speeds lower than the critical value of 53 mph. The results are shown in Fig. 12.5 through Fig. 12.8.

- Figure 12.5 shows how freeway speeds decrease once the critical density of 45 vehicles per lane per mile is exceeded.
- Figure 12.6 shows the entire speed-density relationship. In the example the density of 45 passenger cars per lane per mile separates the regions of uncongested and congested flow. The speed—density data for uncongested flow was calculated from Fig. 12.1.

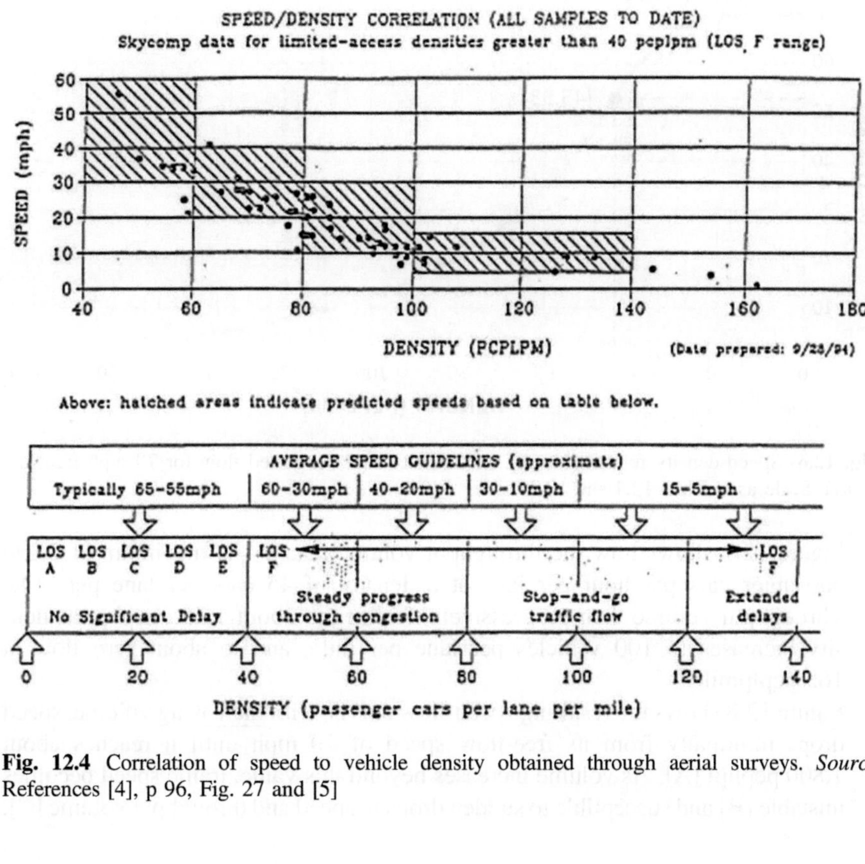

Fig. 12.4 Correlation of speed to vehicle density obtained through aerial surveys. *Source* References [4], p 96, Fig. 27 and [5]

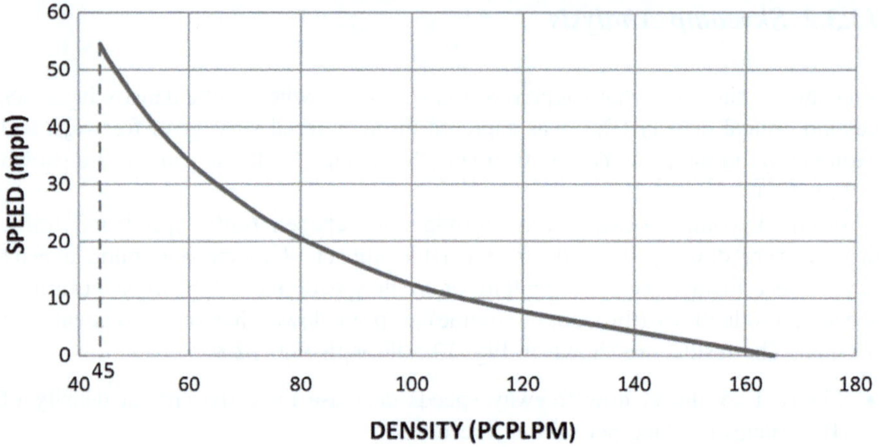

Fig. 12.5 Estimated speed/density relationship for congested freeway flow. *Source* Scale same as in Fig. 12.4

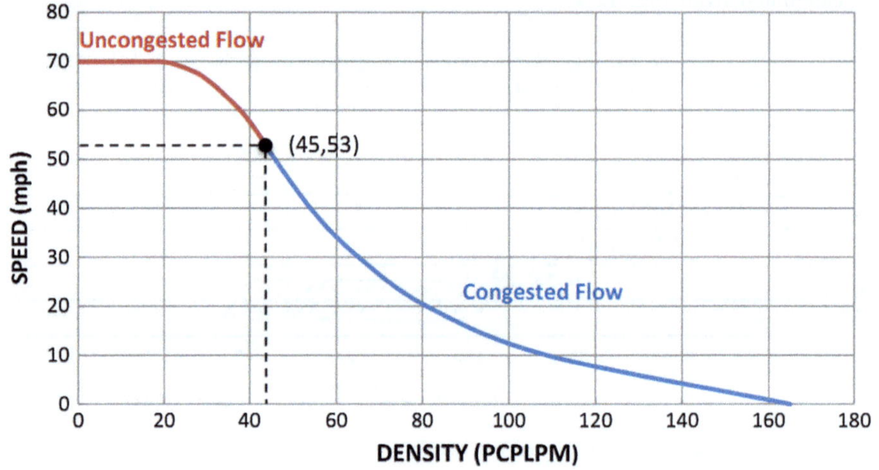

Fig. 12.6 Speed-density relationship for uncongested and congested flow for 70 mph freeway. *Source* Scale as in Figs. 12.4 and 12.5

- Figure 12.7 shows how the throughput volume reaches its maximum of 2,390 passenger cars per hour per lane at a density of 45 cars per lane per mile. Throughput volume then progressively declines to about 1,250 pcphpl as density increases to 100 vehicles per lane per mile, and to about zero flow at 165 pcplpmile.
- Figure 12.8 shows the resulting speed-flow curve. With increasing volume, speed drops minimally from its free-flow speed of 70 mph until it reaches about 1,800 pcphpl [A]. As volume increases beyond this value, traffic speed becomes unstable [B] and susceptible to sudden drops in speed and throughput volume [C].

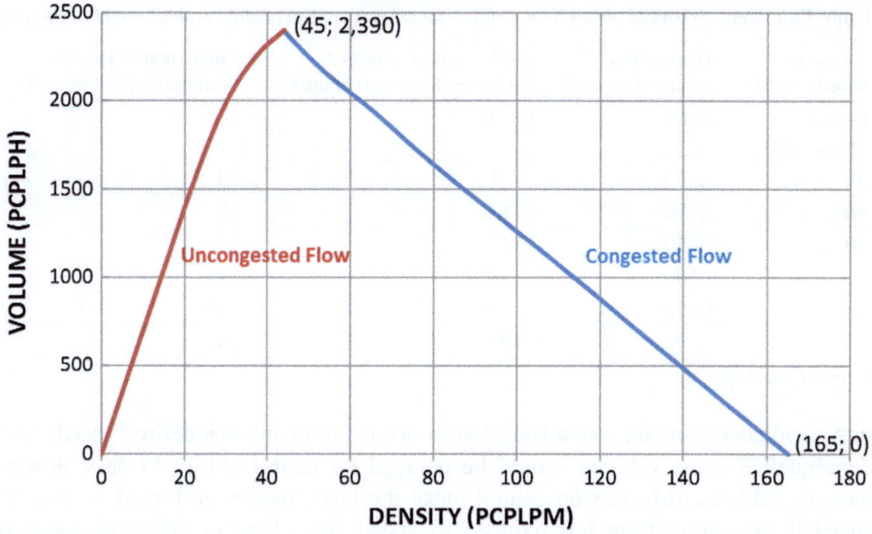

Fig. 12.7 Throughput volume for uncongested and congested flow for 70 mph freeway. *Source* Figure 12.6

The speed volume relationship shown in Fig. 12.8 indicates that: when the V/C = 1.0, the 53 mph traffic speed cannot be sustained for long. At a density of 45 pcplpm, the space between vehicles averages 97 ft [(5,280 ft per mile/45 cars/lane-mile)−20 ft/car]. Many drivers would consider this space too short to merge

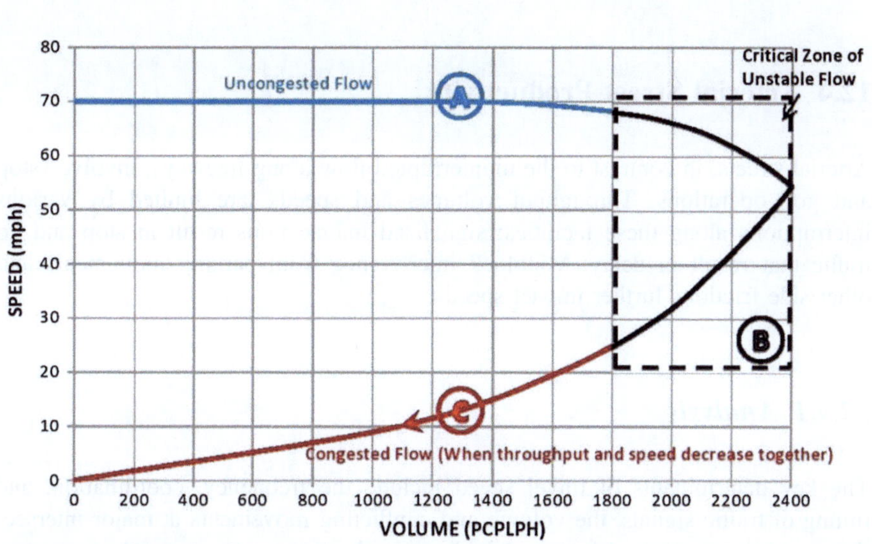

Fig. 12.8 Speed-volume relationship for uncongested and congested flow for 70 mph freeway. *Source* Source scale as in Fig. 12.7

Table 12.2 Percent loss in throughput volume for freeway speeds below critical speed of 53 mph

Average speed (mph)	Throughput volume (pcphpl)	% Loss in throughput volume (from maximum)	Incremental loss in throughput volume (%)
Critical Speed = 53	2,400	0	–
50	2,320	3	3
40	2,150	10	7
30	1,850	23	13
20	1,500	38	15
10	1,050	56	18
0	0	100	44

Source Calculated

into an adjacent lane for a safe lane change needed to maintain a desired speed. As a consequence some vehicles would be delayed by their inability to pass slower-moving vehicles: drivers who cannot make the lane change are forced to drop its speed if the car in front has reduced its speed. This kind of driver response is multiplied in a chain reaction of speed reductions as vehicles come closer to one another. When this occurs the traffic throughput volume of the roadway gets progressively smaller, with a loss rate of 20 vph [2,390/(165–45)] for each unit increase in density above 45 vplm (Fig. 12.8); and when density reaches its maximum (Dj) value, a stop-and-go traffic movement prevails.

The magnitude of the loss in throughput volume for various values of speed below its critical value, is shown in Table 12.2. It may be seen that the incremental loss in throughput increases at a faster rate with each incremental loss in speed.

12.4 Arterial Street Productivity

Arterial streets, in contrast to the uninterrupted flow along freeways, involve "stop and go" operations. Throughput volumes and speeds, are limited by various interruptions along these facilities: signalized intersections result in stop and go traffic that result in delay. Midblock interference from parking movements and other side frictions further impact speeds.

12.4.1 Analysis

The key determinants of travel speed include the frequency, coordination, and timing of traffic signals, the volume and conflicting movements at major intersections, and the number of lanes available on each intersection approach.

The 2010 Highway Capacity Manual (Volume 3) contains detailed procedures for analyzing intersection and roadway performance [1]. The analysis procedures are best suited for evaluating specific locations. Many of the equations are complex and require computerized analysis.

This section, focuses on a general assessment of how traffic speed on arterial roads is impacted by the volumes they carry.

Figure 12.9 shows five types of arterials whose free-flow speed ranges from 40 to 12 mph. Streets with low free-flow speeds (between 20 and 12 mph) are included in the figure, although they are unlikely to function as "arterial streets."

It can be seen that as traffic volume increases, speeds generally decreases until a critical speed is reached and throughput volume reaches its maximum value (V/C = 1.0). The heavy lines are superimposed on the initial analysis to give a working approximation of how speeds decrease as traffic volumes increase. When the volume-to-capacity ratios are less than 0.60, the speeds change very little. As volume-to-capacity ratio increases, there is a sharp decline in speeds.

Table 12.3 summarizes the relationship between free-flow speed (Sf), and the critical speed (Sc) for maximum throughput volume (Vm)—reached when V/C = 1.0, for each class of arterial streets in Fig. 12.9.

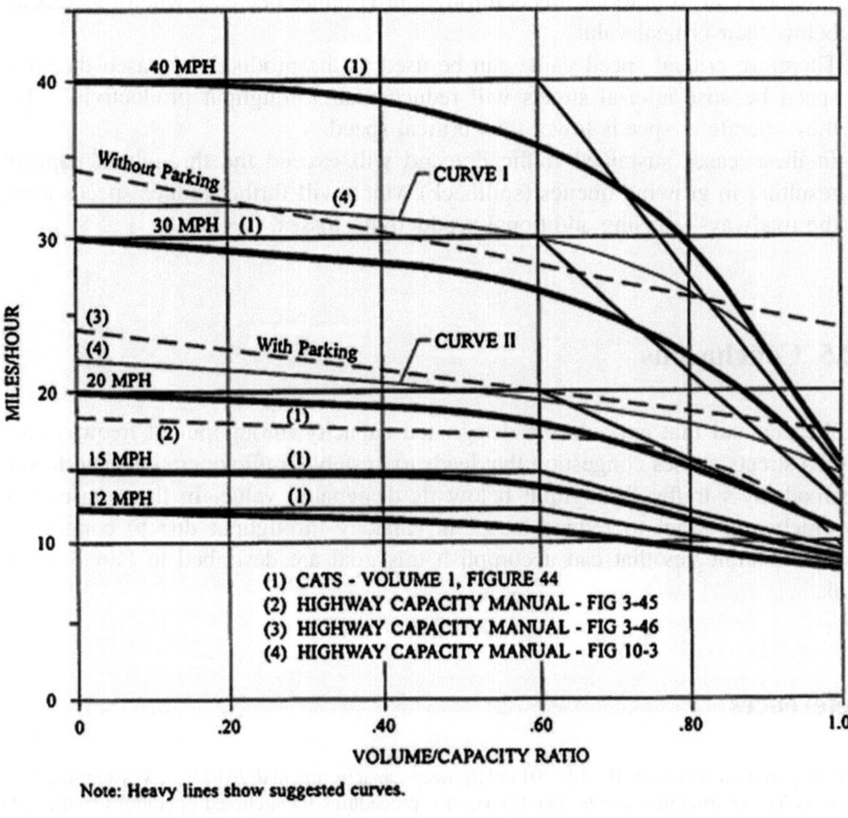

Fig. 12.9 Speed and volume/capacity ratios for arterial streets. *Source* Reference [4], p 30, Fig. 3 and 6

Table 12.3 Free-flow speed and critical speed at maximum throughput volume for arterials of different free-flow speeds

Average free-flow speed (Sf) of arterial streets	Critical speed (Sc) at Vm (V/C = 1.0)
40	17 mph
30	14
20	11
15	9
12	8

Source Figure 12.9

12.4.2 Implications

The preceding speed-flow relationships lead to the following implications:

1. Although Fig. 12.9 does not show what happens to throughput volume when traffic speed drops below its critical value (as shown for freeways), it may be assumed that for arterial streets throughput volumes decrease when speeds drop below their critical value.
2. Therefore critical speed value can be used as the productivity-based threshold speed because arterial streets will reduce their throughput productivity when they operate at speeds lower than critical speed.
3. In these cases sustained traffic demand will exceed the throughput capacity resulting in growing queues (spillback) which will further reduce speeds along the roadway—causing additional productivity losses.

12.5 Conclusions

Traffic demand that exceeds the designated capacity throughput of freeways and arterial streets causes congestion that leads to unstable traffic operations and lowers the roadway's traffic throughput below its designated value. In these cases it is extremely important to reduce losses in capacity throughput due to congestion. Available strategies that can accomplish this goal are described in Part 3 of the book.

References

1. Transportation Research Board (2011) Highway capacity manual 2010. s.n, Washington DC
2. Texas Transportation Institute. Quality control procedures for archived operations traffic data: synthesis of practice and recommendation. http://www.fhwa.dot.gov/policy/ohpi/travel/qc/rec_procedures.cfm

3. Carter EC, Homburger WS (1978) Introduction to transportation engineering. Institute of Transportation Engineers, Washington, DC
4. Board, Transportation Research (1997) NCHRP (National Cooperative Highway Research Program) report 398, user's guide
5. Skycomp Inc. (1992) Arterial Photo Survey for the New Jersey Statewide Traffic and Incident Master Plan Study, Rockville, MD
6. Wilbur Smith and Associates (1915) Urban Truck Road Systems and Travel Restrictions, US

Chapter 13
The Costs and Other Consequences of Traffic Congestion

13.1 Introduction

In an increasingly fast paced and globally oriented economy, the efficient movement of persons and goods is a competitive necessity [1]. Traffic congestion adversely impacts quality of life and economic productivity in metropolitan areas. It increases fuel consumption, the cost of traveler and freight movement, the number of crashes, and tailpipe pollutants harmful to human health.

This chapter sets forth the broad consequences of traffic congestion on these issues of concern.

13.1.1 Congestion Impacts on Travelers

Traffic congestion is a growing concern—especially for those who live in large urban areas and have long commutes. It increases mental stress and disrupts peoples' daily schedules: drivers in congested roads have been reported to experience more stress and aggression. The dissatisfaction with the daily commute has been found to produce undesirable psychological and physiological responses, including elevated blood pressure, increased negative mood states, lowered tolerance for frustration, increased irritability, and more impatient driving behavior [2–4].

Congestion disrupts the daily schedules of business and family activities. Regular commuters who avoid being late for work leave early and often miss breakfast with their children or spouses; in their return trip home if caught in congestion they can miss having dinner with the family, or a child performance at school.

The impact of personal time lost in congestion includes not only the stress and disruption to family life, but perhaps even more also important it includes the opportunity cost of that lost time. For employers it could be the loss in productivity from employees who report late for work.

J.C. Falcocchio and H.S. Levinson, *Road Traffic Congestion: A Concise Guide*,
Springer Tracts on Transportation and Traffic 7,
DOI 10.1007/978-3-319-15165-6_13

Research has shown that congestion may reduce the number of job destination opportunities available: it has been reported that some employers, concerned with work attendance reliability, favor hiring people who live close to the company over hiring those who may be late in getting to work because of congestion [2, 5].

13.1.2 Congestion Impacts on Business Costs

A survey of business leaders in Portland, Oregon [6, 7] reported the following congestion impacts on business production costs:

- Costs of additional drivers and trucks due to longer travel times;
- Costly 'rescue drivers' to avoid missed deliveries due to unexpected delays;
- Loss of productivity due to missed deliveries;
- Shift changes to allow earlier production cut off;
- Increased inventories;

Golob and Regan [8] report that "road congestion is perceived as a more serious problem by managers of trucking companies engaged **in intermodal operations**, particularly private and for-hire trucking companies serving airports and rail terminals, companies specializing in **refrigerated** transport and private companies engaged in **Less-Than Truck-Load (LTL)** operations".

NCHRP Report 463 [9] identifies three types of business costs impacted by traffic congestion:

i. Direct travel costs of all business-related travel, including vehicle operating expenses and the value of time for drivers (and passengers);
ii. Logistics and scheduling costs, including effects on inventory costs such as stocking, perishable items, and just-in-time (JIT) processing;
iii. Reduction in market areas for workers, customers and incoming/outgoing.

Just-in-Time (JIT) Inventory Costs

JIT production techniques have led business to demand a faster, more frequent and more reliable supply of goods to reduce inventory costs. A study reported that "*JIT methods increased deliveries by a factor of two and decreased the size of deliveries by about half*" [9, 10]. This JIT-induced truck volume demand adds to urban traffic congestion (which negatively impacts travel time reliability).

Since travel time reliability is critically important to JIT operations, business will see inventory costs increase as travel time reliability is reduced with increasing congestion.

Transit Operating Costs

Congested roadways slow down transit vehicles. This causes longer round trip cycle times that require more vehicles in service to meet travel demand, additional operator costs, and higher fuel consumption. These conditions increase the cost of operation.

In addition, congested speeds will increase travel time variability with its negative impacts on the level of service experienced by passengers. The resulting consequences include uneven vehicle spacing that produces uneven vehicle loads—with the lead bus in the platoon overcrowded and the following vehicle nearly empty.

13.2 Calculating the Costs of Traffic Congestion

Traffic congestion costs includes the value of the extra time spent in congested traffic, the extra fuel consumed, as well as the additional crashes that result, and the additional amount of air pollutants that are generated.

The assumptions for baseline travel conditions for congestion cost calculations often assign a cost value to the difference between free-flow travel speeds and actual travel speed. This difference is often referred to as "lost" time or travel "delay."

Such "cost of congestion" approaches, when applied in large urban areas, can be misleading because they do not recognize that, in large urban areas, congestion is the outcome of higher land use density—itself the successful result of other urban policies—and a somewhat lower threshold speed is a more realistic approach to quantifying congestion (as was discussed in Chap. 8).

Since travel demand is derived from social and economic activities (which add value to society), the additional travel time from increased demand is the natural consequence of satisfying increasing demand. "Empty cities are not generally considered successful cities; nor should empty roads" [11].

What should be the baseline for calculating congestion costs? The European Conference of Ministers of Transport [11] notes that "the impacts of congestion are not abstract—they must be linked to roadway users' experiences and expectations. Instead of attempting to calculate the "overall cost" of congestion from a theoretical viewpoint, it may be more realistic and more productive to compare current levels to and costs of congestion with past (and expected future) levels." This approach to calculating congestion allows assessing the extent to which congestion is reducing travel time reliability and accessibility to urban facilities and services.

An alternative approach in calculating congestion costs is by establishing baseline congestion threshold speeds similar to the values suggested in Chap. 8, where the concept of user acceptance of threshold congestion speeds were indentified for different size of urban areas, road types, and time of day; and in Chap. 12, where the concept of "critical speed" was discussed.

13.2.1 Components of Congestion Costs

The various components of travel cost in congested traffic are shown in Eq. 13.1:

$$\begin{aligned} &\text{Travel Cost Per Vehicle in Congested Traffic} \\ &= [(\text{Additional Trip Time from Congestion}) \times (\text{Value of Travel Time})] \\ &+ [\text{Additional Fuel Cost From Congestion}] \\ &+ [\text{Cost of Additional Crashes}] \\ &+ [\text{Cost of Additional Air Pollution to Human Health}] \end{aligned} \quad (13.1)$$

The variables of this equation are discussed below in order:

1. Additional Trip Time
2. Value of Travel Time
3. Additional Fuel Consumption
4. Additional Crashes
5. Health Cost of Additional Air Pollutants.

13.2.1.1 Additional Trip Time

a. Based on Average Threshold Speed of Congestion

$$\begin{aligned} \text{Additional Trip Time} &= (\text{Trip Time at Congested Speed}) \\ &- (\text{Trip Time at Threshold Speed}) \end{aligned} \quad (13.2)$$

where:

$$\text{Trip Time at Congested Speed} = (\text{Congested Speed}) \times (\text{Trip Length})\ \text{Trip} \quad (13.3)$$

$$\text{Time at Threshold Speed} = (\text{Threshold Speed}) \times (\text{Trip Length}) \quad (13.4)$$

Trip length is an important variable in the analysis of congestion costs. This is because the cost of congestion is smaller for shorter trips and bigger for longer trips.

b. Based on Time Variability

The travel time experience of most road users is not based on the average value of congested travel times but instead it is based on the worst days when their trip was unexpectedly delayed. Figure 13.1 shows that the 95 percentile travel time rate is almost double the average travel time rate.

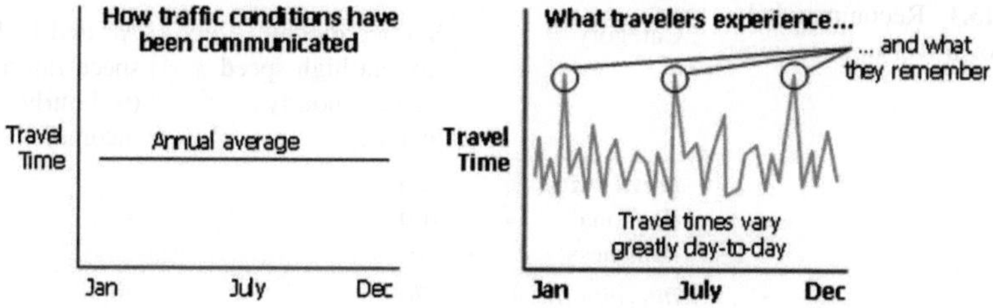

Fig. 13.1 Averages do not tell the full story. *Source* Reference [12], Fig. 1

For example, a traveler on congested roads who experiences variable travel times on the commute to work needs to allow for this variability by starting the trip sufficiently earlier to ensure on-time arrival most of the time.
However, since travel time variability data are not typically measured, they generally are not included by transportation agencies in their calculation of trip times and costs. This is potentially a large underestimate when one considers that in large urban areas between 58 and 67 % of congestion delay on US arterial roads is attributable to random events—a major source of travel time variability (see Table 7.2). Moreover, unexpected delay time is reported at a value of over 2.5 times the value of travel time under normal conditions [2].

c. <u>Travel Time Reliability Impacts on Freight Movement</u> [13]
The reliability of the highway system in enabling goods to get where they need to be when they need to be there is a key performance indicator because **"Unexpected delays can increase the cost of transporting goods by 50–250 %"** [14].
"Studies have shown that reliability is one of the most important factors influencing choices in freight transportation. A 1999 study estimated that carriers on average value savings in transit time at $144–193 an hour and savings in schedule delay at $371 an hour [15]. As in the case of automobile travelers, truckers value time savings in congested conditions more than twice as highly as overall travel time savings." [2].

13.2.1.2 Value of Travel Time

The valuation of travel time reported in the literature represents average values that were aggregated across all travelers and conditions. They are not intended to represent the travel time values experienced by individual travelers. Since the effect of travel time reliability is rarely considered in project assessments, proposed transportation projects that improve travel time reliability may be undervalued [2].

<u>Person Travel</u>
Table 13.1 gives the best single figures for defining the value of travel time savings (VTTS) as a percentage of hourly income. Table 13.2 summarizes a likely range of travel time values for each trip category. The ranges are not necessarily symmetrical

Table 13.1 Recommended values of travel time savings

Category	Surface modes[a] (except high speed rail) (% hourly income)	Air and high-speed rail travel (% hourly income)
Local travel -Personal -Business	50 100	– –
Intercity travel -Personal -Business	70 100	70 100
Vehicle operators	100	100

Source Reference [16]

[a] Surface figures apply to all combinations of in-vehicle and other time. Walk access, waiting, and transfer time should be valued at 100 % of hourly income when actions affect only those elements of travel time

Table 13.2 Plausible ranges for values of travel time savings

Category	Surface modes[a] (except high-speed rail) (% hourly income)	Air and high-speed rail travel (% hourly income)
Local travel -Personal -Business	35–60 80–120	– –
Intercity travel -Personal	60–90	60–90
-Business	80–120	80–120
Vehicle operators	80–120	80–120

Source Reference [16], US Department of Transportation Guidance on the Average Values of Traveler Time Savings (VTTS)

[a] Surface figures apply to all combinations of in-vehicle and other transit. Walk access, waiting, and transfer time should be valued at 80–120 % of hourly income when actions affect only those elements of travel time

about the average estimates of Table 13.1. Therefore, in addition to evaluations based on the most likely estimates of congestion costs, these ranges should be used to reflect potential errors in estimation.

The 2009 US average hourly earnings for determining the value of travel time savings are shown in Table 13.3.

Table 13.4 shows average costs of travel time for US metropolitan areas, as well as nationwide. These estimates are based on the average wage in the region, and they are adjusted by the cost to trucks of delayed delivery of goods

Table 13.3 Recommended hourly earnings rates for determining values of travel time savings

	2009 US dollars per person-hour	
Category	Surface modes[a] (except high-speed rail)	Air and high-speed rail travel
Local travel -Personal -Business	$23.90 $22.90	– –
Intercity travel -Personal -Business	$23.90 $22.90	$45.60 $62.60
Truck drivers	$23.70	–
Bus drivers	$23.60	–
Transit rail operators	$38.90	–
Locomotive engineers	$33.00	$33.00
Airline pilots and engineers	–	$73.30

Source Reference [16], US Department of Transportation Guidance on the Average Values of Traveler Time Savings (VTTS)
[a] Based on Table 13.1

Table 13.4 Cost of travel time congestion in selected areas of the United States (NYMTC 2006)

Source	Cost of delay	Units	Year	Region
2001 Baltimore Regional Transportation Plan *Baltimore Regional Transportation Board*	$12.96	$/vehicle-hour	1996	Baltimore area
2003–2008 TIP Capital District Transportation Committee (Albany, NY)	$8.18	$/vehicle-hour	1991	Albany area
Texas Transportation Institute *Urban Mobility Report* (2004)	$13.45	$/person-hour	2002	Average US urban area
RAND California	$12.85	$/person-hour	N/A	California
United States Department of Transportation	$12.70	$/person-hour	1998	Nationwide
HERS model	$15.59	$/vehicle-hour	1995	Nationwide
IDAS model	$9.63	$/person-hour	1995	Nationwide
Caltrans	$8.16	$/person-hour	2000	California
Washington State Department of Transportation	$6.12	$/person-hour	2000	Washington State

Source Reference [17], Table 1

Table 13.5 HERS value of travel time (1995 dollars) by benefit category and vehicle type

$ per person-hour	Vehicle class						
Category	Small auto	Medium auto	4-tire truck	6-tire truck	3–4 axle truck	4-axle comb.	5-axle comb.
On-the-clock							
Labor/fringe	26.27	26.27	8.02	21.88	18.22	21.95	21.95
Vehicle	1.72	2.02	2.18	3.08	8.08	7.42	7.98
Inventory	0.00	0.00	0.00	0.00	0.00	1.65	1.65
Total	27.99	28.29	20.20	24.96	27.02	31.02	31.58
Other trips							
Percentage of miles (%)	90	59	0	0	0	0	0
Value	12.78	12.78	NA	NA	NA	NA	NA
Weighted average	14.30	14.33	15.08	25.27	27.91	31.64	32.25

Source Federal Highway Administration, The Highway Economic Requirement System: Technical Report (updated 3/97)
Note 1995 dollars

[17]. These costs range from a low of $6.12 per person hour in Washington State in 2000, to $13.45 per person hour for an average US urban area in 2002. To bring these costs to current dollars they should be adjusted by the annual inflation rate.

Commercial Vehicles

The Highway Economic Requirements System (HERS) developed by the Federal Highway Administration [18] divides the value of time for commercial vehicles into two components: on-the-clock trips (trips drivers take as part of their work) and other trips "off-the-clock trips." The suggested weighed average costs are shown in Table 13.5. They range from about $14 per hour for small autos upward to more than $32 per hour for five axle combination

Not all truck drivers perceive the same value of time when they are caught in traffic congestion. "There is a substantial difference in the value of time between truck drivers who are independent operators (and hence internalize the full business costs of their delays) and truck drivers who are paid an hourly wage [19]. Independent operators were found to be more willing to pay tolls to avoid delay than drivers who are paid an hourly wage" [9].

Some Key Factors Affecting the Value of Travel Time

The cost of delay time varies according to trip purpose, amount of delay, time of day, a person's income, and the conditions under which delay time is experienced and thus valued [20].

Trip Purpose

Being late to a job interview is likely to be viewed as more costly than being late to a dinner party.

Amount of Delay

The cost of a 1 min delay may be less than 1/30 the cost of a 30 min delay.

Trip Length

The cost of a 5 min delay for a 30 miles trip is not the same as the cost of 10 min delay for trips of one mile.

Traveler Income

The unit cost of delay is higher for a high-income traveler than for a low-income traveler.

Travel Conditions

The perceived cost of a pre-announced 15 min delay on a section of congested road is not the same as the cost of the same 15 min delay experienced on a road providing no travel time delay information to drivers.

A trip delay of 15 min on a sweltering summer day in a vehicle with a malfunctioning air conditioner, is more onerous than if driving a vehicle with functioning climate control.

Conclusion

Travel time values are to be perceived differently by different road users according to their particular needs, travel conditions of their trip, duration of delay, and if delay information is known or unknown to travelers. These conditions should be recognized in developing time costs for assessing the benefits if congestion relief programs.

13.2.1.3 Fuel Consumption and Traffic Congestion

The Fuel Economy Guide [22] published by US Department of Energy lists the following factors that can lower a vehicle's fuel consumption:

- Aggressive driving (speeding and rapid acceleration and braking) can lower gas mileage by as much as 33 % at highway speeds and 5 % around town.
- Excessive idling, accelerating, and braking in stop-and-go traffic;
- Cold weather (engines are more efficient when warmed up);
- Driving with a heavy load or with the air conditioner running;
- Improperly tuned engine or underinflated tires; Use of remote starters.
- Many short trips taken from a cold start can use twice as much fuel as one multipurpose trip covering the same distance. And let the car idle to warm up doesn't help with the fuel economy: it actually uses more fuel and creates more pollution.

- Small variations in vehicle manufacturing can cause fuel economy variations in the same make and model,
- Some vehicles do not attain maximum fuel economy until they are "broken in" (around 3,000–5,000 miles).
- An extra 100 lbs. can decrease fuel economy by 1–2 %.
- A roof rack or carrier provides additional cargo space and may allow one to meet one's needs with a smaller car. However, a loaded roof rack can decrease fuel economy by 5 %.

Factors influencing fuel economy include:

- Observing the speed limit: each 5 mph one drives over 60 mph can reduce your fuel economy by 7–8 %.
- Idling gets 0 miles per gallon and costs as much as $0.04 per minute.
- Using cruise control on the highway to help maintain a constant speed usually will save gas.
- A car that is noticeably out of tune can decrease gas mileage by about 4 %.
- Keeping tires inflated to the recommended pressure and using the recommended grade of motor oil can improve fuel economy by up to 5 %.
- Replacing a clogged air filter can improve gas mileage on older cars with carbureted engines.

Figure 13.2 shows the average relationship between fuel economy and speed [21]. The higher fuel economy occurs when speeds range from about 35–55 mph. The lowest fuel economy occurs when speeds are less than 15 mph. Thus traffic congestion has an adverse effect on fuel economy.

Excessive idling, accelerating, and breaking in stop-and-go traffic (average speed of 5 mph) increases fuel consumption by over 3 times the amount consumed at highway speed is 45 and 55 mph, and by 2 times the amount consumed at 15 mph on city streets. Table 13.6 illustrates the incremental fuel consumed at slow speeds compared to higher speeds using data extrapolated from Fig. 13.2.

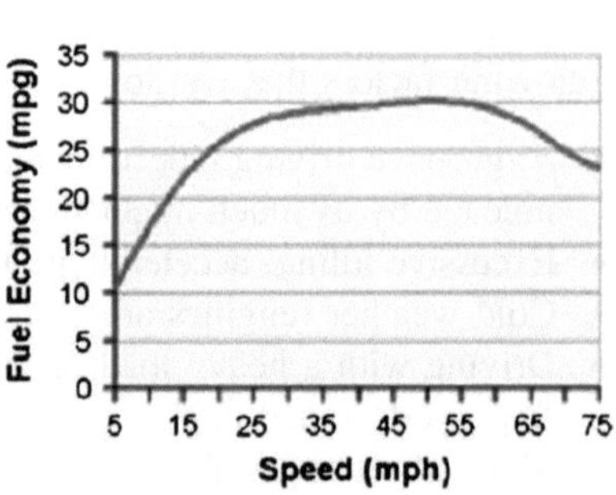

Fig. 13.2 Fuel economy versus speed. *Source* Reference [21], p 42

Table 13.6 Estimated impact of traffic congestion on fuel consumption

Speed (mph)	Fuel consumption (gals./mile) *Source* Fig. 5.3	Additional fuel consumed when speed changes from:	Additional fuel consumed when speed changes from
5	0.100	45/55–5 mph = +0.067 gals./mile (+300 %)	15–5 mph = +0.055 gals./mile (+167 %)
15	0.045	45/55–15 mph = +0.012 gals./mile (+36 %)	–
45–55	0.033	–	–

Source Estimated from Fig. 13.2

13.2.1.4 Some Examples of Reported Congestion Costs

This section gives some examples of the reported costs of traffic congestion.

US Urbanized Areas

The cost of traffic congestion in the US urban areas is reported annually by the Urban Mobility Report (UMR) [22] and is widely quoted by the national press. The UMR defines the time and fuel costs of congestion as:

$$\begin{aligned}\text{Time and Fuel Costs} &= [(\text{Actual Travel Time}) - (\text{Free - Flow Travel Time})] \times [\text{Value of Travel Time}] \\ &\quad + [(\text{Fuel Consumption in Actual Traffic Conditions}) - (\text{Fuel Consumption in Free - Flow Conditions})] \\ &\quad \times [\text{Unit Cost of Fuel}]\end{aligned} \tag{13.5}$$

Using free-flow travel time as the threshold of congestion the following results were documented for 2010:

- In 2010, congestion caused urban Americans to travel 4.8 billion h more and to purchase an extra 1.9 billion gallons of fuel.
- Congestion causes the average urban resident to spend an extra 34 h of travel time and use 14 extra gallons of fuel, which amounts to an average cost of $713 per commuter (Table 13.7).
- The value of wasted time, fuel and truck operating costs amounted to $101 billion total congestion cost, $23 billion of which is due to truck congestion.

The cost of congestion delay time calculated by the Urban Mobility Report (UMR) is based upon the difference between actual travel time and free-flow travel time. However, this assumption over-states the cost of congestion for the following reasons:

(1) In large urban areas, traffic volumes in the rush hours cannot be expected to travel at free-flow speeds because peak period free-flow conditions in large metropolitan areas are practically impossible to achieve.
(2) Increasing traffic volume is the outcome of increasing social and economic activity—a desired social goal. The added travel time from free-flow conditions

Table 13.7 Major findings of the 2011 urban mobility report (439 US Urban Areas)

Measures of …	1982	2000	2005	2009	2010
…individual congestion					
Yearly delay per auto commuter (h)	14	35	39	34	34
Travel time index	1.09	1.21	1.25	1.20	1.20
Commuter stress index	–	–	–	1.29	1.30
"Wasted" fuel per auto commuter (gallons)	6	14	17	14	14
Congestion cost per auto commuter (2010 dollars)	$301	$701	$814	$723	$713
…the nation's congestion problem					
Travel delay (billion hours)	1.0	4.0	5.2	4.8	4.8
"Wasted" fuel (billion gallons)	0.4	1.6	2.2	1.9	1.9
Truck congestion cost (billion of 2010 dollars)	–	–	–	$24	$23
Congestion cost (billion of 2010 dollars)	$21	$79	$108	$101	$101

Note The value of time for the particular year is the same for all urban areas and was estimated for passenger vehicles and trucks. The fuel costs are the per-gallon average price for each state. The value of a person's time is derived from the perspective of the individual's value of their time, rather than being based on wage rate. Only values of truck operating time are included; the value of the commodities is not

Source Reference [22], p 1, Exhibit 1

should not be considered an avoidable "cost" without considering the benefits of congested travel.

(3) A more rational congestion threshold metric is needed: one that is scaled to the size of the urban area and type of road (for example, see Table 8.6).

International Examples

This section summarizes some international examples of congestion cost estimates.

1. VTPI (Victoria Transport Policy Institute) [23] summarizes key congestion cost estimates from the US, New Zealand and central Europe (Table 13.8). However, these data may not be comparable without knowing the threshold of congestion used in each case.
2. Transport Canada [24] research report "calculates recurring and non-recurring congestion costs (including the value of excess delay, fuel use and greenhouse gas emissions). This approach identifies the various congestion threshold baselines that represent the point at which urban-peak speed reductions are considered unacceptable congestion." The 2002 annual congestion costs for various Canadian cities are shown in Table 13.9, for various congestion speed threshold levels of 50, 60, and 70 % of free-flow speed" (Table 13.9).

Numerical Example

The example assumes that population of City 1 is greater City 2, and the population of City 3 is smaller than that of City 2. Although freeways in each city experience a free-flow speed of 60 mph, their threshold congestion speeds increase with

Table 13.8 Congestion cost estimate summary table—selected studies

Publication	Costs	Cost value	2007 USD
Delucchi (1997)	Total US in 1991	$34–146 billion (1991)	$52–222 billion
	Per urban peak mile	$0.07–0.32	$0.11–0.49/ mile
Lee (2006)	US traffic congestion delay costs, relative to free flowing traffic	$108 billion (2002)	$124 billion
	Delay costs based on willingness to pay	$12 billion	$14 billion
TRB (1994)	Congested urban roads per vehicle mile	Average of $0.0 10–0.15*	$0.14–0.21/ mile
Texas Transportation Institute (2007)	Total US in 2005	$78.2 billion (2005)	$83 billion
Winston and Langer (2004)	Total US congestion costs	$37.5 billion (2004)	$41 billion
Land Transport New Zealand (2005)	Benefits of TDM mode shift per km	SI.27–Auckland,	$1.09/mile
		$0.98–Wellington,	$0.84/mile
		$0.09–Cri5tchurch (NZ$2002/Km)	$0.08/mile
FHWA (1997)	Urban Highway Car	$0.062/VMT*	$0.08/mile
	Bus	$0. 128	$0.17
Maibach et al. (2008)	Urban collectors in European centres over 2 million—Car	0.5 €/vkm 2000	$0.89/mile
	Truck	1.25 €	$2.23

* indicates the currency year is assumed to be the same as the publication year
Source Reference [23], pp 5.5-15, Table 5.5.4-1

decreasing city size, producing different conclusions for each city as to the amount of congestion associated with an actual congestion speeds of 30 mph (Table 13.10).

This example shows that important role played by the different threshold value of congestion, in calculating its magnitude. Congestion cost estimates are often developed using different congestion thresholds and different assumptions/methods. When comparing the results of different studies it is vital to state the assumptions used in measuring congestion delay. The need to clarify the assumptions inherent in congestion delay calculations is critical in interpreting its significance in transportation decision making.

13.2.1.5 Traffic Crashes and Traffic Congestion

Many roadway design features and operating practices not only increase congestion, but they also increase traffic crashes. Freeway designs such as short weaving distances, inadequate ramp lengths, and left-hand access points create conflicts and

Table 13.9 Annual (2002) estimated congestion costs (Million $$) in various Canadian cities

Location	Threshold		
	50 %	60 %	70 %
Vancouver	$737	$927	$1,087
Edmonton	$96	$116	$135
Calgary	$185	$211	$222
Winnipeg	$121	$169	$216
Hamilton	$20	$33	$48
Hamilton (old)	$17	$23	$30
Toronto	$1,858	$2,474	$3,072
Ottawa-Gatineau	$100	$172	$246
Ottawa-Gatineau (no rural)	$97	$166	$238
Montreal	$1,179	$1,390	$1,580
Qucbcc City	$73	$104	$138
Total, base	$4,370	$5,596	$6,745
Total, Old Ham./no rural Ottaiva-Gat.	$4,364	$5,580	$6,721

Note Congestion costs were estimated from a combination of sample observations (e.g., floating car method, traffic counts), travel demand models and transportation supply models
Source Reference [23], p 5.5-16, Table 5.5.4-3

Table 13.10 Estimates of congestion speeds for three threshold speeds in canadian cities

	City 1	City 2	City 3
Free-flow speed (mph)	60	60	60
Threshold speed as % of free-flow speed (%)	50	60	70
Threshold speed (mph)	30	36	42
Actual speed (mph)	30	30	30
Congestion speed reduction (mph)	None	6	12

traffic delays that produce crashes. Similarly, complex and offset street intersections, absence of turning lanes, unduly long traffic signal cycles, and on-street parking can create long queues during peak traffic periods that are often the cause of rear-end collisions.

Traffic congestion increases the density of vehicles occupying the road (vehicles per mile of road increases). As vehicles follow each other at close spacing, tend to change lanes more frequently, and merge into crowded lanes to exit or enter the roadway, the risk of crashes increases (Fig. 13.3, [2]).

Research in Colorado [25, 26] also found a relationship between traffic volume/density, speed, and crashes (Fig. 13.4). This figure shows how the crash rate increases with both AADT and traffic density (passenger cars per lane per mile): At 60,000 AADT, the crash rate is 0.64. At 96,000 AADT, it is 0.85., and at about 150,000 AADT, with a “super critical” density greater than 45 pc/mi/ln, the crash rate is 1.56 crashes per million VMT. The authors conclude that better managed Interstate freeways could lead to reduced crash rates.

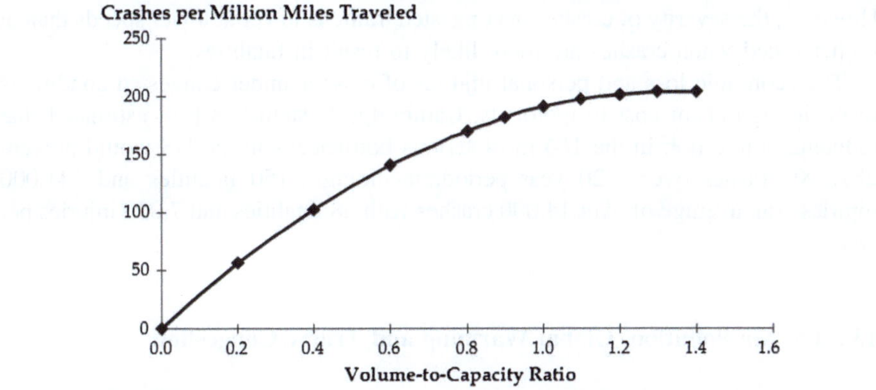

Fig. 13.3 Relationship of congestion and accident rates on urban highways. *Source* Reference [2], the positive impacts of transportation investment, NCHRP Project 8-36, task 22—compilation of working papers. Cambridge systematics, transportation research board. February 2002. Page 4–6, Fig. 2

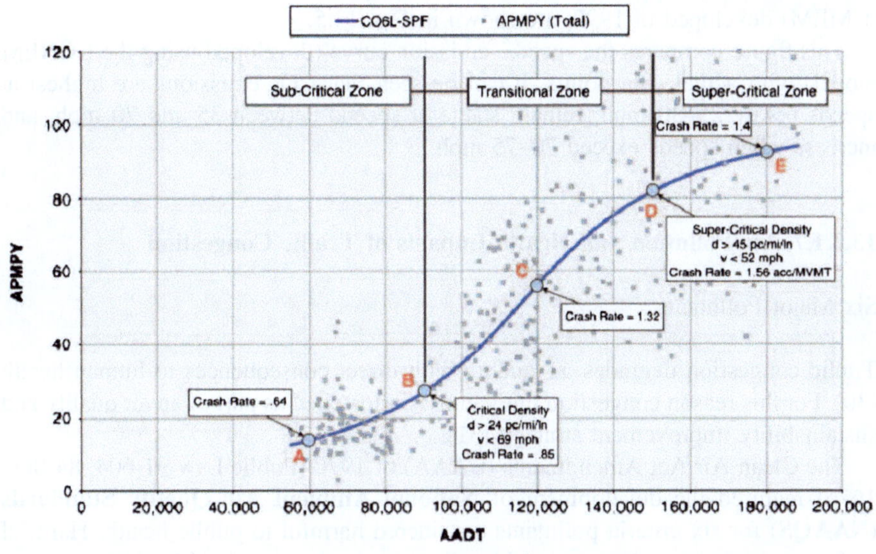

Fig. 13.4 Linking traffic volume, speed, and crashes in urban interstate highways. *Source* Reference [26], p 17. © National Academy of Sciences, Washington, DC, Reproduced with permission of the Transportation Research Board

The greatest effect of incidents on congestion occurs during peak hours. One study [27] using data from the 1970s to early 1980s reports that the presence of a stalled vehicle in a traffic lane can cause a delay of 100–200 vehicle hours to other motorists; and crashes that involve injuries or spills create larger delays. Each crash typically lasts 45–90 min, causing 1,200–2,500 vehicle hours of delay [2].

However, the severity of crashes in congested traffic is lower at lower speeds than at higher speed when crashes are more likely to result in fatalities.

The economic loss and personal injuries of crashes under congested conditions result in significant cost to motorists. Cambridge Systematics [28] estimated that reducing congestion in the 166 most serious bottlenecks in the US would prevent 287,200 crashes over a 20 year period, including 1,150 fatalities and 141,000 injuries—an average of over 14,000 crashes with 58 fatalities and 7,050 injuries per year.

13.2.1.6 Air Pollution, Global Warming and Traffic Congestion

The emissions of carbon dioxide (CO_2) have been associated with Greenhouse Gas (GHG) emissions. In 2007, highway travel accounted for almost 80 % of the 28,000 million pounds of CO_2 related to transportation. Roadway traffic is therefore a major contributor to GHG.

The relationship between speed and emissions is important from both congestion and air quality perspectives. Results of a comprehensive modal emission model (CMEM) developed in 1999, are shown in Fig. 13.5.

This figure compares the speed—emission curve (developed using the CCNEM model) with actual data points. It can be seen that CO_2 emissions are highest at speeds below 25–30 mph; remain stable at speeds between 35 and 70 mph, and increase when speeds exceed 70–75 mph.

13.2.1.7 Air Pollution and Health Impacts of Traffic Congestion

Six Major Pollutants

Traffic congestion degrades air quality with direct consequences to human health [30]. For this reason congestion mitigation is often cited as part of an air quality and sustainability improvement strategy [31].

The Clean Air Act Amendments (CAAA) of 1970 (Public Law 91-604, 84 Stat. 1676) required the development of **National Ambient Air Quality Standards (NAAQS)** for six criteria pollutants considered harmful to public health. Harmful emissions include: carbon monoxide (CO), lead, nitrogen dioxide (NO_2), ozone, particulate matter (PM10), and sulfur dioxide (SO_2) [31].

A brief description of each of these six major pollutants, as reported in [32], follows:

1. **Carbon Monoxide**

Carbon Monoxide (CO) enters the blood stream and links to hemoglobin, reducing delivery of oxygen to the body's organs and tissues. The health threat from lower levels of CO is most serious for those who suffer from cardiovascular disease.

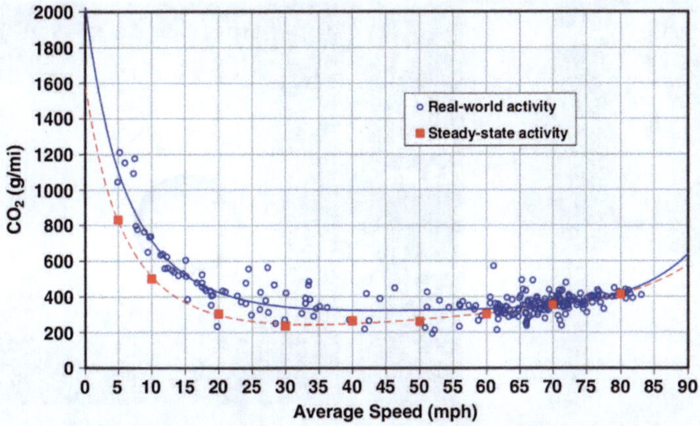

Fig. 13.5 CO_2 emissions as a function of average trip speed. *Source* Reference [29], p 166, Fig. 3

However, impairment of cognitive skills, vision, and work capacity may occur with elevated CO levels in healthy individuals [32].

2. <u>**Nitrogen Oxides**</u> **(NO_x)**
The group of gases composed of only nitrogen and oxygen, including nitrogen dioxide (NO_2) and nitric oxide (NO), which is rapidly converted to nitrogen dioxide (NO_2) after emission from vehicles and other sources. Nitrogen Oxides (NO_x) are gases produced by fuel combustion. Exposures have been associated with lung irritation, emergency department visits and hospital admissions for respiratory conditions. Nitrogen oxides contribute to the formation of ozone.

3. <u>**Volatile Organic Compounds**</u> **(VOC)**
VOCs include a variety of chemicals, some of which may have short- and long-term adverse health effects. VOCs are a major contributor to Ozone.

Ozone at ground-level is created by a chemical reaction between oxides of nitrogen (NO_x) and volatile organic compounds (VOC) in the presence of sunlight.

Ground-level ozone is the primary constituent of smog. Sunlight and hot weather cause ground-level ozone to form in harmful concentrations in the air. As a result, it is known as a **summertime air pollutant**.

4. <u>**Particulate Matter**</u> **(PM)**
Particle pollution or **PM**, is a complex mixture of extremely small solid particles and mists. Particulate pollution is made up of a number of components, including acids (such as nitrates and sulfates), organic chemicals, metals, and soil or dust particles. Smaller particles of 10 micrometers in diameter or smaller (**PM10**, **PM2.5**, and **UFP** are shown in Fig. 13.6 [33]) more easily bypass the natural defenses of the body, and are more easily inhaled deep into lung tissue, where they can cause health problems [30].

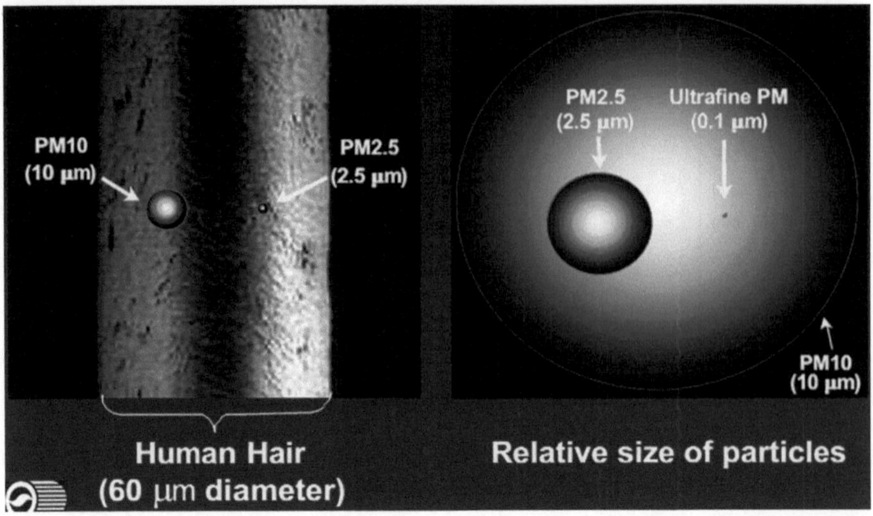

Fig. 13.6 Comparison of PM10, PM2.5, and ultrafine PM. *Source* Reference [33], Slide 2

- **Fine Particles (PM2.5)**
  Fine Particles (**PM2.5**) are small, airborne particles with a diameter of 2.5 micrometers or less. PM2.5 that can penetrate deep into the lungs, causing inflammation of the airways, exacerbating lung and heart disease, increasing hospital admissions and contributing to premature mortality. Sources of PM2.5 include all types of combustion sources; the elemental composition of PM2.5 can vary by source and determine PM2.5 health effects. Studies [34–36] found that motor vehicles contribute up to one-third of ambient PM2.5 in urban areas in the US [30].
- **Ultrafine Particulate Matters**
  Ultrafine particles (**UFP** or **Ultrafine PM,** with diameter <0.1 µm) are another component of motor vehicle emissions, produced by gasoline and diesel engines.

5. <u>**Lead**</u>

Element that exists in fuels and absorption of it by human body leads to higher Blood Lead that brings multiple adverse health results. **However, transportation no longer accounts for a large share of pollution from lead**. Highway vehicles currently account for less than 1 % of total lead emissions, primarily because of the use of unleaded gasoline.

6. <u>**Sulfur Dioxide**</u>

SO_2 is a gaseous pollutant formed by the combustion of fuels containing sulfur (e.g., coal, oil). It is usually measured in winter, since the residual heating oil contributes most of it to the city's concurrent air pollution.

Table 13.11 Contribution of transportation to emissions of air pollutants in the United States, 1999 (millions of short tons)

	Pollutant						
Source category	CO	NO_x	VOCs	PM10	Lead	SO_2	Total
Transportation							
Total	75.1	14.1	8.5	0.8	0.5	1.3	100.3
Highway vehicle share	49.9	8.6	5.3	0.3	0.02	0.4	64.5
Fuel combustion	5.3	10.0	0.9	1.0	0.5	16.1	33.8
Industrial processes	7.6	0.9	8.0	1.3	3.2	1.5	22.5
Miscellaneous	9.4	0.3	0.7	20.6	0.0	0.01	31.0
Total	97.4	25.3	18.1	23.7	4.2	18.9	187.6
Share of total (percent)	–	–	–	–	–	–	–
All transportation	77.0	56.0	47.0	3.0	12.0	7.0	53.0
Highway vehicles	51.0	34.0	29.0	1.3	0.5	2.1	34.0

Source Reference [31], p 43. Table 2-1

Table 13.11 estimates the share of each of the above air pollutants emitted in the US in 1999, by four major sources: *Transportation, Fuels Combustion, Industrial Process, and Miscellaneous.*

The transportation sector accounts for 53 % of the total annual pollutants including 34 % generated by highway vehicles. The impact of highway vehicles' air pollutants on the US population, however, is much is greater than 34 %. This is because most of the US population lives in urban areas where the largest number of highway vehicles is concentrated.

Congestion Impacts on Tail Pipe Emissions

Emissions increase where vehicles spend more time in congestion. It has been reported that stop-and-go traffic congestion "releases as much as three times the pollution as free-flowing traffic" [37].

Vehicle emissions are lowest in moderate speed ranges, when vehicle speeds are more uniform, and rise at higher speeds.

Health Risk from Exposure to Air Pollution

Vehicle-related pollutants such as particular matter and ultrafine particles (soot from gasoline or diesel), nitrogen oxide, and carbon monoxide are highly concentrated immediately downwind from major roadways. Their concentration is highest at the edge of the road and reduces as distance from the roadway increases (Fig. 13.7).

These pollutants are highly dangerous to the health of those who live within a risk zone 1,500 ft from the roadway. Figure 13.8 shows example distances where multiple health effects are found, including adult heart attacks, lung disease, chronic heart disease, and children asthma and lung development [37].

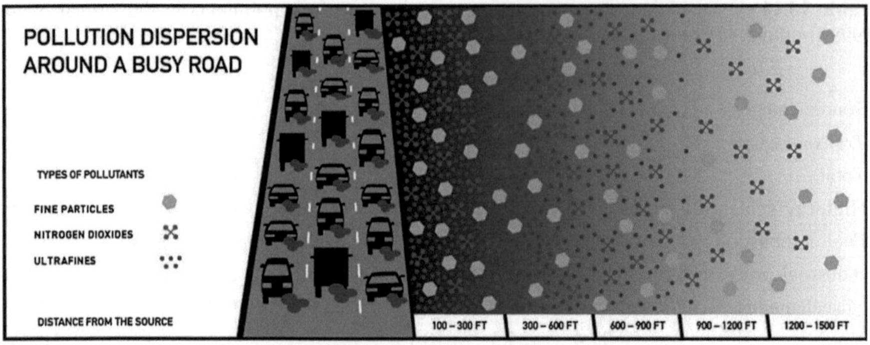

Fig. 13.7 Pattern of air pollutant concentrations emitted by highway vehicles at various distances from a busy road. *Source* Reference [37], p 3. Fig. 1

Fig. 13.8 Health findings for people living near roads with heavy traffic. *Source* Reference [37], p 4. Fig. 2

Some Examples of Pollution Impacts on Large Urban Areas

US Urbanized Areas

Several of the most congested metropolitan areas are also the most polluted. The CMAQ Special Report 264 [31] compared the most congested urban areas in the year 2000, with concurrent air quality status classified by EPA. The results of this comparison are shown in Table 13.12.

Studies reported by the Environmental Defense Fund, 2007 [22] also show the following:

- "Children are especially vulnerable to the effects of traffic-related air pollution; studies show increased prevalence of asthma, respiratory symptoms, and stunted lung development." [38–43]
- "Higher exposure to traffic emissions was associated with increased risk of breast cancer among women in Erie and Niagara Counties in New York State" [44].
- "A Los Angeles Study found that if researchers more accurately estimate exposure, based on localized rather than ambient air pollution levels, estimates of risk of death from heart attacks triple" [45].

Table 13.12 Air quality status of congested urban areas

	Rank on TTI congestion rating			
Urban area	Populallon (thousands)	TRI	TTI	Air quality status (Areas in nonattainment)
Los Angeles, CA	12,600	1	1	Extreme ozone; serious CO; serious PM10
San Francisca–Oaklaric, CA	4,025	2	3	Ozone (unclassified)
Seattle–Everett, WA	1,905	3	2	–
Washington, DC–MD–VA	3,490	4	4	Serious ozone
Chicago, IL–Northwestern IN	3,065	5	7	Severe ozone
San Diego, CA	2,700	5	9	Serious ozone
Boston, MA	3,020	7	4	–
Portland–Vancouver, OR–WA	1,490	a	a	–
Atlanta, GA	2,860	9	10	Serious ozone
Las Vegas, IW	1,260	9	16	Serious CO
Denver, CO	1,360	11	11	Serious CO; moderate PM_{10}
Houston, TX	3,130	12	11	Severe ozone
New York, NY–Northeastern NJ	16,430	13	16	Severe ozone; moderate CD
Miami-Hialali, FL	2,100	13	14	–
Detroit, MI	4,020	15	13	–

Note EPA classifications of nonattainment areas as of July 20, 2000. *TRI* travel rate index; *TTI* travel time index (see text for definitions); *CO* carbon monoxide; *PM10* particulates between 2.5 and 10 μm and less than 2.5 μm in mean aerodynamic diameter. Population data are for urbanized areas; only developed land with a density of greater than 1,000 persons per square mile is included in the boundary

Source Reference [31], p 61, Table 2-3

- "Another study from Worcester, Massachusetts found a 5 % increased risk of acute heart attack for each kilometer closer a subject lived to a major roadway" [46].
- "In New York City, more than 2 million people live within 500 ft of major roadways: in Manhattan this accounts for 75 % of the population Table 13.13. Shows the population living within the 500 foot risk zone in each of the city's five boroughs".

International Experience

The same Environmental Defense report [37] identifies similar findings by European researchers:

- "A key study from 2005 that found that the risk of asthma increased 89 % for each quarter mile closer children lived to a major roadway; the follow-up 2007 study found decreased lung air flow function for children living within about 1,500 ft of a major roadway" [43].
- "A study in Stockholm found a 40 % increase in lung cancer risk for the group with the highest average traffic-related NO_2 exposure" [47].

Table 13.13 New York city population living in the traffic pollution risk zone

New York City borough	Total population in risk ZDne	18 years and younger	65 years and over	Minority
Bronx	218,900	44,700	24,900	143,200
Brooklyn	595,700	158,800	41,300	399,100
Manhattan	1,217,700	204,000	149,200	515,400
Queens	379,300	85,200	51,800	211,700
Staten Island	36,300	9,300	4,100	6,200
Total	2,447,900	521,000	291,300	1,274,600

Source Reference [37], p 5, Table 1

- "Multiple studies have found serious health effects from exposure to heavy-duty diesel trucks, including increased mortality rates. Diesel emissions on busy roads have been associated with triggering asthma attacks, and may play a role in the initial onset of asthma" [48].

Beijing

Guo et al. [49] estimated the heavy traffic impacts on air quality and health of **Beijing**. The annual rate of motor vehicle growth has steadily accelerated from about 300,000 in 1989 to 3,500,000 in 2008.

This rapid increase in vehicle registration has caused severe traffic congestion. In 2008 the average traffic speed in the morning and afternoon peak periods were reported at 15.4 mph (24.7 kmph) and 13.9 mph (22.3 kmph) respectively.

Road transport-related air pollution was attributed to 3,413 deaths 2004. It also accounted for 16,030 episodes of acute bronchitis, 4,900 cases of chronic bronchitis, 598 cardiovascular hospital admissions, and 19,159 cases of asthma attack.

13.3 Implication

Growing motor vehicle use in metropolitan areas has increased traffic congestion. The consequences of this congestion include increases in the cost of doing business; increases in crashes and fuel consumption, high levels of air pollution harmful to human health, and a degradation in the quality of life. Reducing congestion, therefore, will reduce travel costs, crashes, fuel consumption, and will improve air quality and human health.

References

1. NGA Report
2. The Positive Impacts of Transportation Investment (2002) NCHRP Project 8-36, Task 22, Final Report, Working Paper 4—Compilation of Working Papers. Cambridge Systematics, Transportation Research Board

3. Wiesenthal DA, Hennessy DL (1999) Traffic congestion, driver stress, and driver aggression, s.l. Aggressive Behavior 25
4. Novaco RW, Stokols D, Milanesi L (1990) Objective and subjective dimensions of travel impedance as determinants of commuting stress. s.l. Am J Commun Psychol 18(2):231–257
5. Cambridge Systematics, Inc. (1994) Impact of urban congestion on business, repaired for national cooperative highway research program project 2–17(5). Transportation Research Board, Washington, DC
6. The Cost of Congestion to the Economy of the Portland Region. s.l. (2005) Portland Business Alliance, Metro, Port of Portland and Oregon Department of Transportation
7. Centre for International Economics (2006) Business costs of traffic congestion. s.l.: Victorian Competition and Efficiency Commission
8. Golob TF, Regan AC (2001) Impacts of highway congestion on freight operations—perceptions of trucking industry managers. s.l. Transp Res Part A 35(2000):577–599
9. Weisbrod G, Vary D, Treyz G (2001) NCHRP Report 463: economic implications of congestion. Transportation Research Board, Washington, DC
10. Minahan T (1969) How buyers changed it. Purchasing Sept 5, 1996
11. European Conference of Ministers of Transport (2007) Managing urban traffic congestion. s.l. OECD/ECMT, Paris
12. Produced by Texas Transportation Institute and Cambridge Systematics, Inc. (2006) Travel time reliability: making it there on time, all the time. s.l. Published by Office of Operations, Federal Highway Administration, US Department of Transportation, Washington, DC
13. US Department of Transportation, Federal Highway Administration (2006) Freight performance measurement travel time in freight—significant corridors. Dec. 2006
14. White paper prepared for FHWA by HLB Decision Economics (2001) Creating a freight sector within HERS. Department of Transportation (USDOT), Federal Highway Administration (FHWA), Washington, DC
15. Small Kenneth, Noland Robert, Chu Xuehao, Lewis Divid (1999) NCHRP Report 431: valuation of travel-time savings and predictability in congested conditions for highway user-cost estimation. National Academy Press, Washington, DC
16. Guidance on the Average Values of Traveler Time Savings (VTTS). s.l. US Department of Transportation
17. New York Metropolitan Transportation Council (NYMTC) (2006) Update of the congestion management process procedures. Nov 16, 2006
18. Federal Highway Administration. The Highway Economic Requirement System (HERS). s.l. Technical Report, (updated 3/97)
19. Beaton P, Chen C, Meghdir H (1997) Stated choice for transportation demand management models: a study in predictive validity using a disaggregate truth set. s.l. Transp Res Rec 1598:1–8
20. Office of Transportation Policy, US Department of Transportation (2011) Revised departmental guidance on valuation of travel time in economic analysis
21. Fuel Economy Guide, Model Year 2012 (2012) s.l. US Department of Energy, office of energy efficiency and renewable energy. US Department of Energy, Washington, DC
22. Schrank D, Lomax T, Eisele B (2011) TTI's 2011 urban mobility report—powered by INRIX traffic data. Texas Transportation Institute, the Texas A&M University System, College Station
23. Litman TA, Doherty E (2009) Transportation cost and benefit analysis: techniques, estimates and implications, s.l. Victoria Transport Policy Institute, 2 Jan 2009
24. Itrans Consulting Inc. (2006) Cost of non-recurring congestion in Canada- final report. Transport Canada Economic Analysis, Ottawa
25. Kononov J, Lyon C, Allery BK (2011) Relating flow, speed and density of urban freeways to functional form of a safety performance function. Transp Res Rec 2236 (Safety Data, Analysis, and Evaluation 2011, 1:11–19)
26. TR News, No. 278, January–February 2012
27. Grenzeback LR, Woodle CE (1993) The true costs of highway congestion. ITE J

28. Cambridge Systematics, Inc. Unclogging America's Arteries
29. Boriboonsomsin, Matthew Barth and Kanok (2008) Transportation Research Record 2058: Real-World Carbon Dioxide Impacts of Traffic Congestion. s.l. Transportation Research Board
30. Levy JI, Buonocore JJ, von Stackelberg K (2010) Evaluation of the public health impacts of traffic congestion: a health risk assessment. Environ Health Glob Access Sci Sour 9
31. Special Report 264: The Congestion Mitigation and Air Quality Improvement Program: Assessing 10 Years of Experience (2002) Transportation Research Board, Washington, DC
32. National air quality and emissions trends report (1999). s.l. US Environmental Protection Agency, March 2001. EPA 454/R-01-004
33. Presentation: Ultrafine Particulate Matter: Public Health Issues And Related Research (2003) Air Resources Board, California Environmental Protection Agency, Jan 31, 2003
34. Rizzoa MJ, Scheff PA (2007) Fine particulate source apportionment using data from the USEPA speciation trends network in Chicago, Illinois: comparison of two source apportionment models. Atmos Environ 41(29)
35. Hammond DM, Dvonch JT, Keeler GJ, Parker EA, Kamal AS, Barres JA, Yip FY, Brakefield-Caldwell W (2008) Sources of ambient fine particulate matter at two community sites in Detroit, Michigan. Atmos Environ 42(4)
36. Lee S, Liu W, Wang YH, Russell AG, Edgerton ES (2008) Source apportionment of PM2.5: comparing PMF and CMB results for four ambient monitoring sites in the southeastern United States. Atmos Environ 42(18)
37. All Choked Up: Heavy Traffic, Dirty Air and the Risk to New Yorkers (2007) Environmental Defense
38. Gauderman WJ, Avol E, Lurmann F, Kuenzli N, Gilliland F, Peters J, McConnell R (2005) Childhood asthma and exposure to traffic and nitrogen dioxide. Epidemiology 16:6
39. Kim JJ, Smorodinsky S, Lipsett M, Singer BC, Hodgson AT Ostro B (2004) Traffic-related Air Pollution near busy roads: the east bay children's respiratory health study. Am J Respir Crit Care Med 170(5)
40. Janssen N, Brunekreef B, van Vliet P, Aarts F, Meliefste K, Harssema H, Fischer P (2003) The relationship between air pollution from heavy traffic and allergic sensitization, bronchial hyperresponsiveness, and respiratory symptoms in dutch schoolchildren. Environ Health Perspecter 111:1512–1518
41. Venn AJ, Lewis SA, Cooper M, Hubbard R, Britton J (2001) Living near a main road and the risk of wheezing illness in children. Am J Respir Crit Care Med 164(12):2177–2180
42. Gauderman WJ, Vora H, McConnell R, Berhane K, Gilliland F, Thomas D, Lurmann F, Avol E, Kunzli N, Jerrett M, Peters J (2007) Effect of exposure to traffic on lung development from 10 to 18 years of age: a cohort study. Lancet 368:571–577
43. Hoffmann B, Moebus S, Stang A, Beck EM, Dragano N, Möhlenkamp S, Schmermund A, Memmesheimer M, Mann K, Erbel R, Jöckel KH (2006) Residence close to high traffic and prevalence of coronary hear disease. Eur Heart J 27:2696–2702
44. Nie J, Beyea J, Bonner MR, Han D, Vena JE, Rogerson P, Vito D, Muti P, Trevisan M, Freudenheim JL (2005) Environmental exposure to traffic polycyclic aromatic hydrocarbons (PAHs) and risk of breast cancer. American Association for Cancer Research Annual Meeting, Anaheim, CA
45. Jerrett M, Burnett MRT, Ma R, Pope III CA, Krewski D, Newbold KB, Thurston G, Shi Y, Finkelstein N, Calle EE, Thun MJ (2005) Spatial analysis of air pollution and mortality in Los Angeles. Epidemiology 16:727–736
46. Tonne C, Melly S, Mittleman M, Coull B, Goldberg R, Schwarts J (2006) A case-control analysis of exposure to traffic and acute myocardial infarction. Environ Health Perspect
47. Nyberg F, Gustavsson P, Jarup L, Bellander T, Berglind N, Jakobsson R, Pershagen G (2000) Urban air pollution and lung cancer in Stockholm. Epidemiology 11(5):487–495
48. Schwartz J (2004) Air pollution and children's health. Pediatrics 113(4S):1037–1042
49. Guo XR, Cheng SY, Chen DS, Zhou Y, Wang HY (2010) Estimation of economic costs of particulate air pollution from road transport in China. Atmos Environ 44(28)

Part III
Congestion Relief Strategies

Chapter 14
Overview of Congestion Relief Strategies

14.1 Introduction

Keeping congestion manageable is essential, since excessive congestion can adversely affect a community's livability and economy. A basic objective of congestion-relief actions is to reduce congestion to manageable levels since its complete elimination is neither practical nor cost-effective in large urban areas.

There are two basic categories of traffic congestion that should be addressed: nonrecurring and recurring.

- Nonrecurring congestion is characterized by the occurrence of unexpected events that reduce road capacity e.g., vehicle crashes and other vehicle breakdowns, inclement weather, work zones), emergencies, or by surges in traffic demand that temporarily exceed roadway capacity.
- Recurring congestion results from expected increases in demand at locations where it exceeds roadway capacity during known time periods such as the am and pm peak hours of travel. It also results from reductions in capacity at known locations where there is an imbalance in lane continuity (e.g., three lanes merging into two lanes).

This chapter overviews the various ways to better manage congestion. It provides the context for various congestion relief measures that are presented in the chapters that follow. They give more details on the applicability, user benefits, user costs, and other effects, as well as public acceptance of capacity enhancement/expansion strategies, and travel demand reduction strategies.

Congestion relief actions should be keyed to the needs, opportunities, and resources of each urban area. Relevant considerations include:

- the location and extent of congestion
- the character of the development pattern and its transportation connectivity
- the number, size, and complexity of the various governmental jurisdictions and public agencies involved

J.C. Falcocchio and H.S. Levinson, *Road Traffic Congestion: A Concise Guide*,
Springer Tracts on Transportation and Traffic 7,
DOI 10.1007/978-3-319-15165-6_14

- likely existing and future resources available for congestion relief
- community attitudes pertaining to congestion relief versus other values and needs.

Within each urban area agency responsibility should be established and agencies should coordinate their congestion management and relief actions. Those arrangements will vary, depending on the size, complexity, and congestion problems of each area. Coordination of tasks among responding agencies is especially important to reduce response times to nonrecurring congestion events.

14.2 Framing Strategies for Managing Nonrecurring Congestion

Reducing the intensity, duration, and extent of congestion created by nonrecurring events involves the application of supply (adaptation) and demand management (mitigation) strategies/actions that are implemented in response to the type of event that reduces capacity or increases demand. These are fully described in Chap. 15.

One critical element of strategies for managing nonrecurring congestion is the application of ITS technologies (such as sensors and real-time information and communication) whose key function is to provide early detection of random events and to inform responding agencies about the location of the event and traffic flow impacts. Rapid recognition, response, and recovery are essential. A critical component in expediting response times requires responding agencies to coordinate their functions.

14.3 Framing the Strategies for Managing Recurring Traffic Congestion

Congestion management actions and strategies for recurring congestion should address real problems and needs. They should be perceived as being reasonable in terms of costs, benefits, and impacts. And they should be acceptable to the impacted communities.

Key considerations include the size and structure of the urban area, location of deficiencies on existing streets and highways, and community policies and attitudes regarding highways, transit and land use.

Congestion develops when demand exceeds capacity. Therefore, balancing roadway supply and demand is essential. The approach to congestion relief in this book is framed by strategies that focus on enhancing roadway capacity and strategies that reduce traffic demand.

Enhancing roadway capacity consists of two basic approaches: (1) maximize the throughput of existing roadways through operational improvements including the application of ITS technology, and (2) adding new capacity (e.g., removing geometric bottlenecks by widening roadways, adding lanes, or building new roadways). These strategies are described in Chaps. 15–17.

Reducing traffic demand consists of strategies that reduce automobile use or that offer mobility alternatives to private motor vehicles. These include: congestion pricing, road pricing, parking pricing, regulatory restrictions on auto use, employer-based programs, improved public transportation, and land use patterns that encourage alternatives to auto travel. These strategies are described in Chaps. 18–23.

14.3.1 Overview

Reducing congestion in a growing metropolitan area is typically achieved by implementing a combination of strategies that maximize available capacity, add new capacity, reduce automobile use, and reduce the growth in motor vehicle traffic. These strategies might involve increasing the choice in the number of routes and travel modes, or the creating traffic lanes that involve the payment of a toll for higher-speed and higher travel time reliability.

Table 14.1 gives the nine possible states of capacity and/or traffic demand that impact on traffic congestion.

The capacity-demand states (0–8) identified in Table 14.1 are briefly summarized below using Fig. 14.1.

0 Existing condition: This state denotes the congestion level being experienced and serves as a reference point from which the benefits and costs of changes in either roadway capacity, traffic demand, or both, can be evaluated.

1 Traffic demand increases (D+) and capacity remains the same (So): By keeping capacity constant (So), an increase in traffic demand over time (D+) increases congestion (lower speed) and increases vehicle-miles traveled (VMT). An example of this case is where population, car ownership, and economic activities grow while roadway capacity does not change. This condition prevails in large urban areas where transportation investment levels have not kept pace with demand growth. Demand mitigation strategies are most appropriate in large urban areas where roadway capacity increases are least feasible.

2 Traffic demand decreases (D−) and capacity remains the same (So): By keeping capacity constant (So), a decrease in traffic demand over time (D−)

Table 14.1 Possible combination of traffic demand and roadway capacity strategies

Traffic Demand	Roadway capacity		
	Increase	Same	Decrease
Increase	5	1	6
Same	3	0	4
Decrease	7	2	8

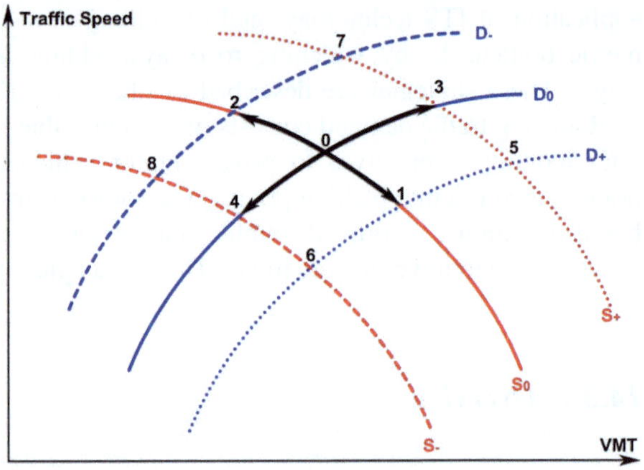

Fig. 14.1 Changes in average traffic speed and road use (VMT) resulting from strategies that change capacity and/or demand

produces a reduction in VMT and a reduction in congestion. This is likely to exist in areas where population and employment decrease.

3 Traffic demand remains the same (Do) and capacity is increased (S+): Increasing capacity (S+) while traffic demand remains the same (Do) reduces congestion and increases VMT. This is likely to happen where the capacity increase is implemented in a short time period (through operational improvements), while population and employment remain constant. Most of the VMT increase on the improved road results from traffic diverted from other routes and or time periods.

4 Traffic demand remains the same (Do) and capacity decreases (S−): This condition increases congestion and reduces VMT. Roadway capacity decreases in high density areas, especially in the middle of the day, where street space is shared with pedestrians, delivery vehicles, utility repairs, etc. In these cases congestion will increase and VMT will decrease.

5 Capacity increases (S+) in response to increasing demand (D+): This strategy aims at managing congestion to be within acceptable levels by balancing expected demand growth with additional capacity. In this case VMT increases while congestion is managed at acceptable levels. This condition is easiest to implement in smaller metropolitan areas.

6 Traffic demand increases (D+) and roadway capacity decreases (S−): This condition results in higher congestion and in VMT growth. It is typically found in the megacities of developing economies where population and employment growth increases land use density while transportation capacity growth does not keep pace with growth in private vehicles use. Consequently the presence of too many people and vehicles crowding the road space they share *reduces* the vehicle throughput (capacity) of the road network.

7 Traffic demand decreases (D−) and roadway capacity increases (S+): This condition is unlikely to be found in large metropolitan areas.

8 Traffic demand decreases (D−) and roadway capacity decreases (S−): This condition would reduce VMT. Its impact on congestion would be determined by the relative proportional decreases in demand and capacity. Actions that reduce demand would consist of increasing the cost of private vehicle use and improving access to alternative modes to the private car. Actions that reduce roadway capacity would involve reducing the number of roadway lanes used by private vehicles and dedicating them for public transport vehicles.

14.4 Strategies That Address the Causes of Congestion

Chapters 4–7, classified the causes of traffic congestion into four categories: (1) peaking of travel demand; (2) concentration of activities; (3) area-wide traffic demand growth exceeding capacity growth over time; and (4) recurring and non-recurring bottlenecks. Strategies that address each of these causes are briefly outlined below.

14.4.1 Cause: Recurring Peaking of Travel Demand

When travel demand exceeds available roadway capacity (usually at least twice a day—the morning peak demand and the evening peak demand) the duration of stop-and-go congestion is longer than the period when the demand exceeds capacity.

Example

Consider a bridge with an average directional capacity of 4,000 vph; a peak hour demand volume of 6,000 vph; and demand volumes below capacity before and after the peak hour.

Figure 14.2 shows the volume-capacity comparison before during, and after the morning peak hour (8–9 am).

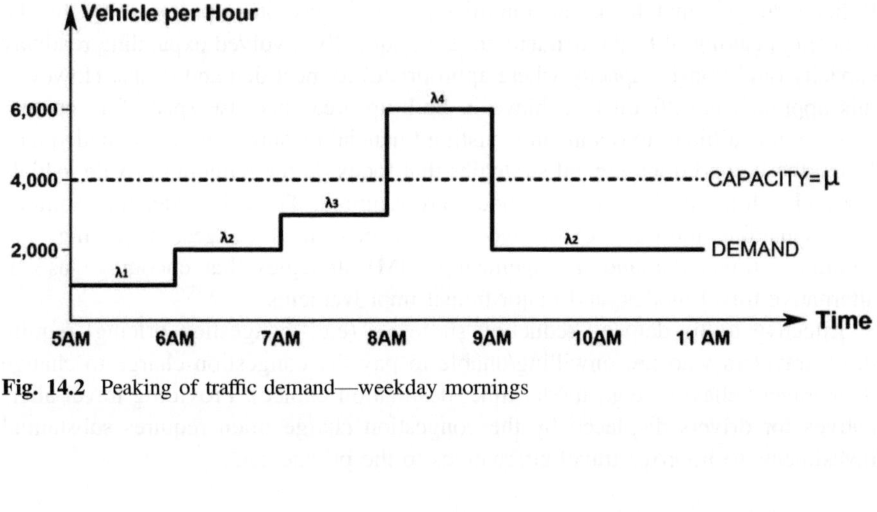

Fig. 14.2 Peaking of traffic demand—weekday mornings

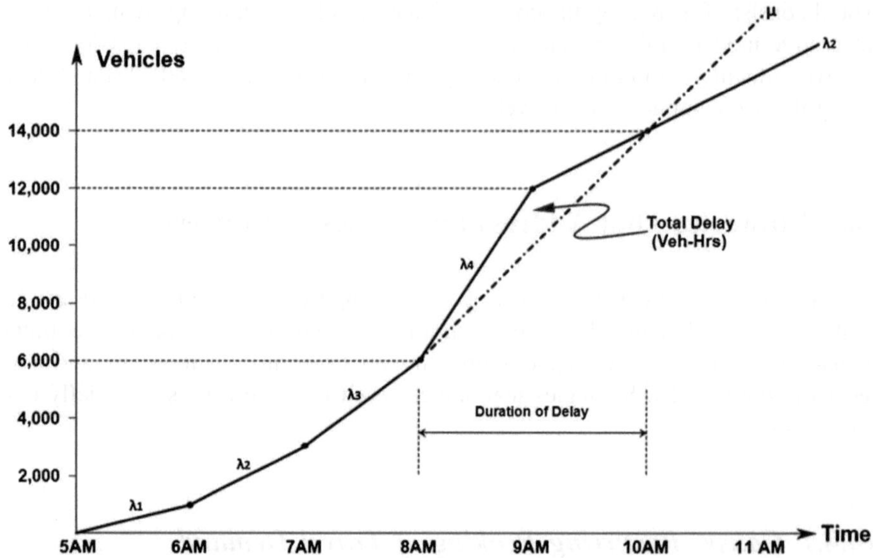

Fig. 14.3 Extent and duration of traffic delay when peak hour demand exceeds roadway capacity

Although traffic demand exceeds capacity for only 1 h, Fig. 14.3 shows that congestion delay on the approach roadways lasts 2 h (from 8 to 10 am), adding an average trip delay of 15 min to each of the 8,000 (14,000–6,000) vehicles delayed with a maximum trip delay of 30 min experienced by those vehicles arriving at 9 am.

14.4.1.1 Possible Strategies

Relief strategies that focus on minimizing the congestion impacts caused by the recurring peaking of traffic demand have traditionally involved expanding roadway capacity (and transit capacity where appropriate) to meet demand peaks. However, this approach is difficult to achieve in built-up areas because space for capacity expansion is difficult to obtain and existing financial resources are often inadequate. When these conditions prevail strategies that focus on the reducing private vehicle demand volume during the peak hours are required. These include, for example, peak spreading through flexible work hours, peak hour congestion pricing, and commuter travel demand management (TDM) strategies that encourage use of alternative travel modes, and major transit improvements.

Effective traffic demand reduction strategies (e.g., congestion pricing) require those travelers who are unwilling/unable to pay the congestion charge to change their travel behavior (e.g., mode, time, destination choice). Providing travel alternatives for drivers displaced by the congestion charge often requires substantial investments to improve travel alternatives to the private car.

Implementing effective travel demand reduction strategies requires strong stakeholder and political support through:

- Strong political leadership;
- a comprehensive planning and public information process that addresses the social, economic, environmental, and equity concerns of the stakeholders community; and
- addressing the short-term costs to those who are forced to change their travel behavior (by making alternative modes of travel available) against the long term benefits that will accrue to society in the form of sustainable mobility and environmental health.

14.4.2 Cause: Concentration of Activities

Concentrations of activities tend to generate peak vehicle traffic demands that exceed the capacities of roads serving these areas.

Megacenters in suburban areas are primarily accessible by private vehicles that cause severe traffic congestion in the peak hours on expressways and arterial roads leading to the megacenters.

Megacenters located in centers of large cities (e.g., Manhattan, Chicago, Boston, Philadelphia central business districts) consist of high-density commercial and residential developments where transit access predominates, with a high volume of pedestrian trips, and a low per capita share of private vehicle use. These locations typically face excessive traffic congestion as well.

14.4.2.1 Possible Strategies

Congestion relief strategies on the roadways serving suburban megacenters involve proactive strategies aimed at increasing transportation capacities and/or reducing peak hour traffic loads by limiting the growth of these centers to the capacity of their access roads, better access management, and coordinating the center's land use development/design to facilitate customer access by transit or as pedestrians. Coordinating land use decisions (under local control), land developers decisions (under private control) and transportation decisions (under State control) is a challenging task that is not always achievable since it involves the financial interests of private developers and the importance of tax revenues to the local towns' operating budgets.

Congestion relief strategies for large city centers typically require a combination of strategies involving operational and physical improvements to increase multi-modal transportation capacity, as well as travel demand management and parking management policies.

Strategies that reduce automobile use in peak period travel include transit service improvements that divert auto users into transit modes (e.g., new or extended rapid transit lines), flexible work hours to spread the peak demand outside the peak hour, road congestion pricing, downtown parking supply pricing and parking supply constraints within the congested areas, and offering outlying park-and-ride facilities connected to downtown by transit.

Adoption of parking pricing policies in city centers can relieve congestion by:

- influencing the choice of alternative modes for accessing the area
- increasing the availability of parking spaces in congested parts of the city at different times of the day
- reducing the incremental VMT added when searching for a parking space.

Parking supply constraints in large city centers served by off-street public transport can also help reduce congestion by:

- encouraging public transportation access to the city center
- concentrating park-and-ride facilities along outer stations along the transit line
- reducing commuter VMT on radial express highways leading to the city center by intercepting motorists at outlying park-and-ride facilities.

14.4.3 Cause: Area-wide Demand Growth Exceeding Capacity Growth

Population, job growth, and rising incomes increase traffic demand growth that, over time, exceeds growth in roadway capacity.

14.4.3.1 Possible Strategies

Congestion relief strategies in areas experiencing/expecting population and economic growth entail (1) increasing the capacity of the roadway network as well as other modes of transportation, and (2) reducing growth in private vehicle use by (a) coordinating land use growth policies that reduce the need to drive with investments in alternative modes of transportation, and (b) road pricing policies that increase the cost of driving.

Increasing Transportation Capacity

This policy involves expanding roadway and transit capacity to serve trips that are destined to destinations outside the range of the walk/bicycle modes. It also involves constructing a network of accessible bicycle lanes and pedestrian paths that connect to land use activities located within the range of these modes.

The challenge of finding space to build needed facilities in a built environment is often significant in light of public resistance and financial constraints.

The debate on how to increase transportation funding to serve increasing travel demand and to sustain the existing transportation system is a major issue for the US Congress and State legislatures and it is often fraught with conflicting visions and values among the stakeholders.

Coordinating Land Use Growth and Transportation Decisions

Locating different types of activities in the same zone with high development density will reduce the per capita use of private motor vehicles and will enable the use of public transit/walking/biking.

Land use policies that create zones of high density and activity mix, however, require extensive coordination among local zoning boards, and state, and federal transportation officials, that is time consuming and lacks political support.

Policy decisions required to coordinate transportation and land use planning are difficult to achieve in light of:

- the life-style choices for low-density living;
- existing institutional practices that involve separate governmental jurisdictions;
- the short life cycle of elected officials; and
- competing societal needs—education, police and fire protection, health care, etc. —for the limited financial resources available.

Pricing Strategies that Discourage Private Vehicle Use

These strategies include congestion pricing, road pricing, parking pricing, and/or regulatory constraints on the use of private vehicles in certain areas. They are more common in major cities located in Europe, Asia, and South America.

14.4.4 Cause: Bottlenecks (Operational, Physical, Incident-Induced, Weather, Special Events, Work Zones)

Bottlenecks happen where/when the upstream demand volume exceeds the throughput volume of the downstream roadway. They occur along freeways and city streets.

Congestion delays resulting from physical bottlenecks on the roadway system are of a recurring nature: delays happen every day during the same time periods on the same roads. However, delays resulting from incidents, weather, special events, and road maintenance can happen at any time, on any road. These events bottlenecks create non-recurring congestions.

14.4.4.1 Possible Strategies

Relief strategies that focus on removing/reducing bottlenecks produce the most immediate congestion relief considering that 65 % of congestion delays nationwide are attributable to bottlenecks—40 % of delay is from recurring bottleneck locations and 25 % from incident-induced bottlenecks (Table 7.1 and Ref. [1]).

Strategies that reduce the impacts of nonrecurring congestion delays involve the ability to detect the event as soon as it happens and to restore the roadway to full capacity as soon as possible, direct drivers to reduce speed during inclement weather or when the roadway is being repaired, or inform drivers about the location and times of special events that are likely to generate surge in traffic demand on the impacted roads.

Strategies that reduce recurring congestion delays at physical bottlenecks involve grade separation of intersecting traffic streams; road widening at bottleneck locations; the addition of merging and turn lanes; and reconfiguration of entrance and exit ramps at freeways and expressways. However, to prevent new traffic attracted to the improved roadway from nullifying the travel time benefits of bottleneck removal, strategies that reduce bottleneck congestion in highly congested roads should be coupled with strategies that control traffic demand on these roads (e.g., ramp metering).

In urban areas under 1–1.5 million people, recurring congestion can be reduced in intensity, extent, and duration. In larger urban areas, however, congestion relief is generally manifested in reducing its duration.

Strategies involving the removal of operational bottlenecks are relatively easy to implement from an operational standpoint when they are the responsibility of one transportation agency. However, where the owners of the transportation infrastructure are agencies typically controlled by different units of government, coordination of congestion management strategies among these units may be time consuming and not always easily achievable—especially when the limited funding available for this purpose may not be transferrable between agencies.

14.5 Summary

Most urban areas experience more congestion, poorer pavement and bridge conditions, and less public transportation services than required. A variety of solutions is needed for different metropolitan areas, their cities, job centers and shopping areas, and neighborhoods [1] Some of these will involve constructing new road capacity, others will require improving the throughput of existing roads, adding travel alternatives to the automobile, the development of diversified land use patterns, or the redevelopment of neighborhoods that encourage use of non-motorized transportation.

Table 14.2 FHWA congestion reduction toolbox strategies

Improve service on existing roads	**Chapter in book**
• Traffic Incident Management	15
• Traffic Signal Timing	16
• Arterial Management	16
• Access Management	16
• Freeway Management and Traffic Operation	16
• Road Weather Management	15
Pricing	
• Value Pricing	19
Add Capacity	
• Add Capacity/Easing Bottlenecks	17
• Public Private Partnerships	17
Better Work Zones	
• Work Zone Management	15
Travel Options	
• Travel Demand Management	18, 19, 20, 21, 22, 23
• Planned Special Events Traffic Management	15
Traveler Information	
• 511 Traveler Information Telephone Services	15, 16
• Travel Time Message Signs for Travelers	15, 16
• National Traffic and Road Closure Information	15, 16
• Real-Time Travel Time Information	15, 16
• Freight Shipper Congestion Information	15, 16

The various congestion reduction strategies set forth in the Federal Highway Administration's "Congestion Reduction Toolbox" [2] are shown in Table 14.2 This table also shows the book chapters where these strategies are described and assessed for their likely congestion reduction effects from their applications in the United States.

References

1. Shrank D, Eisele B, Lomax T (2012) TTI's 2012 Urban Mobility Report, powered by INRIX Traffic Data. s.1. Texas A&M Transportation Institute, The Texas A&M University System
2. Federal Highway Administration. Congestion Reduction Toolbox, US Department of Transportation, Washington, DC

Chapter 15
Managing Nonrecurring Congestion

15.1 Introduction

Nonrecurring congestion accounts for over half of all traffic delays in the United States [1] and accounts for up to 2/3 of traffic delays in urban areas larger than one million population [2]. Therefore preventing and reducing the impact of nonrecurring congestion is a key strategy for improving traffic conditions.

Various strategies for addressing traffic delays from nonrecurring events have emerged in recent years, and they are now receiving the same level of priority given to strategies that address recurring congestion.

These nonrecurring strategies are keyed to the type and causes of delay—whether delays are caused by incidents (e.g., crashes, vehicle breakdowns), work zones, special events that generate surges in traffic demand, inclement weather, or construction zones, and major evacuations.

To effectively minimize the adverse road user impacts of nonrecurring congestion requires the use of accurate real-time information that enables transportation agencies to proactively and quickly respond to changes in traffic conditions caused by incidents, adverse weather, road maintenance, or random surges in traffic demand from special events. These changes can consist of traveler advisory information in maintaining safe speed, lane changes, or dynamic traffic control policies that minimize delay, and that inform travelers about the location and extent of congested conditions. Real-time information is a major benefit of Intelligent Transportation Systems (ITS) technology that enables the emerging practice of Active Transportation and Demand Management (ATDM) [3].

J.C. Falcocchio and H.S. Levinson, *Road Traffic Congestion: A Concise Guide*,
Springer Tracts on Transportation and Traffic 7,
DOI 10.1007/978-3-319-15165-6_15

15.2 Incidents

Traffic incidents reduce roadway capacity. Estimates of the amount of freeway capacity reduction, as a function of number of lanes blocked by the incident, are provided in Table 15.1. This table shows that an incident reduces freeway capacity even when it is located at the shoulder of the roadway, and it is not physically blocking a lane. A roadway incident reduces the capacity of the roadway to various degrees depending on the number of lanes on the roadway, the number of lanes blocked by the incident, and the duration of the incident.

The factors involved in determining the amount and duration of delay resulting from an incident are illustrated in Fig. 15.1. They are described to show the intensity (amount of delay), duration (hours of congested conditions), and extent (number of vehicles delayed) of congestion caused by the incident. It can be seen that the duration of delay to the impacted traffic can substantially exceed the duration of the incident.

Definition of terms in Fig. 15.1:

t_0 = time when incident occurs;
t_1 = time when incident is detected;
t_2 = time when incident is reported to the traffic management center (TMC);
t_3 = arrival time of first responders with the means to restore roadway capacity;
t_4 = time when capacity is restored (all lanes are reopened);
t_5 = time when traffic flow is restored to conditions prevailing before the incident with no demand reduction;
t_6 = time when traffic flow is restored to conditions prevailing before the incident but traffic demand is reduced by ramp closings or by diverting traffic to other roads via dynamic traffic information devices;
$(t_1–t_0)$ = time elapsed after incident occurs and when it is detected;
$(t_2–t_1)$ = time elapsed after the incident is detected and when it is reported to the TMC;

Table 15.1 Percentage of freeway section capacity available as a function of incident condition and number of freeway lanes

Number of freeway lanes in each direction	Shoulder disablement	Shoulder accident	Lanes blocked		
			One	Two	Three
2	0.95	0.81	0.35	0	N/A
3	0.99	0.83	0.49	0.17	0
4	0.99	0.85	0.58	0.25	0.13
5	0.99	0.87	0.65	0.40	0.20
6	0.99	0.89	0.71	0.50	0.25
7	0.99	0.91	0.75	0.57	0.36
8	0.99	0.93	0.78	0.63	0.41

Source Reference [1], pp 1–9, Table 1.2

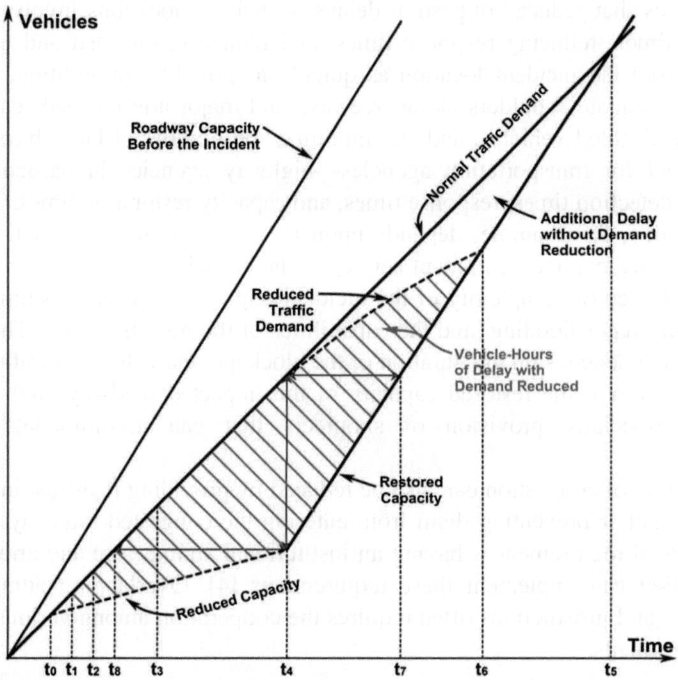

Fig. 15.1 Factors impacting on congestion delay from a roadway incident

$(t_3 – t_2)$ = time elapsed after the incident is reported and when the incident team arrives;
$(t_4 – t_3)$ = time needed to restore capacity (all lanes reopened);
$(t_5 – t_0)$ = duration of delay if traffic demand continues unabated;
$(t_6 – t_0)$ = duration of delay if traffic demand is reduced;
$(t_7 – t_4)$ = delay time to last vehicle in longest platoon (Q_{max}) before it resumes normal speed;
$(t_4 – t_8)$ = maximum delay time before capacity is restored;
$(t_4 – t_0)$ = time interval when roadway operates at reduced capacity.

The impact of an incident on delay depends upon:

- The severity of the incident (number of lanes blocked) and the volume of traffic on the roadway at the time of the incident;
- The time the incident is detected;
- The time it is reported to the responding agency;
- The agency response time;
- The time elapsed before the roadway capacity is completely restored;
- The volume of upstream traffic approaching the incident location.

Strategies that reduce congestion delays at incident locations involve reducing detection times, reducing response times, and removing disabled and emergency vehicles from the incident location as quickly as possible. In addition, the availability of adequate shoulders along freeways and major arterial roads can provide refuge for disabled vehicles, and can minimize the number of lanes blocked.

The goal for transportation agencies—highway agencies in particular—is to minimize detection times, response times, and capacity restoration times. The initial response time, for example, depends upon the location and coordination of fire, police, ambulance, and equipment services. The capacity restoration time depends upon the degree of complexity of the incident (e.g., crash severity, simple vehicle breakdown, major flooding, and the capabilities of the response team. The incident duration time depends on the duration of the blockage, the volume of traffic demand (5 am or 5 pm?), the restored capacity of the impacted roadway, and its design features—especially provision of shoulders that can accommodate disabled vehicles.

The extent of congestion can also be reduced by providing real-time information to drivers and/or preventing them from entering the congested roadways.

An critical requirement is having an institutional architecture and arrangements in place that can implement these requirements [4]. Overlapping administrative boundaries and jurisdictions often requires the cooperation among various agencies and governments:

- Responses to non-recurring events should be mainstream rather than ad hoc
- Coordinated approaches require that barriers to institutional change be overcome
- Institutional change can be driven by experiences that expose the weakness of existing institutional structures and by reconfiguring the institutional framework to better meet incident response requirements

Traffic incident management should be a planned and coordinated process to detect, respond to, and remove traffic incidents and restore traffic capacity as safely and quickly as possible. This coordinated process involves a number of public and private sector partners, including: Law Enforcement, Fire and Rescue, Emergency Medical Services, Transportation, Public Service Communications, Emergency Management, Towing and Recovery, Hazardous Materials Contractors, and Traffic Information Media.

Creating congestion management centers at strategic locations in urban areas can provide for a reduced response time to incidents. This helps to reduce impact duration, intensity, and extent.

Specific supply and demand strategies that can reduce incident delay are briefly identified below.

15.2.1 Supply Strategies

Key supply strategies include:

- Identifying incidents more quickly, reducing response times, and clearing incident scenes more rapidly will dramatically reduce incident delays—especially in peak hours.
- Stationing of response vehicles at strategic locations to minimize the time needed to remove incidents
- Training emergency responders with practices that minimize their presence at the incident site
- Providing traveler information to divert traffic away from the incident location.

Incident response performance can be improved and response times reduced by widespread deployment of surveillance and detection technologies.

15.2.2 Demand Reduction Strategies

Demand reduction strategies include:

- diverting traffic away from incident bottlenecks by:
  - providing real-time information to drivers on incident locations and congested roadways and
  - advising travelers of available alternative routes.
- Closing on-ramp to prevent additional traffic from entering the congested highway section

Advance variable message signs along freeways can help to divert motorists away from congested roads impacted by an incident.

15.3 Special Events

Special events account for up to 5 % of congestion delay [3].

- Special events (e.g., conventions, football games) are time and location-specific. Ideally the starting and ending times of these events should be scheduled outside the peak travel hours.

15.3.1 Supply Strategies

Supply strategies include:

- Pre-event coordination and planning
- Applying traffic engineering actions such as (1) parking bans on major streets serving the event, police control of traffic signals, and reversible travel lanes on streets leading to and from the event
- Developing traffic control plans in response to surges in traffic volume at the start and end of events
- Coordinating operations during the event
- Providing traveler information

15.3.2 Demand Reduction Strategies

Demand reduction strategies include:

- Metering upstream flow to maintain a better balance with downstream roadway capacity (e.g., real-time traffic signal timing; controlling the exit rate from parking lots/garages.
- Possibly closing various entry ramps downstream of the ramps where the event traffic enters the roadway

15.4 Inclement Weather

Adverse weather conditions have a major impact on traffic congestion, safety and operations. Weather affects driver behavior, vehicle performance pavement friction, and roadway infrastructure. Although weather events and their impacts are non-recurring, they are generally predictable. Inclement weather usually has an area wide impact but its severity is sometime concentrated in specific parts of the urban area where it accounts for up to 10 % of total traffic delay [5].

15.4.1 Supply Strategies

Supply strategies include:

- Better prediction and detection of rain, snow, ice, and fog on specific roads to assist transportation managers in roadway treatment strategies
- More effective treatment of roadway surface to improve traction and prevent ice bonding (sand, salt, anti-icing chemicals)
- Posting fog warnings on Dynamic Message Signs (DMS)

- Listing flooded routes on websites
- Reducing speed limit with Variable Speed Limit (VLS) signs and modifying traffic signal timing

15.4.2 Demand Strategies

Demand strategies include actions intended to give guidance to drivers on the safe use of roadways during adverse weather conditions:

- Issuing travel advisory information to drivers about weather events (snow, ice, fog) in specific areas or roadways to minimize crashes.
- Listing flooded routes on web sites to direct drivers to alternate routes
- Establishing control strategies that permit or restrict traffic flow and regulate roadway capacity

15.5 Work Zones

Work zones account for up to 19 % of delay in large urban areas and up to 27 % of delay in small urban areas. The degree to which work zones impact traffic speed depends upon the length of time the work zone is in place, the number of lanes closed or detoured by the work zone, and the hours when the work is performed, and driver awareness of the work zone.

15.5.1 Supply Strategies

Supply strategies include:

- reducing the time to complete the work
- scheduling the work during off peak hours
- maintaining roadway capacity (e.g., temporary contra-flow lanes)
- providing breakdown (shoulder) lanes
- maintaining safe work practices
- providing alternate routes

15.5.2 Demand Strategies

Demand strategies include:

- posting safe speed limits
- real-time monitoring of traffic speeds

- effective enforcement of speed limit
- providing traveler information on work zone locations and times

15.6 Information Technology (IT)

Each metropolitan area needs to establish an IT architecture that enables the monitoring of nonrecurring events, and provides real-time information to travelers on (1) incidents that cause major bottleneck delay (intensity, extent, and duration); (2) work zones; (3) road weather; (4) alternate route/travel modes.

15.6.1 Active Traffic and Demand Management (ATDM)

Application of Active Traffic and Demand Management (ATDM) strategies enables the system operator to manage incident impacts by a number of strategies aimed at real-time coordinated management of traffic controls, lane assignments, traveler information, etc. ATDM comprises a series of strategies (e.g., traffic sensors, traffic management centers, managed lanes, rapid incident response) to dynamically manage roadways and corridors in response to non-recurring sources of congestion [6]. These include:

>Queue warning display systems: to warn drivers of the presence of congestion downstream;

> Dynamic routing and traveler information: the use of dynamic message signs to display rerouting instructions in response to non-recurring congestion events; and

> Dynamic lane markings: the delineation of lanes to manage traffic flow patterns created by the above strategies.

15.7 Examples of Best Practice

Over the last 15 years, many states in the US, have built transportation management centers, deployed intelligent transportation systems (ITS) over critical segments of their road networks, deployed safety service patrols (using HELP vehicles), and developed interagency arrangements to both incident management and traveler information.

15.7.1 Institutional Best Practices

Examples of institutional best practices adopted by several states are summarized in Table 15.2.

Table 15.2 Examples of institutional best practices in operational activities

• An increasing number of states have quick clearance laws to support the removal of stopped vehicles from obstructing the road. Florida DOT (FDOT), for example, carried out an aggressive statewide campaign of signage, radio spots, billboards, and brochures to inform the public about the law and its benefits
• Both the FDOT Rapid Incident Scene Clearance (RISC) program and Georgia DOT Towing and Recovery Incentive Program (TRIP) are public–private partnerships that use both incentive payments and disincentive liquidated damages to ensure shortened clearance times for heavy vehicle wrecks; these programs have reduced the average clearance times by 100 %
• Oregon DOT has used a set of unique contractor requirements (staged tow trucks, traffic supervision, and public advisories) as part of effective work zone traffic control
• Detroit metropolitan area transportation agencies are part of a regional multiagency coalition that tracks and manages weather problems and treatment strategies, including flexible interjurisdictional boundaries for efficient operations
• The 16-state I-95 Corridor Coalition has supported an operations academy, which is a 2-week residential program designed to provide middle and upper managers in state DOTs with a thorough grounding in various aspects of SO&M state of the practice
• The Maryland DOT Coordinated Highways Action Response Team (CHART) is a formal, multiyear budgeted ITS and operations program with an advisory board that provides oversight and strategic direction. It is chaired by the deputy administrator/chief engineer for operations and includes district engineers, the director of the Office of Traffic and Safety, the director of the Office of Maintenance, the Maryland State Police, the Maryland Transportation Authority, the Federal Highway Administration, the University of Maryland Center for Advanced Transportation Technology, and various local governments
• Washington State DOT (WSDOT) has formalized interactions among units and managers involved in its SO&M program. TMC managers from around the state meet every 6 weeks to coordinate with regional Incident Response Program managers, who in turn meet quarterly for operations coordination with the state patrol. TMC managers and incident response managers coordinate activities and issues by meeting with the statewide traffic engineers group and the maintenance engineers group
• The Oregon Transportation Commission moved some capacity funding to the operations program to create an Operations Innovation Program that awards funding to projects selected on a competitive basis for their potential to demonstrate innovative operations concepts related to congestion mitigation and freight mobility
• Virginia DOT has reorganized its senior management to include a deputy director for operations and maintenance responsible for all SO&M activities, as well as maintenance resources
• WSDOT has made a strong and transparent commitment to performance measurement as evidenced by the quarterly Gray Notebook, which tracks performance based on five WSDOT legislative goals, including mobility/congestion, and includes regular updates on progress in the application of operations strategies such as incident management and HOT lanes

Source Reference [4], p 14, Table ES.7

15.7.2 Regional Cooperation in Managing Nonrecurring Events

Examples of regional operational collaboration are given in Table 15.3.

Table 15.3 Examples of regional collaboration in operational activities

Name and location	When started	Members	Operational activities
TRANSCOM	1986	• Metropolitan Transportation Authority – NYC Transit – MTA Bridges and Tunnels • Connecticut Department of Transportation • New Jersey Department of Transportation • New Jersey State Police • New Jersey Transit Corporation • New Jersey Turnpike Authority • New York City Department of Transportation • New York City Police Department • New York State Bridge Authority • New York State Department of Transportation • New York State Police • New York State Thruway Authority • Port Authority of New York and New Jersey • PATH	*Operations Information* • Collects and disseminates real-time regional incident and construction information • During major incidents, construction, and special events, helps marshal regional resources for incident response, including its member agencies' variable message signs and highway advisory radio • Provides services under contract to the I-95 Corridor Coalition *Construction Coordination* • Maintains a long-term database of all construction projects planned or under way *Special Events* • Assists with interagency coordination for special events *ITS* • TRANSMIT: Vehicles equipped with transponders for electronic toll collection are used as probes on road-ways for real-time determination of travel times and speeds and for the detection of incidents • TRANSCOM Regional Architecture: Integrates member agencies' ITS, allowing for the electronic sharing of information among the agencies' operations centers • Trips123: Website with real-time information and transit trip planning services for the general public • Interagency Remote Video Network (IRVN): Enables the sharing of member agencies' CCTV feeds

(continued)

Table 15.3 (continued)

Name and location	When started	Members	Operational activities
TranStar	1993	• The Texas Department of Transportation • Harris County • The Metropolitan Transit Authority of Harris County • The City of Houston	*Incident Management* • Monitors traffic incidents with more than 600 regional closed-circuit television cameras (CCTVs) • Dispatches vehicles to remove debris or hazardous materials • Communicates with emergency vehicles about the most direct routes to an accident scene • Motorist Assistance sends tow trucks to stalled Vehicles • Dynamic message signs (DMS) • Synchronized traffic signals, speed sensors • Highway Advisory Radio • Ramp meters and other devices • Transit authority is a partner, and there are HOT lanes, but TranStar does operations on all roads
Freeway and arterial system of transportation (FAST)—Las Vegas	2003	• RTC • Clark County • NDOT • City of Henderson • City of Las Vegas • City of North Las Vegas	• Operates the TMC • Ramp meters • DMS • Signal timing • Lane-use control signs • Each entity (e.g., city, county) maintains the physical equipment and power for traffic signals, while FAST is responsible for timing, traffic signal synchronization, and the communication network
MTC-BATA	1997	• Part of the Metropolitan Transportation Commission (MTC), San Francisco Bay Area	• The Bay Area Toll Authority (BATA) administers programs and allocates all toll and other rev-enues (except the $1 seismic surcharge) from the seven state-owned toll bridges. BATA funds the day-to-day operations, facilities maintenance, and administration of the bridges

(continued)

Table 15.3 (continued)

Name and location	When started	Members	Operational activities
TSSIP (Denver)	2003	• Denver Regional COG, Colorado DOT, 28 local governments	• Works with the Colorado DOT and local governments to coordinate traffic signals on major roadways in the region • Facilitates the implementation of a regional vision for transportation operations using both technology and regional partnerships
NITTEC	1995	• Buffalo and Fort Erie Public Bridge Authority • City of Buffalo • City of Niagara Falls, New York • City of Niagara Falls, Ontario • Erie County • Ministry of Transportation, Ontario • New York State Department of Transportation • New York State Thruway Authority • Niagara County • Niagara Falls Bridge Commission • Niagara Frontier Transportation Authority • The Niagara Parks Commission • Niagara Region • Town of Fort Erie	• Traffic Operations Center (TOC) • Closed-circuit television (CCTV) • Dynamic message signs (DMS) • Highway Advisory Radio (HAR) • TRANSMIT: Gathers vehicle travel time information • Road Weather Information System (RWIS) • Skyway Closing System: Advanced warning system that alerts motorists to closures on the Buffalo Skyway • Advanced Traffic Controllers (ATC): Traffic counting stations that transmit real-time traffic information to the TOC
• FAST-TRAC • Road Commission for oakland county	1992	• Oakland County, Michigan • Local governments • MDOT	• Operates a TMC • Website with real-time traffic information • Traffic signal control in response to congestion • Special event management. • Maintains a database of road construction projects • Variable message signs

(continued)

Table 15.3 (continued)

Name and location	When started	Members	Operational activities
CLARUS	Designed 2004–2005; tested 2006	• FHWA • NOAA • A number of states	• Research and development initiative to demonstrate and evaluate the value of Anytime, Anywhere Road Weather Information that is provided by both public agencies and the private weather enterprise to the breadth of transportation users and operators
AZTech	1996	• Led by the Maricopa County Department of Transportation and Arizona DOT—more than 75 public and private agencies • Arizona DOT • Arizona Department of Public Safety • Arizona State University • Maricopa County • Valley Metro • Phoenix • Mesa • Glendale • Peoria • Scottsdale • Eight other cities or towns, local police, and fire departments	• AZTech supports the following efforts along with its public and private partners: • Traveler Information at the Phoenix Sky Harbor International Airport • Regional Emergency Response Team (REACT): Focuses on incidents on arterials • When an incident occurs, one incident commander from each agency reports to the command post. The command post will then implement the correct plan of action • Phoenix International Raceway (PIR) Special Event Management • Coordinate between TMCs so all use the same communication protocols and can easily share information

Source Reference [4], p 80, Table D-1

15.7.3 Road User Benefits

Examples of system operating benefits for various types of incident management strategies are detailed in [4]. Reported benefits include the following:

- Effective traffic incident management reduces incident duration by 30–50 %
- Road weather information systems reduce crash rates from 7 to 80 %
- Dynamic message signs reduce crashes by 3 % and improved on-time performance up to 15 %
- Work zone management reduces system delays up to 50 %
- Active traffic management increases traffic throughput up to 7 %, and decreases incidents up to 50 %

15.8 Summary Assessment of Experiences in Managing/Mitigating Nonrecurring Congestion

A 2008 report by Cambridge Systematics et al. [7] contains a review and synthesis of congestion management practices in US metropolitan areas. The report covered both non-recurring and recurring congestion.

The mitigating impacts of strategies that were applied to nonrecurring congestion (Incidents, Work Zones, Road Weather, and Special Events Traffic) and the dissemination of *Traveler Information* about their effects on traffic congestion were evaluated and the findings are summarized below.

Incident Management was found to be *highly* effective in reducing congestion at the *local and area-wide* scales of impact. It was primarily applied to freeways, and could be implemented in the *short term* at *low to medium cost*, with *minimum* institutional or regulatory barriers. The potential future effectiveness of incident management was rated *extensive* (area-wide impacts).

Work Zone management was *highly* effective in reducing congestion at the *local* scale, was applied in some form in up to 2/3 of the metropolitan areas, and could be implemented in the *short term* at *low cost*, with *minimum* institutional or regulatory barriers. The potential future effectiveness of work zone management was rated *moderate* (limited to work zones only).

Road Weather management had a *medium* effect in reducing congestion at the local scale of application and a *low* effect at the area-wide scale. The extent of application in metropolitan areas was *moderate* (up to 2/3 of metropolitan areas use it), and could be implemented at *low cost,* with *minimum* institutional or regulatory barriers. The potential future effectiveness in congestion reduction of road weather management was rated *moderate* (limited to inclement conditions only).

Planned Special Events Traffic management was *highly* effective at the local scale of application, was used in about to 2/3 of the metropolitan areas, and could be implemented at *low cost* with a *medium amount* of institutional or regulatory barriers. The potential future effectiveness in reducing congestion was rated *moderate* (limited to special event locations only).

Traveler Information management was found to be of *low to medium* effectiveness in congestion reduction at both the local and area-wide levels, had a *moderate* to *extensive* application in metropolitan freeways networks, could be implemented at *low to medium cost* within a *short term* timeframe, and with a *minimum* amount of institutional or regulatory barriers. Its potential future effectiveness in reducing congestion impacts on road users was rated *extensive*.

References

1. Freeway Management Handbook (1997) Federal highway administration. Report No. FHWA-SA-97-064, Washington, DC
2. Lockwood S (2006) NCHRP Project 20–24: The 21st century operation oriented state DOTs. American Association of State Highway and Transportation Officials, Washington, DC
3. Federal highway administration (2012) Managing demand through travel information services s.l. Department of Transportation, Federal highway administration, 11/15/2012. www.ops.fhwa.dot.gov/publications/manag_demand_tis/traveinfo.htm
4. Brinkerhoff P with Delcan, Philip J Tarnoff (2012) SHRP2 Report S2-LO6-RR-1: Institutional architectures to improve system operations and management. Transportation Research Board, Washington, DC
5. Lockwood S (2006) The 21st century operation oriented state DOTs, NCHRP Project 20–24. Transportation research board, American Association of State Highway and Transportation Officials, Washington, DC
6. Traffic Technology International (2012) The ongoing war against traffic congestion. s.n. www.traffictechnologytoday.com
7. Cambridge Systematics and Resource Systems Group (2008) NCHRP 20–24A Task 63: Effective management practices for congestion management: final report. s.l. requested by: American Association of State Highway and Transportation Officials (AASHTO), Washington, DC

Chapter 16
Adaptation Strategies for Managing Recurring Congestion—Operational Improvements

16.1 Introduction

Increasing roadway capacity has been the traditional method to cope with and manage traffic congestion. Capacity enhancement strategies generally improve mobility of all road users. Sometimes, however, they are designed to improve mobility of special users—such as high-occupancy vehicles (e.g., carpools and public transport vehicles).

There are two types of capacity enhancement strategies: (1) those that restore lost capacity by eliminating operational inefficiencies and bottlenecks in the existing roadway system; and (2) those that add new road capacity. This chapter describes strategies that increase the operational efficiency of the existing system. Chapter 17, discusses strategies that add new roadway capacity.

This chapter shows how operational strategies can reduce congestion by making better use of existing streets and highways. Getting the "most" from existing streets and highways has been a long-established traffic engineering practice to increase capacity, reduce delay, and improve traffic safety in the short term at modest costs and with minimum adverse community impacts.

16.2 Scope

The various sections of this chapter describe the strategies that improve traffic operations, give application guidelines for each, suggest relevant analysis methods, and set forth likely effectiveness of the various strategies. In this respect the suggested analysis methods complement the various capacity, delay, and level of service procedures of the 2010 Highway Capacity Manual [1].

Traffic operational improvements that increase capacity include a combination of traditional strategies such as: parking restrictions, signal timing and signal

J.C. Falcocchio and H.S. Levinson, *Road Traffic Congestion: A Concise Guide*,
Springer Tracts on Transportation and Traffic 7,
DOI 10.1007/978-3-319-15165-6_16

coordination improvements, turn restrictions at key intersections, one-way streets, reversible commuter lanes, movable median barriers during peak periods, better lane striping, managing road access to and from traffic generators, etc. They also include the application of advanced technologies that use real-time information of traffic conditions to implement dynamic traffic control strategies that optimize traffic flow and provide travelers with real-time information on traffic conditions.

16.3 Analysis Overview

Congested roadways and intersections generally can be identified by observations and from travel time and delay studies. Demand-to-capacity analyses are useful in identifying problem locations. As a general guide, improvements should reduce the demand-to-capacity ratio to less than 0.85 for achieving stable flow. Reducing the red times on each approach at signalized intersections can also reduce delays.

The 2010 Highway Capacity Manual [1] contains detailed procedures for estimating capacity, delays, and facility performance. Additional guidelines are found in the various editions of "Traffic Engineering" handbooks that are published by the Institute of Transportation Engineers [2] and the Manual of Uniform Traffic Control Devices for Streets and Highways [3]. These procedures can be used to estimate change in service levels resulting from traffic operational improvements.

The sections that follow give descriptions, applications, guidelines, and travel time savings for the following operational strategies:

- Traffic Signal Timing and Coordination
- On-street Parking and Loading Zone Management
- Intersection Turn Controls and Management
- One-way Streets
- Changeable Lane Assignment
- Ramp Controls
- Access Management
- Intelligent Transportation Systems
- Traveler Information Systems
- Roadside Electronic Screening Programs for Commercial Vehicles
- Integrated Corridor Management

16.4 Traffic Signal Timing and Coordination

Effective traffic signal timing and coordination is one of the most basic and effective strategies to reduce congestion. The Federal Highway Administration (FHWA) indicates that there are more than 240,000 traffic signals in the United States. The agency estimates that poor traffic signal timing accounts for 5–10 % of all traffic delays or almost 300 million vehicle hours of delay on major roadways alone. Over 75 % of the signals in the US could be improved by adjusting and updating equipment or by adjusting timing and coordination plans. The FHWA recommends that jurisdictions assess retiming signals at least once every 3 years.

16.4.1 Signal Location and Spacing

Traffic signals should be installed at intersections that meet the national Manual on Uniform Traffic Control Devices recommendations [3] and state standards pertaining to minimum volume and annual crashes. Too many signals along a roadway generally should be avoided, since adding signals generally reduces travel speeds and increases delay (see Table 16.1).

Typically signals are located almost every block in the central business district and other major commercial centers. In cities, average spacing should not be closer than one quarter mile; while wider signal spacing is desirable in suburban areas.

16.4.2 Cycle Length

The cycle lengths should (1) accommodate conflicting traffic volumes at intersections without excessive delay on each approach, (2) provide enough green time on side streets to let pedestrians cross safely, and (3) permit effective coordination of signals along a roadway and. Within this context, cycle lengths should be as short as possible. When cycle lengths are too long, delays can be excessive. Although cycle lengths depend upon the number of phases, they generally should not be longer than 2 min. Shorter cycles are desirable where transit vehicles operate along the streets.

Table 16.1 Percentage increase in travel times and increasing signal density

Signals per mile	Percent increase in travel times (compared with 2 signals per mile)
2.0	0
3.0	9
4.0	16
5.0	23
6.0	29
7.0	34
8.0	39

Source Reference [4]

16.4.3 Cycle Phases

The number of phases should be kept at a minimum. Two-phase operation is desirable wherever possible. As the number of phases increases the red time on each approach increases and so does delay. A third phase, and sometimes a fourth phase, is needed to serve left turns or accommodate a third street entering the intersection. Additional pedestrian-only phases generally should be limited to locations with heavy turning movements and heavy pedestrian crossings.

Design and operating strategies that restrict or divert left turns and simplify intersection geometry can reduce both cycle lengths and delay.

16.4.4 Coordination

Coordinating traffic signals along a roadway minimizes stops and delays during both peak and off-peak hours. Coordination requires signals to operate on common cycle lengths. In addition, uniform spacing of signals is desirable.

Signals should be coordinated on city streets (where they are spaced close together), and on suburban roads where their spacing does not exceed one mile.

The traffic speeds that result from coordination depend upon both the signal spacing and the cycle length. Higher speeds require longer distances between signals; shorter cycle lengths and closer spacing of signals produce lower speeds [5].

Figure 16.1 shows these relationships.

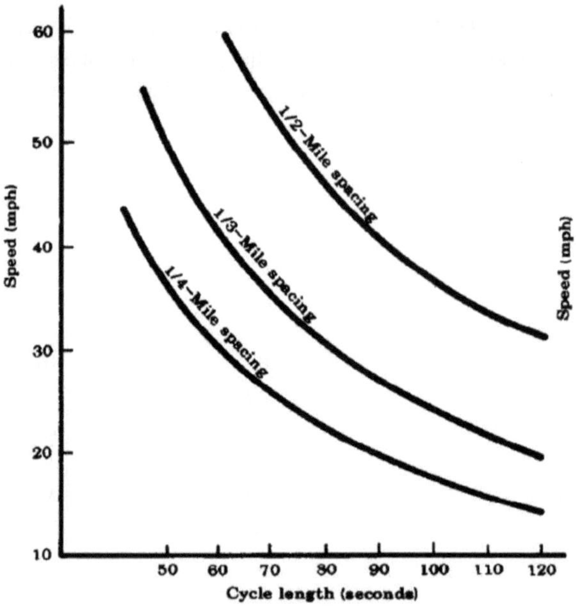

Fig. 16.1 Speed of traffic progression as a function of cycle length and signal spacing. *Source* Reference [5], p 57. Figure 7.1

The maximum average traffic speed that can be achieved through a series of signalized intersections depends also on the spacing of the signals. Uniform, or near uniform, signal spacing is essential in maximizing traffic speed. Where this is the case, the through band, which indicates the amount of traffic that can pass through a series of signals during the green phase, equals the green time provided by the signal. Thus signals placed at quarter mile intervals can permit a through band in each direction, at the designated progression speed (e.g., 30 mph using a 60 s cycle). If a new signal is introduced, say one quarter mile from the next signal, the through band of progression is significantly reduced.

16.4.5 Operating Strategies [5]

Cycle lengths, phase splits, offsets, and progression speeds can vary throughout the day. In some cases, where traffic volumes are unbalanced (e.g., inbound in the morning and outbound in the evening) the signal progression should favor the heavier traffic flow.

16.4.6 Equipment

Signals can be pre-timed or traffic-responsive. Computer control and coordination of signals is a common practice. Fully traffic-actuated signals work best at isolated intersections. Semi-actuated signals are appropriate at minor streets and should be integrated with the coordination pattern.

Traffic adaptive signals represent the state of practice and are increasingly popular in large metropolitan areas. These traffic adaptive controls, used for example in Los Angeles and Manhattan, adjust the signal splits of each cycle based on traffic volume, but can retain the specified background cycles.

16.4.7 Costs

Implementing signal coordination plans and upgrading signal control equipment generally involves considerable costs. Costs also include installing real-time traffic sensors to update signal timing based on actual traffic demand.

Sunkan [6] estimated the cost to retime signals in the range of $2,000–$3,100 per intersection; while the reported costs of computerized synchronization systems in several cities ranges between $30,000 and $60,000 per intersection.

Table 16.2 Congestion management strategies for using existing capacity more efficiently—summary of effectiveness and implementation potential

Type of improvement	% Increase in speed	% Decrease in travel time	% Decrease in delay	% Decrease in stops	Source
Improved timing	0–5	–	–	–	(A)
Coordinating isolated signals	2–25	–	–	–	(A)
Placing isolated signals on computerized system	30	–	–	–	(A)
Placing single dial system on computerized system	15–20	–	–	–	(A)(B)
Placing multiple programmed signals on computerized system	6–10	–	–	–	(A)(B)
California statewide program	15	7	15	10	(A)
Texas statewide program	–	–	19	9	(A)

Notes (A) Reference [7]
(B) Reference [8]
Estimates are adapted in part from: Rowan, N.S., Woods, J.P., Stover, V
(a) Alternatives for improving urban transportation—a management
Overview Prepared for FHWA by TTI, Taxes A&M University, 1977
(b) Technical Memorandum 3, Quantifying Measures of Effectiveness, Tri-State TSM Study, Wilbur Smith and Associates, 1978. *Source* Reference [9]

16.4.8 Reported Time Savings

Reported travel time savings resulting from traffic signal timing improvements, operational treatments along arterial streets, and improved freeway operations are shown in Tables 16.2 and 16.3, respectively. These tables, assembled from various sources, provide a basis for estimating travel time savings from similar congestion reduction treatment elsewhere.

16.5 Curb Parking and Loading Zone Management

On-street parking and loading zones are found throughout the metropolitan area. Curb parking and loading/unloading activities are usually concentrated in older commercial districts and in densely populated residential areas where there is limited off-street parking space available.

On-street parking or standing along major streets often takes away street space that could be used for the movement of motor vehicles. These events frequently create congestion especially during the peak travel times and sometimes throughout the day. A short stretch of curb parking can reduce the capacity of an entire street.

Table 16.3 Reported benefits of traffic signal timing improvements

Action	% Increase in speed	% Decrease in travel time	Unit travel time savings min/miles	Other	Source
Curb parking restriction	20–30	+25	0.24–2.4	–	(B)(A)
Left turn prohibition	15–35	+10	–	–	(B)(A)
Left turn lanes	10–20	–	–	–	(B)
Reversible lanes (heavy direction)	20–50	–	–	–	(B)
One-way streets	20–40 25–50	–	–	–	(B) (C)
Special one-way streets					
Manhattan					
Fifth avenue	+60	−37	2.4	–	(C)
Madison avenue	+21	–	–	–	(D)
3rd Lexington 7th, 8th, ave of the Americas	–	22		65 % reduction in stops	(D)
Denver					
Broadway-Lincoln	+9	–	–	–	(A)

Notes (A) Reference [7]
(B) Reference [8]
Estimates are adapted in part from: Rowan, N.S., Woods, J.P., Stover, V
(a) Alternatives for Improving Urban Transportation—A Management Overview Prepared for FHWA by TTI, Taxes A&M University, 1977
(b) Technical Memorandum 3, Quantifying Measures of Effectiveness, Tri-State TSM Study, Wilbur Smith and Associates, 1978
(C) Reference [10]
(D) Reference [11] *Source* Reference [9]

16.5.1 Curb Parking Restrictions

Curb parking restrictions can provide congestion relief quickly with minimum costs. They can apply to one or both sides of the street. Restrictions are normally applied during busy traffic periods, but they can sometimes extend throughout the day.

16.5.2 Installation Guidelines

Curb parking restrictions are generally appropriate when the following conditions apply:

- Roadways operate at Service Level E or F
- At intersection approaches, at transit stops and at fire hydrant

- Curb bus lanes are needed during peak hours (i.e., 40 buses per hour)
- Where curb parking is prohibited throughout the day, it is essential to provide access to impacted facilities—either on opposite sides of a street or off-street. It is sometimes desirable to provide "access windows" during morning and evening off-peak periods when good delivery and pickups are permitted

16.5.3 Congestion Relief Benefits of Curb Parking Management

Congestion is relieved by:

- Providing additional travel lanes during busy traffic periods
- Implementing dynamically variable parking rates where curb parking demand exceeds curb parking supply thus ensuring parking space availability [12–14]
- Restricting curbside loading and unloading to midblock locations (Fig. 16.2)
- Effective enforcement of parking and loading and loading restrictions.

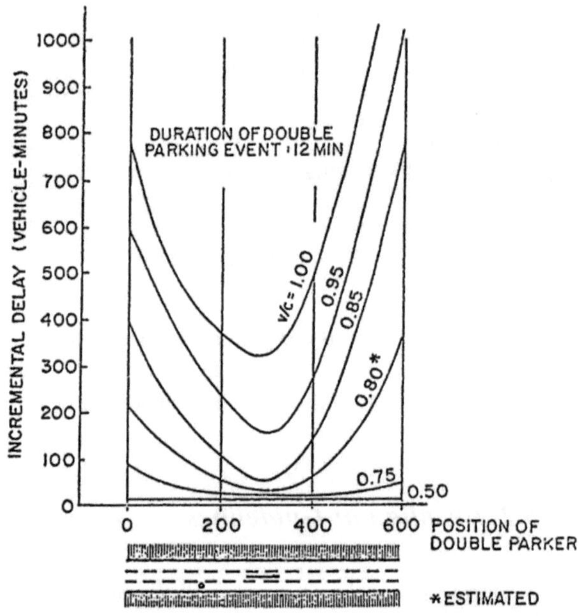

Fig. 16.2 Illustration of delay caused by Double Parker. *Source* Reference [14]

16.6 Intersection Turn Controls and Management

Left and right turns impede traffic flow at many locations. The "right-turn problem" is usually critical in areas of heavy pedestrian activity commonly found in the city center and in older high-density neighborhoods.

Left turns, however, create intersection delays throughout the street system. They conflict with opposing through traffic, can also block vehicles behind them and complicate traffic signal phasing.

To reduce intersection delay many urban areas restrict left and right turns. These restrictions preserve capacity and reduce congestion, and can be in effect all day (7 a.m. to 7 p.m.) or just during rush hours.

Managing left and right turns at busy intersections by restriction or rerouting, is a long established and cost-effective way to reduce delays and congestion at busy intersections. Prohibiting left turns permits fewer phases, shorter cycle lengths, and longer "green times" for the through movement. This translates into more capacity and less delay.

16.6.1 Managing Right Turns

Right turn restrictions are sometime needed on intersection approaches where both right turns and pedestrian volumes are heavy. These restrictions result into shorter cycle lengths and reduce delays.

The travel times gained by restricting right turns can be estimated by applying the following equation [16]:

$$\Delta t = r \cdot p \cdot \frac{t_s}{L} \tag{16.1}$$

where

Δt green time to be gained per cycle
r right turns/cycle (peak 15 min)
p conflicting pedestrians per cycle (peak 15 min)
t_s time loss per pedestrian (e.g., 3–4 s), and
L number of pedestrian channels in crosswalk

Using the above equation, the amount of time lost due to right turns and pedestrian conflicts can be estimated. The estimated times lost per signal cycle for a range of conflicting right turns and pedestrian volumes is shown in Table 16.4.

For example, if there are 300 pedestrians per hour conflicting with 240 right turns per hour (@ 60 cycles/h: 5 pedestrians and 4 right turns per cycle), and 3 s lost per conflict, about 20 s per cycle would be lost, assuming 3 pedestrian channels, or 20 = (4) × (5) × (3)/3.

Table 16.4 Estimated time lost per cycle by conflicting right turns and pedestrian volumes

Typical values of *R/Nc* and *P/Nc*	Time loss per cycle at 3 s per pedestrian channels (lanes)			
	1 Lane	2 Lanes	3 Lanes	4 Lanes
4	12	6	4	3
8	24	12	8	6
12	36	18	12	9
16	48	24	16	12
20	60	30	20	15
24	72[a]	36	24	18

Notes For a 60 s cycle, time loss should not exceed 25 % of cycle or 15 s. Thus, values below the boldface lines are not acceptable, and turns should be prohibited
Source Reference [17]
R = right turns per hour
Nc = number of cycles per hour
P = pedestrians per hour
[a] = Excess cycle length

Therefore, if right turns were prohibited in this case the curb lane would then gain an additional 20 s of additional green time per signal cycle.

Sometimes it is possible to provide curb lanes for moving traffic by removing curb parking, by restriping the roadway, or by minor widening.

16.6.2 Managing Left Turns

Delays from left turn movements (for right-hand driving) are common at intersections along most streets and roads. Depending upon specific circumstances, delays created by left turns generally can be reduced by prohibiting, better accommodating, diverting, or separating left turn movements. Table 16.5 summarizes the various ways of addressing the left turn issue and suggests where each possible improvement applies [5].

16.6.3 Left-Turn Treatment Options

This chapter describes how to minimize the left-turn conflicts at intersections through operational changes. While the next chapter describes how left turn conflicts can be separated through the redesign of the intersection.

(a) Prohibiting Left Turns at Intersections
The prohibition of left turns is common at heavily traveled intersections, especially where it is not possible to provide protected left turn lanes. The

Table 16.5 Treatment of left turns at intersections and driveways

Option	Condition	Application considerations
Prohibit	Full time	Requires alternate routes
	Peak periods only	Requires alternate routes
Provide	Share lane	Limit to minor roads or places where R/W is not available for left-turn lane
	Left-turn lane	Protected or permissive phasing
	Dual left-turn lane	Protected phasing only
Divert	Jug-handle	Divided highways at minor roads (signalized junctions only)
	Modified jug-handle	6-lane divided highways
	Michigan "U"	Divided highways with wide median—allows two-phase signals
Separate	Directional design	Very heavy turns in one direction
	Left-turn flyover	Very heavy turns in one direction
	Through lane flyover	Major congestion points

Source Reference [5], p 70. Table 8.1

prohibition simplifies traffic signal phasing, reduces queues, and improves traffic flow.

(b) Allowing Left Turns at Intersections

Allowing left turns from a lanes used by through traffic (shared lanes), should be avoided along major roadways. Left turns from shared lanes can reduce lane capacity by about 50 %, increase delay to through vehicles, and increase the number of crashes. For these reasons, protected left turn lanes should be provided where space permits.

Sometimes left turn lanes can be provided by adjusting the center median, or by reducing the width of through lanes. Providing five 10-foot lanes on a 50 foot road out of four 12.5-foot lanes, would enable the extra lane to be used as a protected left turn lane.

Where left turn lanes are provided at signalized intersections, they are sometimes given a special phase. However, the green time provided for the left turns can both increases the cycle length and the delays to opposing through traffic.

Two-way left turn lanes are common in suburban settings where one finds roads with frequent driveways serving roadside development. Left turn lanes remove left turning vehicles from the through lanes and reduce delays to through traffic. Conventional left turn lanes are usually provided at signalized intersections in suburban areas.

16.6.3.1 Separating Conflicting Left Turns

Figure 16.3 shows a typical signalized access drive connecting to an arterial. Note that three signal phases are necessary.

Separating conflicting left-turn movements entering and leaving major developments along a roadway is generally desirable. The separation reduces intersection conflicts both along the public road and along the internal circulation roadways of the development. Figure 16.4 shows how separating the left turn movements

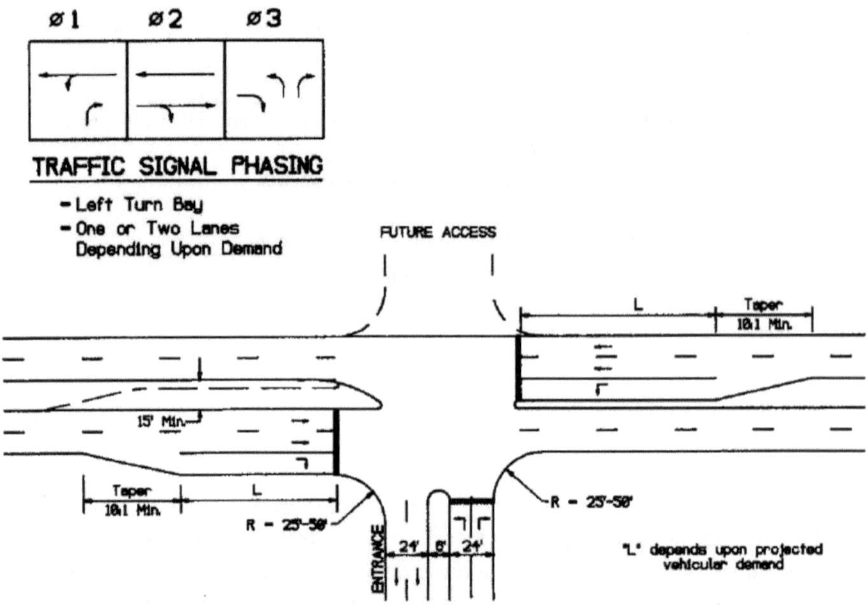

Fig. 16.3 Typical access design. *Source* Reference [5], p 76. Figure 8.18

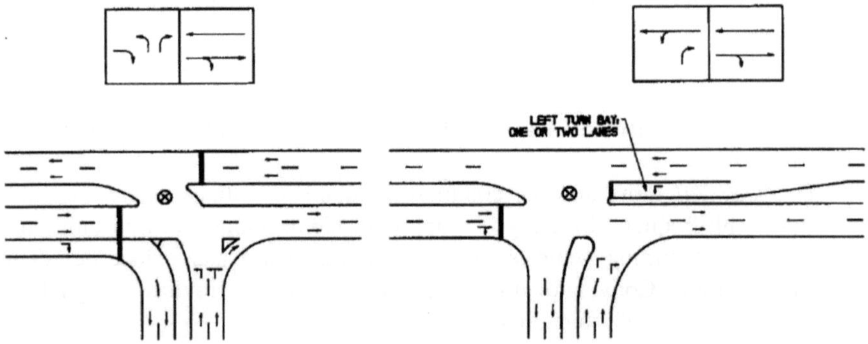

Fig. 16.4 Directional access treatment. *Source* Reference [5], p 76. Figure 8.19

entering and leaving a development site reduces conflicts and simplifies signal operation.

In this example, the left turn movements entering and exiting take place essentially at the same time. This two-phase operation results in shorter "red" times and less delay for motorists on both the arterial and site-access roads.

16.7 One Way Streets

One-way streets have improved traffic speed and reduced congestion since ancient times. They were found in Pompeii where the narrow lanes allowed for the passage of only one lane of vehicles [3].

In the US, one-way streets first emerged on Chestnut Street in Philadelphia and in the Park Row section of Manhattan in 1907. They were progressively implemented in American cities since the 1920s. By 1939, 85 % of the streets in Manhattan operated one-way. Most urban areas in the United States and Canada have one-way streets.

16.7.1 User Benefits

One-way streets are a low-cost strategy that reduces delay, increases traffic speeds, and improves safety. The delay reductions translate into lower emissions and better air quality. One-way street operations increase road capacity and reduce intersection conflicts, travel times and crashes. Case studies from the 1950s indicate that one way conversion from two-way traffic flow increased traffic speeds from 20 to 50 %, with a corresponding reduction in crashes [18, 19].

The changes in traffic volumes, trip times, and number of stops resulting from converting Fifth Avenue in Manhattan to one-way operation are as shown in Table 16.6.

The Madison Avenue-Fifth Avenue one-way couplet in Midtown Manhattan reduced accidents by 27 % and personal injuries 28 % [11].

Table 16.6 Traffic volume and performance changes resulting from two way to one way operation on fifth avenue, Manhattan

Item	% Change
Average daily traffic volume	+19
Average trip time	−31
Average number of stops	−60

Source Reference [20] © National Academy of Sciences, Washington, DC, 2000. Reproduced with permission of the Transportation Research Board

16.7.2 Advantages of One-Way Operations

The reported advantages of one-way streets include the following:

a. Reduced Conflicts at Intersections: Conflicts between through traffic and opposing left turns are eliminated, as is the need for special left-turn phases.
b. Simplified Traffic Signal Phasing: Because the need for special left-turn phases is eliminated, cycle lengths can be shorter resulting in less "red times" and delays.
c. Improved Traffic Signal Progression: Traffic signal offsets along a street can be set at the designed speed since only direction of travel is involved. The through band (or "green wave") for one-way flow is usually greater than that for two-way flow because the green wave for *both* directions of travel is smaller. One-way traffic also simplifies signal coordination within a network.
d. More Opportunities to Add Signals: Traffic signals can be added along a street without any adverse effect on the "through band" and progression.
e. Safer Pedestrian Crossings: There are more traffic "gaps" acceptable to pedestrians since only one direction of travel is involved—making pedestrian crossings easier and safer.
f. Safer Intersections: There are fewer traffic conflicts and eliminating the possibility of head-on collision improves safety.
g. More Travel Lanes: Sometimes, one-way operations permit restriping a roadway to provide an additional travel lane.

16.7.3 Disadvantages of One-Way Operations

Several disadvantages have been cited for one-way streets. Disadvantages include:

a. Longer Travel Distances: Some analysts have found that one-way streets increase the VMT by up to 30 % [20].
b. Increase in Left Turn and Right Turn Volumes: The doubling up of left and right turns can increase pedestrian—vehicle conflicts. In areas of very heavy pedestrian volumes, and additional signal phase may be needed.
c. Possible Confusion to Drivers: Confusion is greatest when drivers are not familiar with the one-way system and/or the system design is not easily understood by drivers.
d. Longer Crossing Distances for Pedestrians: Where medians formerly separated opposing directions of travel, pedestrians have longer distances to walk when medians are removed for one-way operation.
e. Reduce Store Front Exposure: One-way streets can adversely impact store front businesses that depend on pass-by traffic.
f. Adverse Effects on Transit Riders: Transit service works best from a standpoint of passenger perception and identity when it operates in both directions on the same street. The adverse effects of one-way streets on transit service are greater when the one-way streets are far apart.

16.7.4 Applications of One-Way Streets

The use of one-way streets and roads was well expressed by Halsey [21] in his book "Traffic Accidents and Congestion," in which he stated that one-way roads should be used wherever:

- balanced pairs can be found
- the revised routing does not unduly detour traffic too much, and
- streets are too narrow to permit parking on each side, or to allow the effective movement of fire and other emergency vehicles.

16.7.5 Types of One-Way Streets

Within this context, the following types of one-way streets can be developed. Various details are provided in current and earlier editions of the Traffic Engineering Handbook [2].

- Narrow streets that are usually less than 25 feet wide
- Rotary movements around public parks or squares
- Adjuncts or complements to freeway design—such as connections or extensions of freeway ramps or frontage roads that are parallel to the freeway
- Simplification of complex intersections by making a minor street "outbound." Simplification sometimes eliminates a traffic signal phase, allowing more green time added to the remaining two phases (see Fig. 16.5).

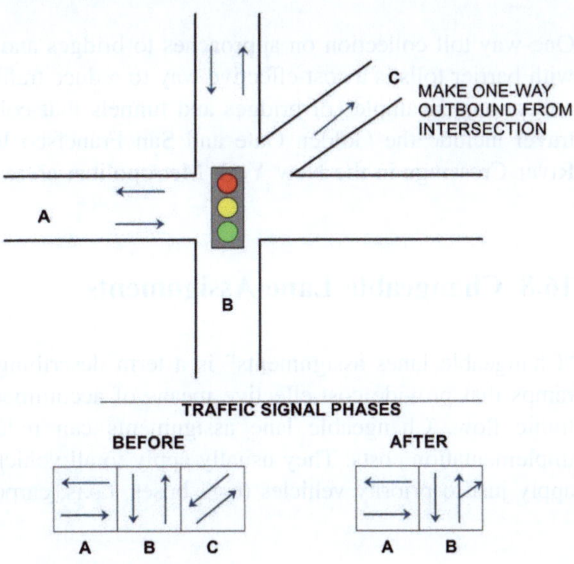

Fig. 16.5 Making the minor street one-way simplifies a complex intersection

- Along pairs of parallel arterial streets, such as: Chestnut Street and Walnut Street, Philadelphia; Broadway and Lincoln Ave., Denver; Washington and Warren Boulevards, Chicago; and most cross town streets and north-south avenues in Manhattan.

16.7.6 Installation Guidelines

The following guidelines underlie the installation of one-way streets:

- One-way pairs should be comparable in width, capacity, continuity and land use
- One-way pairs should be within 600 feet of each other wherever possible
- Provide adequate transition to two-way traffic at the beginning and end of the one-way system
- Provide conflict-free circulation around city blocks
- Avoid interposing a two-way street between a one-way pair
- Coordinate traffic signals along each street (or within a network) to provide progressive movements on common cycle lengths. Avoid long cycle lengths wherever possible
- Manage conflicts between pedestrians and turning vehicles
- Avoid street widths that exceed four moving lanes
- Accommodate public transit service wherever possible

16.7.7 One-Way Toll Collection

One-way toll collection on approaches to bridges and tunnels and along toll roads with barrier tolls is a cost-effective way to reduce traffic delays (and the cost of toll collection). Examples of bridges and tunnels that collect tolls in one direction of travel include the Golden Gate and San Francisco Bay Bridges, on the Hudson River Crossings in the New York Metropolitan area.

16.8 Changeable Lane Assignments

"Changeable lanes assignments" is a term describing reversible lanes, roads and ramps that provide cost-effective means of accommodating the tidal variations in traffic flow. Changeable lane assignments can reduce delay with a minimum implementation costs. They usually apply to all vehicles, although sometimes they apply just to priority vehicles (e.g., buses, taxis, carpools).

16.8.1 Benefits from Applications

This strategy has the potential to achieve significant delay reductions where traffic flows are highly unbalanced. Its implementation can benefit from applications of automated traffic control technologies such as sensors and lane control signals. Delays can be minimized when left turns are prohibited along arterial streets during the hours that the reversible lanes operate.

- Reversible lanes provide an effective means of accommodating tidal variations in traffic flow on roadways connecting the city center or other major activity center with residential areas. Capacity increases in the peak direction of 20–50 % have been reported—with corresponding reductions in travel time. Accident reductions up to 30 % have been reported; however, they probably reflect the drop in congestion-related rear end crashes [22].
- Application of this strategy along Memorial Drive in Atlanta, GA showed that peak travel times in the major-flow direction decreased by 25 % in the am peak, and 24 % in the pm peak [23]
- Reversible lanes along 6 miles of 7th Ave. in Phoenix AZ, increased speeds about 25 % in the AM peak period and 16 % in the PM peak period, but there was a 28 % increase in crashes. The annual cost of crashes and sign installation was $175,000, while the annual travel time savings were valued at about $1,000,000 [18].

Table 16.7 illustrates how the per lane demand-to-capacity ratio of a four-lane, 2-way street can be better balanced through reversible lane operations of lane #3.

16.8.2 Types and Extent of Applications

Reversible lanes are generally applied along radial arterial streets and freeways. They are also used at many bridges/tunnels and toll plazas. Sometimes entire streets are made reversible.

Table 16.7 Illustrative example of the effect of reversible-lane operation on the per lane demand-to-capacity ratio of a hypothetical 4-lane arterial street

		Lane 1	Lane 2	Lane 3	Lane 4
Before	Direction	NB	NB	SB	SB
	Demand volume (VPH)	600	600	200	200
	Capacity (VPH)	650	650	650	650
	Demand/capacity	0.92	0.92	0.31	0.31
After	Direction	NB	NB	NB	SB
	Demand volume (VPH)	400	400	400	400
	Capacity (VPH)	650	650	650	650
	Demand/capacity	0.62	0.62	0.62	0.62

Source Estimated

A 2006 survey for the ITS Deployment Database, 8 of 100 metro areas reported using reversible lanes on 98 miles of *freeways*, representing one percent coverage [23]. And 16 of 106 metro areas reported using reversible lanes on only 126 miles of *arterial streets*. Considerable mileage was found in two metro areas (Fresno CA, and Janesville-Beloit, WI). Other applications of this strategy were found in several freeway bridges/tunnels (e.g., Bay Bridge in San Francisco, Walt Whitman Bridge in Philadelphia, Tappan Zee Bridge in Westchester-Rockland counties, NY, the Long Island Expressway approach to the Queens Midtown Tunnel (NYC), and the Gowanus Expressway Approach to the Brooklyn Battery Tunnel (NYC).

Examples of reversible lanes include:

- I-5, Seattle
- the JFK Expressway, Chicago
- Connecticut Avenue, Washington, DC
- Highland Boulevard, Los Angeles
- North Sheridan Road, Chicago

16.8.3 Strength and Weaknesses

The basic strength of reversible lanes is their ability to serve both directions of travel. In so doing, they reduce congestion by providing a better balance between demand and capacity at a relatively little cost. A possible weakness of reversible lanes is an increase in the number of crashes (especially if left turns are permitted). Therefore, roadways with reversible lanes work best when left turns are banned.

16.8.4 Application Guidelines

Reversible lanes are appropriate where [24]:

- There are pronounced imbalances in the directional traffic volume. The ratio of major to minor movement should be at least 2–1.
- Peak period and peak direction flows are recurrent.
- Peak hour traffic speeds generally are at least 25 % slower than those during the off-peak period.
- Adequate provisions are made to accommodate the traffic in the corridor.
- Adequate capacity is provided at the transitions of the beginning and end points, and parking should be prohibited during the hours that the lanes are in effect.
- Reversible lanes work best when left turns are prohibited at signalized intersections. They are not compatible with the provision of protected left-turn lanes, and they do not work well on streets with median islands.

Reversible lanes on streets can be controlled in various ways:

- Using both curb-mounted and overhead signs are used.
- Overhead lane control signals that conform to the National Manual on Uniform Traffic Control Devices [3].
- Traffic cones and portable barriers.
- Movable physical barriers are typically used on bridges and on freeways.

A common practice is to use physically separated roadways on freeways and to control entry by electronically operate gates. Implementation of reversible flow lanes is applicable for freeways and arterial streets with automated lane control technologies (sensors and lane control signs).

16.9 Ramp Controls

Ramp meters and entrance controls are integral parts of many freeway management programs. They control the rate at which ramp vehicles enter the freeway during peak periods to prevent a speed reduction and a lower volume throughput of the freeway lanes. Metering is most effective when the freeway operates at service levels of "E" or "F".

16.9.1 Applications

Ramp controls work best when the following conditions apply [25]:

- recurring peak period traffic congestion along the freeway (e.g., demand exceeds capacity)
- there are suitable alternative surface routes that can accommodate traffic diverted by the metering strategy (e.g., available continuous frontage road along the freeway)
- the travel time saved by the freeway traffic exceeds the additional delay that diverted traffic imposes on surface routes
- adequate ramp storage for vehicles waiting to enter the freeway (to avoid spillback onto surface streets)

16.9.2 Control Types and Methods

There are several ways to apply entrance ramp controls. Methods include:

(a) permanent or part-time closure
(b) pre-timed or traffic responsive ramp metering

(c) traffic responsive merge control
(d) integrated system controls that apply to a series of entry ramps along the freeway

Metered freeway ramps that operate in a single lane should have a full width shoulder to allow for emergency passing. A common practice is for buses and car pools to bypass the ramp controls. In these cases an additional full-width bypass lane should be provided [2, 26].

16.9.3 Benefits

Typical benefits reported from freeway ramp metering include [23]:

- in the Minneapolis-St. Paul region system-wide ramp metering has increased freeway throughput volume by 30 % and peak period speeds by 60 %, with travel time decreases of 14–27 %
- ramp metering along freeway corridors in other cities have found that freeway travel times decreased by 7 % (Portland, Detroit), 27 % (Denver), 38 % (Austin), and 48 % (Seattle).
- ramp metering, however, creates queues at entry ramps, that can cause traffic to divert to local streets, resulting in increased delay on local streets.
- in Minneapolis-St. Paul, however, when the ramp delays were included, the total delay was still substantially reduced.

16.10 Access Management

Access management is a proactive strategy that balances the need to minimize traffic delay and crashes along arterial roadways with the need to provide access to land development activities adjacent to the roadway. It applies to both existing and new roadways. It extends the concept of access control to these roads by defining access spacing standards, providing suitable geometry, and establishing the necessary legislative authority to implement the desired standards.

The primary goals of access management are to improve safety and to maintain desirable traffic speed and capacity along arterial roads and streets. These goals are achieved by coordinating and consolidating the number and location of curb cuts to adjacent land development, and by controlling the spacing of traffic signals along arterial roads. As noted earlier, each traffic signal per mile added to the roadway reduces speed about 2–3 mph.

There is a large repertory of access management techniques [7] including: separating and physically restricting left turns; restricting curb cuts and direct access to driveways; establishing access spacing requirements; separating obvious conflict areas; eliminating parking along roadway; constructing frontage roads to serve local business traffic and enhancing nearby intersections. NCHRP Report 420, Impact and Access Management Techniques provides a comprehensive list and assessment of these techniques [4]. The Access Management Manual [27] gives further guidance on the use and benefits of access management in traffic operations.

16.10.1 Basic Principles

Several basic principles underlie access management as a strategy for reducing congestion and improving traffic safety. They are briefly described below:

- Limit (or prohibit) direct property access along higher type roads
- Provide a specialized roadway system in which different roads serve different travelers and goods movement needs relative to accessibility and mobility. Figure 16.6 shows an example of the hierarchy—freeways emphasize high speed travel and are designed with complete control of access (to the freeway); local roads emphasize low speed movement whose functions are designed to provide access to land and buildings; while arterials must serve both movement and property access.
- Preserve traffic signal coordination—locate signals only where they fit in the traffic signal coordination plan
- Locate access drives away from intersections in order to minimize traffic conflicts and crashes

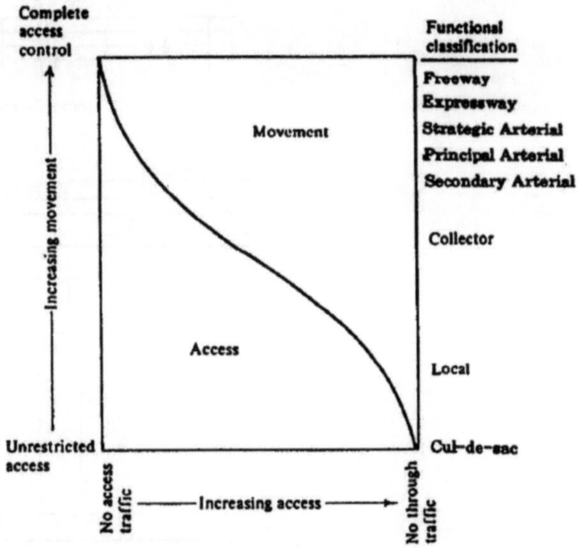

Fig. 16.6 Functions and access control of various road types. *Source* Reference [5], p 50. Figure 6.1

16.10.2 Access Control Methods

Access control is mainly implemented through the police power of eminent domain. It also can be achieved through the geometric design of roadways and access connections to land uses. The development of an access classification system (e.g., Fig. 16.6) can prove useful in determining when, where, and how access can be provided.

16.10.3 Access Design Concepts

Access planning and design should coordinate the three aspects of the access system—the public roadway, private driveway, and private development requirements/needs. All three should be treated as an integral part of the overall access system. In this way access can be provided and congestion minimized.

Road connections to land developments should have adequate turning radius and storage space. The conditions, shown in Fig. 16.7 (top) should be avoided. But, unfortunately, the problem of inadequate throat length is found along many roads.

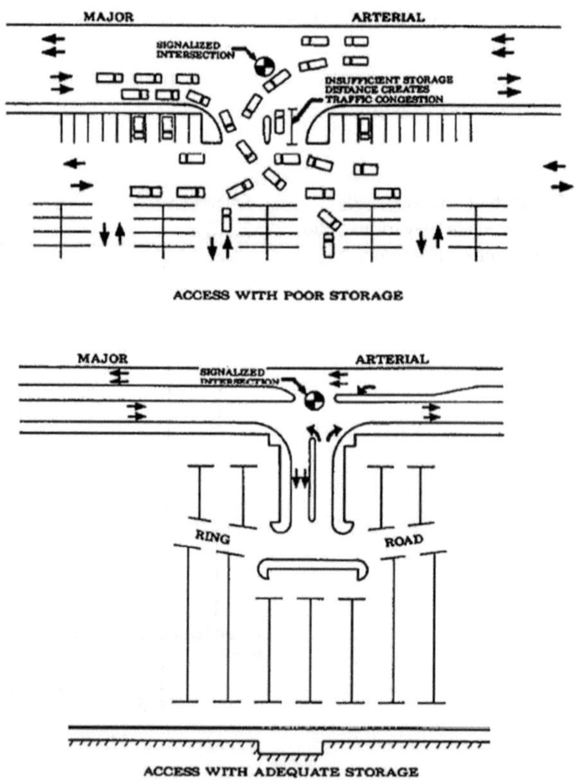

Fig. 16.7 Driveway improvements to increase storage. *Source* Reference [5], p 86. Figure 8.31

Where it exists, this condition can (and should) be corrected by increasing storage space on the connecting driveway, and by separating the left turns into and out of the development. Figure 16.7 (bottom) shows that by increasing the storage space and by separating conflicting left-turn access, conflicts are greatly reduced and congestion can be reduced or eliminated. Additional examples of best access practices are given in [5].

16.10.4 Traffic Speed and Safety Impacts of Access Management Practices

The effects of access and traffic signal frequency on traffic speeds have been well documented.

- The 1994 Highway Capacity Manual [28] reported a speed reduction of 0.24 mph for every access point along unsignalized highways, and up to a 10 mph speed reduction for sections with 40 or more access points
- Each traffic signal per mile added along an arterial roadway reduces traffic speed by about 2–3 mph
- The application of access management techniques has resulted in significant safety and level of service improvements, as shown in Table 16.8 [29].
- Table 16.9 summarizes access management benefits in traffic speeds and safety from eight case studies. Speeds and safety improved at each location.

16.10.5 Conclusions

It is clear that access management is a desirable strategy from a congestion-reduction perspective. A growing number of state and local transportation agencies have implemented access management activities. However, others have not. This is because "controlling or managing access along arterials is a difficult task facing local officials and transportation engineers. This difficulty comes from a time-honored tradition, and in some cases a legal right, for land owners abutting a road to have access to their land" [7].

However, there is a growing recognition of the importance of access management in congestion mitigation, and of the need to better integrate access management with corridor traffic management and land development. Development density and design, which greatly influence levels of transit service and pedestrian access, will likely receive more emphasis in the future, as the "green" goal of VMT reduction is increasingly gaining popularity.

Table 16.8 Summary of research on the effects of access management techniques

	Treatment	Effects
1.	Add continuous TWLTL	• 35 % reduction in total crashes • 30 % decrease in delay • 30 % increase in capacity
2.	Add non-traversable median	• 35 % reduction in total crashes • 30 % decrease in delay • 30 % increase in capacity
3.	Replace TWLTL with a non-traversable median	• 15–57 % reduction in crashes on 4-lane roads • 25–50 % reduction in crashes on 6-lane roads
4.	Add a left-turn bay	• 25–50 % reduction in crashes on 4-lane roads • up to 75 % reduction in total crashes at unsignalized access • 25 % increase in capacity
5.	Type of left-turn improvement (a) painted (b) separator or raised divider	• 32 % reduction in total crashes • 67 % reduction in total crashes
6.	Add right-turn bay	• 20 % reduction in total crashes • Limit right-turn interference with platooned flow, increased capacity
7.	Increase driveway speed from 5 to 10 mph	• 50 % reduction in delay per maneuver; less exposure time to following vehicles
8.	Visual cue at driveways, driveway illumination	• 42 % reduction in crashes
9.	Prohibition of on-street parking	• 30 % increase in traffic flow • 20–40 % reduction in crashes
10.	Long signal spacing with limited access	• 42 % reduction in total vehicle-hours of travel • 59 % reduction in delay • 57,500 gallons fuel saved per mile per year

Source Reference [29]

16.11 Emerging Congestion Management Strategies

Application of new detection, information, and communications technologies in transportation increases efficiency of the system and provide better levels of service for travelers in terms of lower traffic congestion, reduced delays, and safer roads. The application of these advanced technologies in transportation is popularly known as "Intelligent Transportation Systems," or ITS.

16.11.1 Intelligent Transportation Systems (ITS)

ITS strategies entail use of real-time traffic information that allows dynamic traffic signal controls, better traveler information, roadside electronic screening programs, integrated corridor management, and vehicle-infrastructure integration (VII).

Table 16.9 Summary of access management benefits reported in the case studies

Case study		Reported benefits	
Location	Description of improvements	Speeds	Safety
A. Arapahoe Road Denver. CO	Access managed roads with physical medians. Limited turns 1/2 mile	40 mph in pm peak hour compared	4 Acc/million VMT on arapahoe and 7 Acc/million VMT
A. Parker Road Denver. CO (52 miles)	Traffic signal spacing	With 15–20 mph on non-access managed arterials	On Parker compared with up to 13 on other arterials
B. Oakland Park Blvd. Ft. Lauderdale 17 Florida (2.2 miles)	Physical median extended across unsignalized driveways	30 % less delay	Accident rate declined about 10 % injury rate declined 28 %. 30 % fewer mid-block median maneuvers after improvements
C. Jimmy Carter Blvd. Atlanta, Georgia (3.0 miles)	Two-way left lanes on 4-lane road replaced by physical median; 6 through lanes and protected left turn lanes	Speeds reportedly increased	32 % drop in accidents and 40 % drop in accident rate with interim New Jersey barrier median
C. Memorial Dr. Atlanta, Georgia (4.3 miles)	Two-way left lanes on 6-lane road replaced by physical median; 6 through lanes and protected left turn lanes		40 % drop in accidents and 37 % drop in overall accident rate. 64 % drop in left turn accident rate
G. Route 47—Vineland. New Jersey (1.8 miles)	Four narrow lanes replaced by two through lanes plus protected left turn lane	PM peak hour speeds declined from 35 to 32 mph	39 % decline in total accidents. 86 % decline in left turn accidents
G. Route 130—New Jersey (43 miles)	Median openings closed and left turn lanes installed		45 % decline in accident rate
G. Route 23—New Jersey (3.9 miles)	Jug handles built and road cut through two rotaries		34 % decline in accidents

Source Reference [30]

Examples of these strategies are described below:

- ITS traffic control strategies include a range of applications, including metering traffic, dynamic timing of traffic signals in response to changing traffic demands, non-stop toll collection, metering flow entering the freeway, managing the response to traffic incidents, or providing real time information to travelers about traffic delays, expected travel times, and alternative routes/modes.

- Traveler Information Systems—provide travelers with real-time accurate information on roadway traffic congestion and advice on alternative routes if any are available. Real time traffic information reduces trip delays and increases mobility. Various types of traveler information are available in US cities. They include: traffic radio reports, "511" traveler information numbers, on-route Dynamic Message Signs (DMS) posting expected travel time, and through private providers such as NavTeq/Traffic.com and INRIX. The increasing popularity of "smart phones" provides extensive access to real time traffic information—generally for a fee.
- Roadside Electronic Screening/Clearance Programs for Commercial Vehicles: These actions enable credentialed motor carriers to bypass weigh stations. This reduces the volume of trucks at weigh stations and the delay time while being processed in addition to the time spent in queues waiting to be processed. It also can eliminate queues upstream of weigh stations that create mainline congestion and safety hazards.

• In Oregon, the "Green Light Commercial Vehicle Operations (CVO)" project was expected to prescreen 7.2 million trucks between 2002 and 2012, and save 360,000 h and $25 million to the trucking industry [23].

• The Pre-Pass system is a transponder-based electronic system that enables enrolled trucks to bypass weigh stations that have been retrofitted with Pre-Pass infrastructure: provided to the states free of charge but paid for by participating motor carriers who fund the system with monthly service charges. In 2011, the Pre-Pass system handled approximately 50 million trucks for an aggregate saving of over 4 million hours and over $433 million (http://www.prepass.com).

16.11.2 Integrated Corridor Management (ICM)/Active Traffic Management (ATDM)

Integrated Coordinated Management (ICM) of traffic controls, lane assignments, traveler information, comprises a series of strategies (e.g., traffic sensors, traffic management centers, managed lanes, rapid incident response) to dynamically manage roadways and corridors in response to recurring and non-recurring sources of congestion.

- ICM is a relatively new concept. The US Department of Transportation has implemented eight demonstration projects across the US.
- ICM is a relatively new concept. The US Department of Transportation has implemented eight demonstration projects across the US. An example of the successful application of this approach has been achieved in the metropolitan areas of Dallas (Route 75), Alameda and Contra Costa Counties in northern California (I-80), San Diego (I15), and Seattle (I-5) [31].

- it is anticipated that congestion relief benefits from the ICM strategy will be greater than the benefits of individually applied strategies as demonstrated by the application of a similar approach in Europe (Active Traffic Management) that uses:
  - Speed harmonization: reducing speeds in advance of a major bottleneck to minimize the impact of the congestion event and increase overall throughput;
  - Temporary shoulder use: using the shoulder of the roadway in conjunction with speed harmonization, to increase capacity during peak periods;
  - Queue warning display systems: to warn drivers of the presence of congestion downstream;
  - Dynamic merge control: the selective closing or metering of ramps based on traffic demand to maximize throughput, with priority given ramps with higher volume;
  - Dynamic routing and traveler information: the use of dynamic message signs to display rerouting instructions in response to non-recurring congestion events; and
  - Dynamic lane markings: the delineation of lanes to manage traffic flow patterns created by the above strategies.
  - Transit-Traffic Integration: the coordination of transit and traffic operations in both arterial streets and freeway corridors to reduce delay for all travelers.

16.12 Conclusions

The various operational (e.g., traffic engineering) strategies described in this chapter, where effectively and sensibly applied, will reduce congestion delays in most urban areas.

A general description of various strategies, and an assessment of their application, effectiveness, and implementation issues is given in Table 16.10. This table is adapted from a national research project conducted by Cambridge Systematics [23].

The resulting benefits of operational strategies that reduce recurring congestion, however, will probably not meet the long term needs of many cities. Growth in person travel—motorized travel in particular—in the rapid growing areas of the United States will require additional congestion relief strategies and actions. Accordingly, Chap. 17 describes roadway capacity expansion strategies and Chaps. 18–23 describe strategies that reduce traffic demand.

Table 16.10 Congestion management strategies for using existing capacity more efficiently—summary of effectiveness and implementation potential

				Effectiveness		Extent of application		Implementation issues		
		Strategy	Substrategy	Local	Area wide	Current	Potential future	Cost	Noncost barriers	Time frame
1.	a	Traffic signal timing and coordination	Traffic signal optimization and interconnection	High	Medium to high	Moderate—extensive (25–60 % signals optimally timed)	Extensive (all nonfreeways)	Low	Low	Short-term
	b		Centralized, actuated control systems	High	Medium in to high?	Limited (major arterials networks)	Moderate (major arterials. networks)	Medium	Medium	Mid-term
2.		Changeable lane assignments		Medium	?	Limited (<250 miles in US)	Limited	Low to medium	High	Short-term
3.		Congestion pricing	Variable pricing on existing tolled facilities	Varies	Varies	Limited (~10 facilities in US)	Moderate (any tolled facility)	Low/ revenue-generating?	Low to medium	Mid-term
			Area wide pricing systems	High	High	None in US	Unknown	High/ revenue-generating	High	Mid-to long-term
4.		Ramp metering	Real time metering based on mainline volumes/ speeds	High	Medium to high	Moderate (1/3 fwys in 25 metro areas)	Extensive (all major metro area fwy systems)	Low	Medium to high	Short-term

(continued)

Table 16.10 (continued)

				Effectiveness		Extent of application		Implementation issues		
		Strategy	Substrategy	Local	Area wide	Current	Potential future	Cost	Noncost barriers	Time frame
5.		Roadside electronic screening/clearance programs for commercial vehicles		Medium	Low	Moderate—extensive (35 f stales)	Extensive (all states)	Low to medium	Low to medium	Short-term
6.		Loading zone management		Medium	Low	Extensive (most local zoning ordnances)	Extensive	Low	Medium	Mid-term
7.		Access management		High	Medium—high	Moderate (∼29 percent of urban arterials)	Moderate (arterials. especially new or reconstructed)	Low to medium	Medium	Mid-to long-term
8.	a	Port operations	Time-of-day pricing	High	?	Limited (two mayor ports) Limited	Limited (major US ports)	Low/ revenue-generating	Medium	Short-term
	b		Alternative modes (rail, barge)	?	?		Limited (major US ports)	High	Medium to high	Mid-term
9.		Border crossing improvements		High	Not applicable	Limited (prescreening at 55/100 locations)	Limited (all border crossings)	Low to medium	Low to medium	Short-term

(continued)

Table 16.10 (continued)

				Effectiveness		Extent of application		Implementation issues		
		Strategy	Substrategy	Local	Area wide	Current	Potential future	Cost	Noncost barriers	Time frame
10.		Integrated corridor management/ active traffic management		High	Medium	Limited (pilot sites)	Moderate (major travel corridors)	High	Medium	Mid-term
11.		Vehicle infrastructure integraton		High	High	None	Extensive	High	High	Long-term
12.		Traveler information		Low to medium	Low to medium	Moderate/ extensive (fwy network in most metro areas)	Extensive	Low to medium	Low	Short-term

Source Reference [23], p 3-3. Table 3.1

References

1. Highway Capacity Manual. Transportation Research, Washington, DC (2010)
2. Traffic Engineering Handbook. Institute of Transportation Engineers, Washington, DC
3. Manual of Uniform Traffic Control Devices for Streets and Highways, 2009 Edition with Revision 1 and 2. s.l.: US Department of Transportation, Federal Highway Administration (2012)
4. Gluck J, Levinson HS, Stower V (1999) NCHRP Report 420: impacts of access management techniques. Transportation Research Board, Washingto DC
5. Koepke FJ, Levinson HS (1992) NCHRP Report 348: access management guidelines for activity centers. Transportation Research Board, Washington, DC
6. Sunkan S (2004) The benefits of retiming traffic signals. ITE Journal. April 2004.
7. Meyer MD (1997) A toolbox for alleviating traffic congestion and enhancing mobility. Institute of Transportation Engineers, Washington, DC
8. Levinson HS, Cleveland DE, Kostyniuk LP, Waissi GR (1985) Traffic engineering for public transportation—a manual for practice draft 1985
9. Levinson HS, Mentor I (2005) Operational improvements—an overview. s.l.: UTRC2 Research News
10. Homburger WS, Hall JW, Loutzenheiser NC, Reilly WR (1995) Fundamentals of traffic engineering, 14th edn. Course notes. University of California, Berkeley
11. Bruce JA (1967) One-way major arterial streets. Highway Research Board Special Report: improved street utilization through traffic engineering, pp 24–36
12. Falcocchio JC, Jose D, Prassas ES (1995) An inquiry on the traffic congestion impacts of parking and pricing policies in the Manhattan, CBD. s.l.: prepared for New York City Department of Transportation and the University Transportation Research Center, Region 2
13. Shoup D (2005) The high cost of free parking. s.l.: American Planning Association
14. New York City Department of Transportation (2013) Measuring the street: new metrics for 21st century streets
15. Polytechnic Institute of New York (1975) NCHRP Report 3-18(2) traffic control in over saturated street network. Transportation Research Board, Washington, DC
16. Kittelson & Associates, Inc., Herbert S. Levinson Transportation Consultants, DMJM + Harris (2007) TCRP Report 118, bus rapid transit practitioner's guide. Transportation Research Board, Washington, DC
17. Levinson HS, Zimmerman S, Clinger J, Gast J, Rutherford S, Bruhn E (2003) TCRP Report 90: bus rapid transit. Volume 2: implementation guidelines. Transportation Research Board, Washington, DC
18. Homburger WS, Hall IW, Luotzenhouser RC, Reck WR (1996) Fundamentals of traffic engineering. Institute of Transportation Studies, Berkeley
19. Matson TM, Smith WS, Hurd FW (1951) Traffic engineering. McGraw-Hill Book Company, New York
20. Walker GW, Kulash WM, McHugh BT (2000) Downtown streets—are strangling ourselves on one-way networks? In: Transportation research circular E-C019: urban street symposium conference proceedings, Dec 2000, p F-2
21. Halsey M (1941) Traffic accidents and congestion. Wiley, New York, p 315
22. LaPlante JN (1967) Directional flow improvements on urban streets. Highway Research Board Special Report 93: improved street utilization through traffic engineering, pp 17–23
23. Prepared by: Cambridge Systematics and Resource Systems Group (2008) NCHRP 20-24A Task 63: effective management practices for congestion management: final report. Requested by: American Association of State Highway and Transportation Officials (AASHTO), Washington, DC, Nov 2008
24. Titus LT (1976) Speed regulations and other operational controls. Transportation and traffic engineering handbook, Chapter 18. Prentice Hall, Englewood Cliffs

25. Capelle DG, Bau S (1987) Transportation and traffic engineering handbook, 2nd edn. Chapter 25: freeway surveillance and control. Prentice Hall, Englewood Cliffs
26. Freeway Management Handbook. s.l.: Federal Highway Administration, US Department of Transportation, Report No. FHWA-SA-97-064, August 1997
27. Committee on Access Management (2003) Access management manual. Transportation Research Board, Washington, DC
28. Highway Capacity Manual. Transportation Research Board, Washington, DC (1994)
29. SK Transportation Consultants Inc. (2000) Access management location and design: participant notebook for NH Course 133078. s.l.: National Highway Institute, Federal Highway Administration
30. Levinson HS, Koepke FJ (1993) Case studies in access management. s.l.: Prepared for Transportation Research Board
31. O'Connor AT (2014) Development of a framework for the deployment of intelligent transportation systems in a multi-agency environment. Ph.D. dissertation, s.l.: NYU Polytechnic School of Engineering, May 2014

Chapter 17
Adaptation Strategies for Managing Recurring Congestion—Adding New Capacity

17.1 Introduction

Most operational improvements provide gains in capacity. Important as they are, they produce short-lived travel time reductions in areas with a growing population and employment. In these cases the extra efficiency and travel time reductions gained through operational improvements are soon lost as the growth in traffic demand reaches the capacity of the improved roads: "only so much extra efficiency can be squeezed out of an already—stressed highway system" [1].

Adding new roadway capacity is an effective strategy in reducing congestion. As shown in Fig. 17.1, areas that were more successful in adding road capacity experienced a fraction (1/3–1/4) of the congestion growth found in areas less successful in adding road capacity [2].

The reasons underlying the need for capacity expansion can be generalized as follows:

- Many urban roads are congested—even those where operational improvements have been implemented
- Long term growth will exceed the capacity gains resulting from operational strategies
- Newly developing areas will need improved roadway infrastructure
- New and expanded major developments in existing cities and suburbs will need new transportation infrastructure

Adding new roadway capacity to reduce existing and anticipated congestion can be done in various ways. New roadways and lanes can be provided for all vehicles, or they can be restricted to specific types of vehicles.

Examples of capacity strategies that apply to all vehicles include:

- Bottleneck Reduction/Removal
- Intersection Improvements
- Street Connectivity/Continuity Improvements

J.C. Falcocchio and H.S. Levinson, *Road Traffic Congestion: A Concise Guide*,
Springer Tracts on Transportation and Traffic 7,
DOI 10.1007/978-3-319-15165-6_17

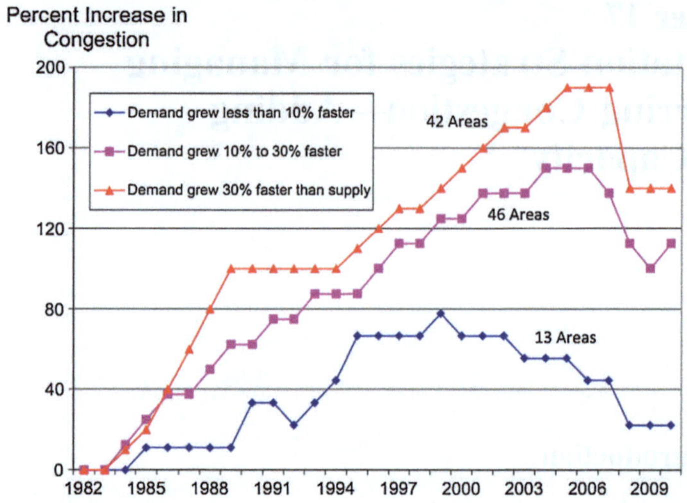

Fig. 17.1 Growth rate in traffic versus growth rate in roadway capacity (1982–2010) and its impact on traffic congestion. *Source* Reference [3], p 16, Exhibit 11

- Roadway Widening and Reconstruction
- New Roadways
- New Toll Roads

Examples of strategies that apply to priority vehicles include:

- Managed Lanes (HOV, HOT, Express Toll/Value Pricing)
- Truck-only Lanes
- Intermodal Access Roads

A summary assessment of these various strategies, developed by Cambridge Systematics and Resource System Group [2], is shown in Table 17.1.

The remainder of this chapter describes new capacity enhancement strategies that apply to all vehicles and priority vehicles. It also sets forth some consequences in the application of these strategies such as the "induced traffic" issue of added capacity.

17.2 Capacity Expansion for All Vehicles

The traditional approach to capacity expansion has been to eliminate congestion for all roadway users by highway bottleneck removal, intersection reconstruction and expansion, widening existing roads, and building new roadways. Collectively these capacity expansion actions have generally reduced congestion and increased mobility.

Table 17.1 Congestion management strategies involving added capacity/physical improvements—summary of effectiveness and implementation potential

		Strategy	Substrategy	Extent of application: Current	Extent of application: Potential future	Effectiveness: Local	Effectiveness: Areawide	Implementation issues: Cost	Implementation issues: Noncost barriers	Implementation issues: Time frame
1.		New roads and roadway widening		Moderate	Moderate	High	High	High	High	Mid-to long-term
2.		New toll roads		Moderate	Moderate	High	High	High/revenue-generating	High	Mid-to long-term
3.	a.	Managed lanes (HOV, HOT, express toll)	Construction of new HOV, HOT, or express toll lane	Limited (four percent of metro fwy miles)	Moderate	High	High	High	Medium to high	Short-term
	b.		Conversion of existing general purpose lanes to HOV, HOT, or express toll	Limited	Limited	Low or negative	Low or negative	Low to medium/revenue-generating	High	Short-term
	c.		Conversion of existing HOV lanes to HOV or express toll	Limited (four applications in US)	Limited (existing HOV lanes)	Medium	Medium	Low to medium/revenue-generating	Medium	Short-term
4.		Truck-only lanes		Limited (three interstate miles)	Limited (high- truck-traffic facilities)	Varies	Varies	High/revenue-generating	High	Mid-to long-term
5.		Bottleneck relief		Moderate	Moderate	High	High	Medium to high	Medium	Mid-to long-term
6.		Intersection improvements		Moderate	Moderate (new/reconstructs)	High	Medium to high?	Low to medium	Low to medium	Mid-to long-term

(continued)

Table 17.1 (continued)

		Strategy	Substrategy	Extent of application		Effectiveness		Implementation issues		
				Current	Potential future	Local	Areawide	Cost	Noncost barriers	Time frame
7.		Intermodal access roads		Limited (near intermodal terminals)	Limited (near intermodal terminals)	Varies	Low	Medium to high	Medium	Mid-to long-term
8.		Access management		Moderate	Moderate especially new arterials	High	Medium to high	Low to medium	Medium	Mid-to long-term
9.		Street connectivity		Moderate (most older urban location)	Moderate (area of new development)	?	?	Low or cost savings	Low	Long-term

Source Adapted from Ref. [2]

17.2.1 Bottleneck Reduction/Removal

A "bottleneck" results along heavy traveled roads where there is a severe imbalance between traffic demand and roadway capacity. Its impact on delay can vary widely in intensity, extent, and duration.

17.2.1.1 Causes

Bottlenecks along freeways and arterial streets are usually caused by "lane drops" and by the convergence of major roadways on approaches to bridges and tunnels. They also result from inadequate freeway geometry such as closely spaced freeway ramps, left-hand freeway entry and exits, short weaving distances, and inadequate interchange design.

Along arterial streets, bottlenecks are usually created by complex intersections with high traffic demand resulting in short "green" times, and long "red" times for each traffic movement.

17.2.1.2 Consequences

Recurring bottlenecks account for 40 % of congestion delays. Creating additional capacity is an essential component of a comprehensive congestion relief program to relieve the system-wide congestion impacts of bottlenecks.

A study prepared for the American Highway Users Alliance estimated that "improvements to the 166 most serious bottlenecks nationwide including traffic operations, demand management, and capacity expansion could significantly reduce delays, crashes, and air pollution and result in significant cost savings" [4].

17.2.1.3 Bottleneck Relief Strategies

The corrective strategies will vary individual circumstances, including location, type, and severity of congestion. They include freeway widening and reconstruction of the main travel lanes and interchanges, and ramp widening, metering, or ramp closure.

Improvements along arterial streets include intersection widening and simplification of intersecting flows, grade separation, and new controlled-access bypass routes around congested business centers.

17.2.1.4 Implications

The congestion benefit of bottleneck removal, however, must be evaluated for the entire roadway system—not just for the bottleneck location. Bottlenecks create delays for traffic upstream of the bottleneck, but they also meter traffic demand at downstream roadways. Removal of a bottleneck, therefore, may transfer the delay from upstream to downstream roadways.

For example, consider a bridge carrying traffic to and from the central business district (CBD) of a large city, where the morning commuter traffic demand destined to the CBD and the evening traffic demand exiting the CBD exceed the capacity of the bridge. In this case the bridge meters the morning traffic demand to the CBD and increasing its capacity will increase the traffic congestion on the roadways within the CBD. On the return trip, however, increasing the capacity of the bridge enables traffic demand exiting the CBD to be processed at a faster rate resulting in lower congestion on CBD streets.

The benefits of a bottleneck removal, therefore, need to be evaluated from the standpoint of its impact on total trip delay and the possible transference of congestion impacts from one area to another. Increasing outbound bridge capacity would lower the intensity and duration of congestion on the CBD streets benefiting its businesses, residents and visitors; while increasing inbound capacity would increase the intensity and duration of CBD congestion.

In this example if the goal is to improve traffic conditions in the CBD, the strategy should be to remove the bridge bottleneck for outbound traffic only.

17.2.2 Intersection Improvements

Intersections are a common source of traffic congestion when they process a large number of conflicting movements. As discussed in the previous chapter, often congestion can be reduced by reducing the conflict points by providing turning lanes within the existing roadway width, restricting turning movements, and/or making streets one-way.

In many cases, however, adding new capacity is needed to reduce conflicts and reduce delay. This strategy entails the expansion and reconfiguration of intersections by (1) intersection widening to provide auxiliary turn lanes, (2) reducing conflicts—e.g., using "jughandles" and "indirect" left turns, (3) replacing traffic signals with roundabouts where traffic signals are inefficient in controlling conflicting traffic flows—especially where more than two streets intersect, and (4) separating the grades of conflicting flows.

17.2.2.1 Auxiliary Through Lanes [5]

Auxiliary lanes consist of roadway widening to increase the capacity of arterial roads in suburban areas. They are provided upstream and downstream of the intersection to serve left turn, through, or right turn volumes at heavily traveled suburban intersections. Illustrative sketches of such treatments are shown in Fig. 17.2.

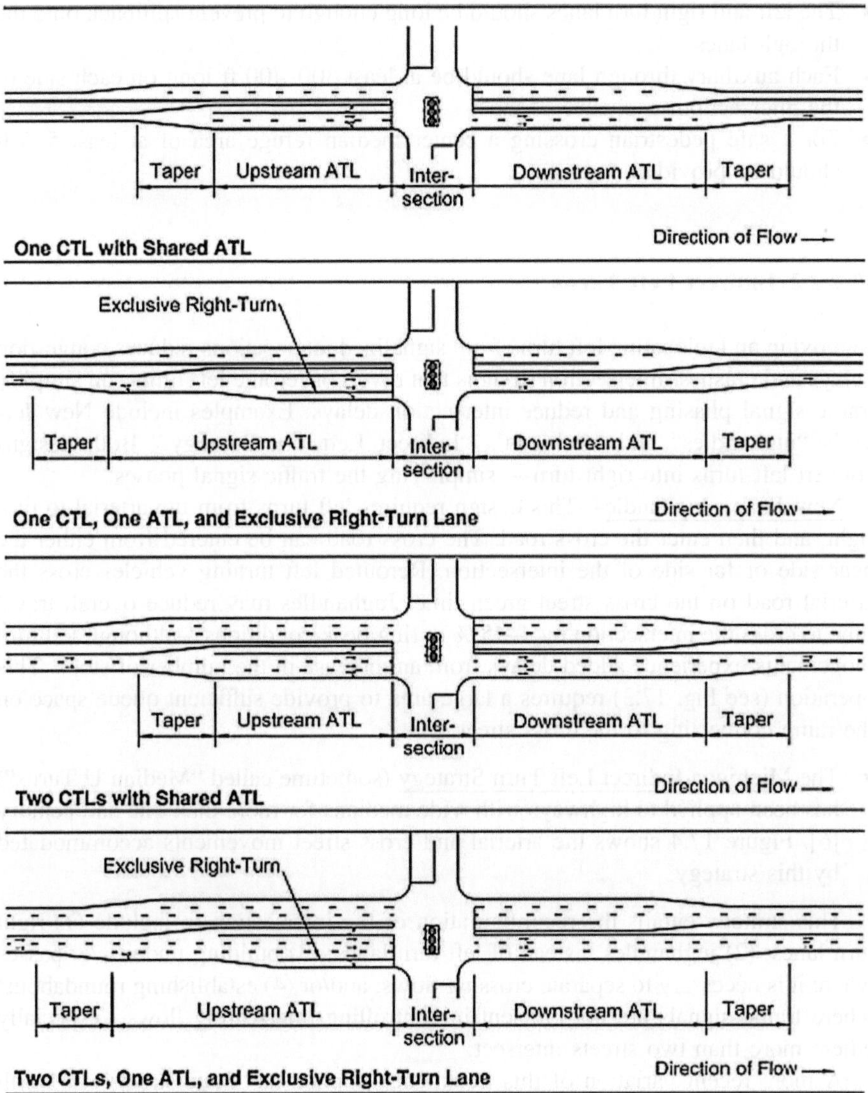

Fig. 17.2 Auxiliary through-lanes configurations. *Source* Reference [5]

Each of the configurations consists of a right-hand addition upstream of the intersection, and a right-hand merge downstream of the intersection. It should be noted that:

- Auxiliary through lanes are most effective in reducing congestion at signalized intersections when through traffic volume on the normal through lanes exceeds capacity of the intersection

- The left and right turn lanes should be long enough to prevent spillback onto the through lanes
- Each auxiliary through lane should be at least 300–400 ft long on each side of the intersection, exclusive of taper
- For a safe pedestrian crossing a center median refuge area of at least 5–6 ft should be provided.

17.2.2.2 Indirect Left Turns

Removing and relocating left turns from signalized intersections reduces congestion delays and crashes. Intersection designs that divert or reroute left turns can simplify traffic signal phasing and reduce intersection delays. Examples include New Jersey's "jughandles" and Michigan's "Indirect Left-turn Strategy." Both designs convert left turns into right turns—simplifying the traffic signal phases.

New Jersey Jughandle—This design requires left turns from the arterial to turn right, and then enter the cross road. The cross road can be entered from either the near side or far side of the intersection. Rerouted left turning vehicles cross the arterial road on the cross street green time. Jughandles may reduce overall travel time to cross the intersection by 4–45 % during peak conditions—although left turn movements experience added delays from an increase in the number of stops. This operation (see Fig. 17.3) requires a large area to provide sufficient queue space on the ramp connecting to the cross street.

- The Michigan Indirect Left Turn Strategy (sometime called "Median U Turns") has been applied to highways with wide medians for more than one half century [6]. Figure 17.4 shows the arterial and cross street movements accommodated by this strategy.

This strategy entails the reconfiguration of the intersection to include (1) right turn lanes; (2) jughandles instead of left-turn lanes; (3) building under/over passes where it is necessary to separate crossing flows; and/or (4) establishing roundabouts where traffic signals are not efficient in controlling intersecting flows—especially where more than two streets intersect.

A more recent variation of this treatment—applied as "Superstreets" in North Carolina—works well when the arterial road has the dominant flow, (where there are and no pedestrians). This treatment, sometimes called the "Restricted Crossing U Turn (RCUT)" is shown in Fig. 17.5.

All the cross street through traffic entering the Superstreet must turn right, then proceed several hundred feet before making a U turn like the Michigan treatment. Side street intersections are signalized where they enter the arterial, and sometimes the U turns are signalized as well.

Both concepts allow two-phase traffic signals along the main roadway, with the arterial getting most of the green time. The safety and congestion relief benefits of

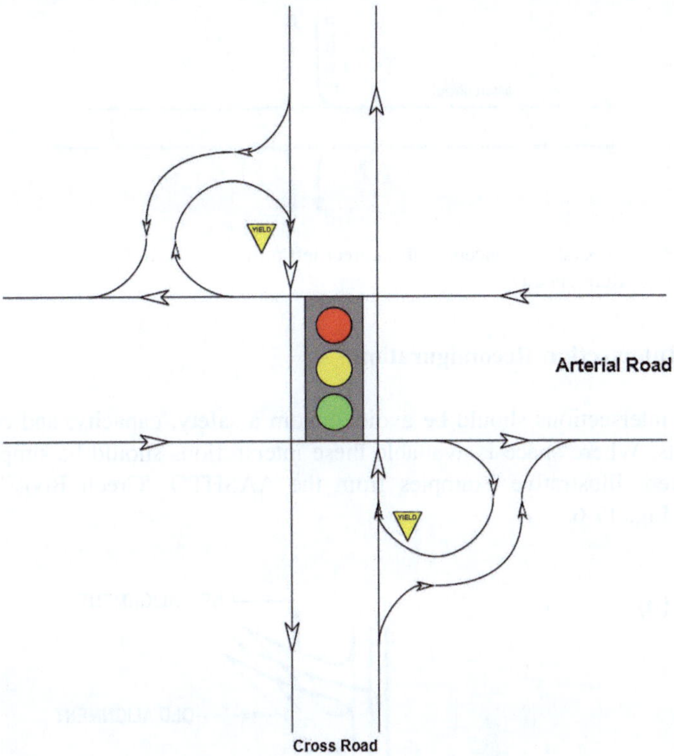

Fig. 17.3 New Jersey "Jughandle" concept

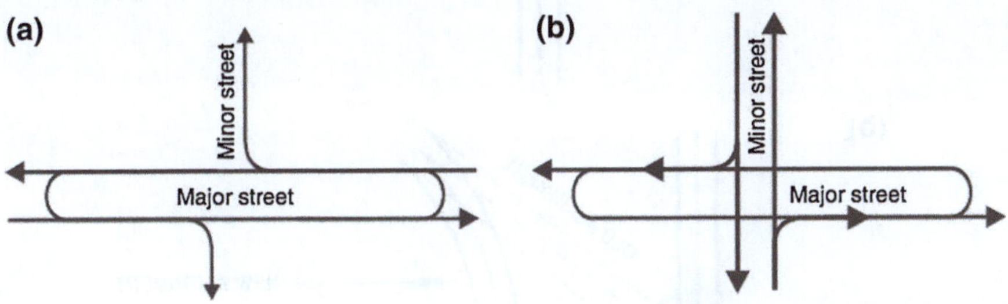

Fig. 17.4 Michigan U turn left turn movements. *Source* Reference [6], p 47, Fig. 13. Primary source: Reference [7]

the directional median crossovers are well documented. Directional crossovers experience one-third the crash rates of two-way left turns and about two-thirds the rate of bi-directional crossovers [8]. A study by Koepke and Levinson [9] found a gain of 14–18 % in capacity as compared to conventional intersection designs with dual left turn lanes.

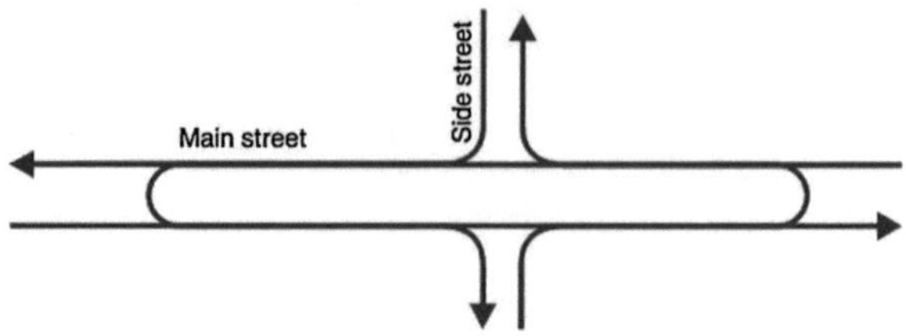

Fig. 17.5 Basic superstreet concept with indirect left turns. *Source* Reference [6], p 48, Fig. 15. Primary source: Reference [7].

17.2.2.3 Intersection Reconfiguration

Multi-leg intersections should be avoided from a safety, capacity, and congestion standpoints. Where space is available these intersections should be simplified and reconfigured. Illustrative examples from the AASHTO "Green Book" [10] are shown in Fig. 17.6.

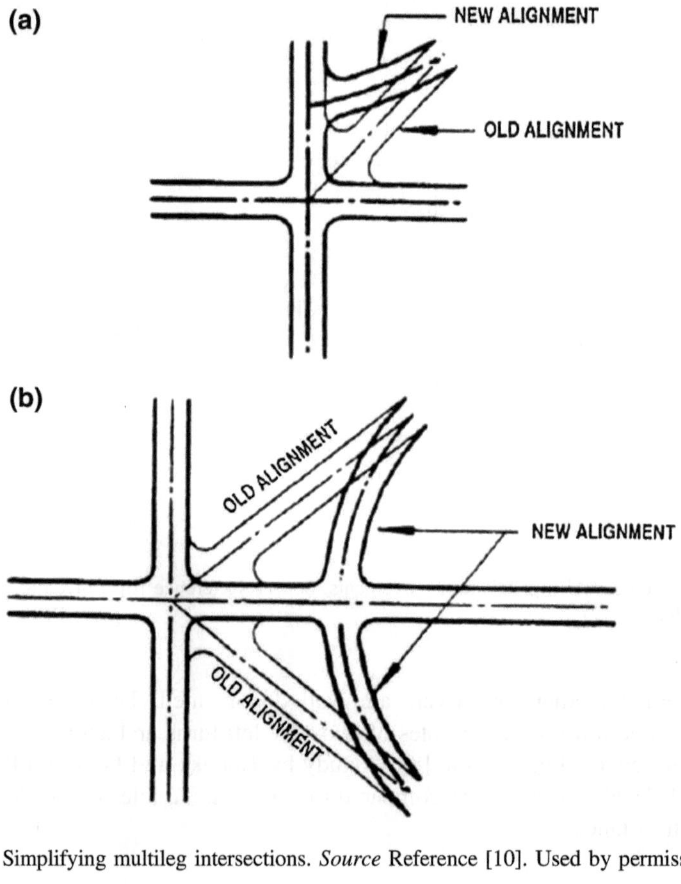

Fig. 17.6 Simplifying multileg intersections. *Source* Reference [10]. Used by permission.

In each example the diagonal roadway entering the intersection is re-aligned to create a simple right angle intersection with the main arterial. The distance between the existing and newly created intersection should be set to avoid vehicle spillback and should allow independent and coordinated operations with the main intersection.

17.2.2.4 Roundabouts

Roundabouts emerged in the United States in the 1990s after several decades of successful implementation in Great Britain, France, and Australia [11].

They differ from the earlier traffic circles and rotaries in several respects: (1) the diameter of the turning circle is much smaller, (2) vehicles entering the roundabout yield to circulating vehicles, and (3) the entering traffic is sometimes slightly deflected.

Design Principles of Roundabouts

NCRP 672 [11] sets forth the following basic design principles for roundabouts:

- Provide slow entry speeds and consistent speeds throughout the roundabout by using deflection
- Provide appropriate number of lanes and lane assignments to achieve adequate capacity, lane volume balance, and lane continuity
- Provide smooth channelization that is intuitive to drivers and results in vehicles naturally using intended lanes
- Provide adequate accommodation for the design vehicles
- Design to meet needs of pedestrian and cyclists
- Provide appropriate sight distance and visibility for driver recognition of the intersection and conflicting users.

Features, Types, and Dimensions

The key features, types, and dimensions of modern roundabouts are shown in Fig. 17.7.

The types include: (1) mini-roundabouts, (2) urban compact roundabouts, (3) urban single-lane roundabouts, (4) urban double-lane roundabouts, (5) rural single-lane roundabouts, and (6) rural double-lane roundabouts.

The roadway dimensions vary with the types of roundabouts and the design vehicles. Some key roundabouts entry speeds and design vehicle-related dimensions are shown in Table 17.2.

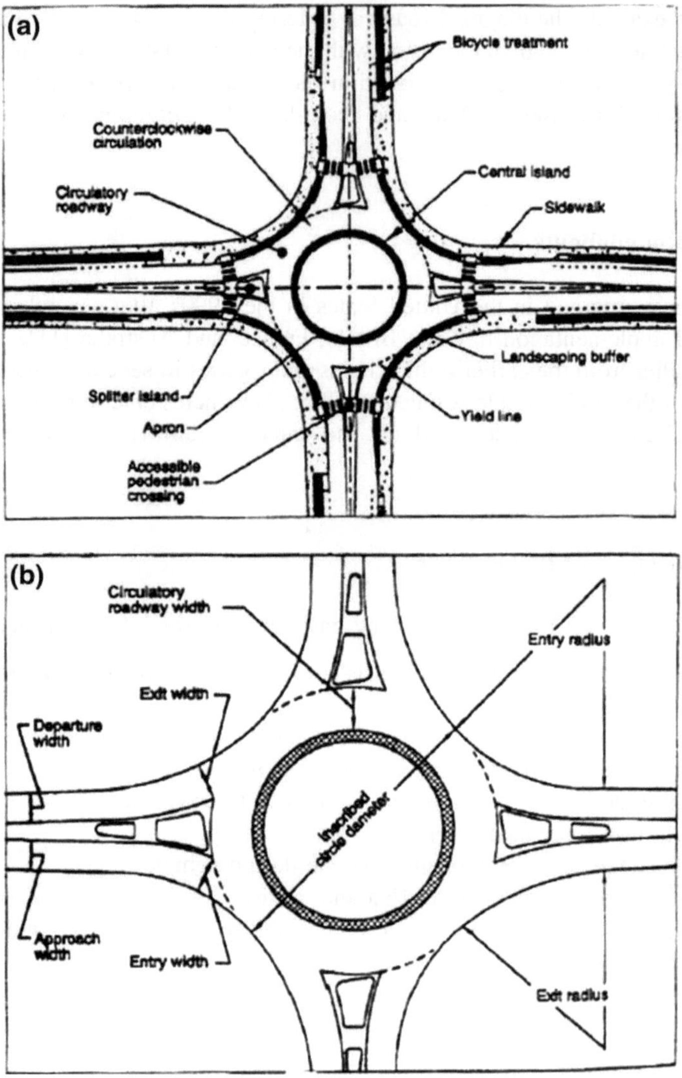

Fig. 17.7 Key features of the roundabout. *Source* Reference [11]

Key dimensions for roundabouts pedestrian crossings are as follows:

- Pedestrian refuge width on splitter islands = 6 ft
- Pedestrian crossings are located 25–50 ft before the yield sign.

The pedestrian walkway should be located at street level, and should include a detectable warning surface.

Table 17.2 Roundabouts design entry speeds and typical inscribed circle diameters

Type of roundabout	Recommended maximum entry speed (MPK)	
Mini-roundabout	25	
Urban compact	25	
Urban single lane	35	
Urban double lane	40	
Rural single lane	40	
Rural double lane	50	
Type of roundabout	Circle typical design vehicle	Common inscribed circle diameter range (ft)
Mini-roundabout	SN-30	45–90
Single-lane roundabout	WB-40	90–150
	WB-50	105–150
	WB-67	130–180
Multi-lane roundabout 2 lanes	WB-50	150–220
	WB-67	165–220
Multi-lane roundabout 3 lanes	WB-50	200–250
	WB-67	220–300

Source Reference [10]

Effectiveness

Single-lane roundabouts have been reported to carry up to 25,000 vehicles per day. Two-lane roundabouts have been reported to carry 40,000–50,000 vehicles per day [6].

A growing body of literature indicates that roundabouts can reduce delay, increase capacity, and improve safety. Cambridge Systematics [2] reports that:

- in 38 cases reviewed, modern roundabouts reduced delay, increased capacity, and increased safety.
- in three roundabouts in Kansas, delay decreased from 13 to 23 %.
- in 10 intersections with ADTs between 14,000 and 41,000 with signal control, could have experienced lower average peak hour delays between 17 and 92 %, had they been constructed as roundabouts.

17.2.2.5 Arterial Connections to Freeways

In many urban areas congestion is common along arterial roadways in the vicinity of ramps to and from freeways. This is the result of (1) heavy traffic volumes on the arterial street and (2) heavy left turns (sometimes overlapping along the roadway)

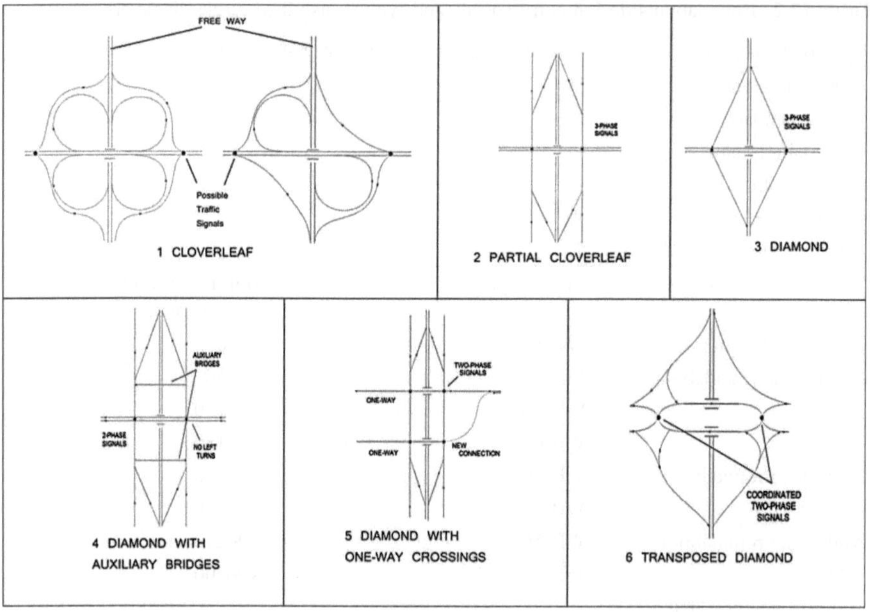

Fig. 17.8 Possible interchange connections that reduce congestion. *Source* Made by authors

combined with complex phasing. Corrective actions include changing the ramp configurations and building new connecting roadways.

Typical freeway-arterial street interchanges, with possible congestion—reducing improvements are shown in Fig. 17.8.

1. The cloverleaf intersection—once the solution to traffic congestion—has been superseded by other ramp configurations because of the space required and the weaving on the freeway travel lanes. Sometimes two-phase traffic signals are installed along the intersection of the arterial roadway and the exiting ramp traffic.
2. The partial cloverleaf is increasingly used since it eliminates weaving on the freeway travel lanes. Ramp terminals at the interchange arterial street are controlled by two-phase traffic signals. There are no left turns from the arterial roadway to the freeway.
3. The diamond interchange is widely used because it requires less land and can be integrated into the city street system where sometimes continuous one-way frontage roads run parallel the main freeway. However, the arterial roads often become congested at the interchange because of heavy overlapping left turns, and because they need special traffic signal phases to accommodate them. Single (and sometimes dual) left-turn lanes are provided along the arterial roadway.

To address this diamond interchange issue several treatments have been applied that include (a) building additional crossings over or under the freeway lanes and (b) reconfiguring the arterial roadway, and its ramp connections.

- One solution—applied along the I-494 freeway in the Detroit metropolitan area —was building two one-way bridges over the freeway spaced 400–600 ft from the arterial road. These bridges form a one-way couplet to accommodate left turns from both the freeway exit ramps and the intersecting arterial roadway. Left turns are prohibited at frontage road intersections with the arterial thereby simplifying the traffic signal phasing.
- A similar possibility is to build a new bridge over the freeway and connect it to the existing street system to form a one-way couplet with the existing bridge across the freeway.
- The most far reaching treatment of diamond interchanges is what is called the "displaced diamond" interchange -"DDI"- or "double crossover" intersection. This approach is amply documented in the TRB publication "Design and Operation Performance of the Double Crossover Intersection and Diverging Diamond Interchange." [12]. This treatment simplifies the traffic signal phasing but it requires more space along the crossroad.

The interchange designs of new freeways should be tested for their congestion impacts, and the necessary modifications made to minimize traffic congestion.

17.2.2.6 Separating Grades

Grade separations (flyovers and underpasses) have been built at heavily traveled and complex road junctions in many cities to avoid severe congestion. The higher volume movement is removed from the at grade intersection thereby resulting in reduced delay—both on the main arterial and cross street.

Some applications of intersection grade separations are given in Table 17.3.

- Most treatments carry the main arterial under or over the intersecting streets
- Left-turn flyovers are found in Chicago and Miami Beach
- A "three-level diamond" at the intersection of Telegraph Road and Eight mile Road in metropolitan Detroit separates the high volume through movements on both six to eight lane roadways; the middle level of the interchange is signalized to accommodate the turning movements on both roadways
- The Route 4 and 17 cloverleaf interchange in New Jersey separates grades on two major roadways with expressway characteristics
- Connecticut Avenue in Washington, DC originally had two levels below the street: there was a short two-track street car tunnel on the upper level that was closed when street car service was abandoned; and a longer lower level roadway tunnel.

Table 17.3 Some examples of intersection grade separation

Location		Type			
		Flyover or fly-under	Left turn flyover	3-level diamond	Interchange
Boston	Cambridge Massachusetts Ave at Commonwealth, Huntington, Memorial Drive	○			
Chicago	Western at Belmont and Clybourn	○			
	Archer at Ashland		○		
Detroit metro area	Telegraph Rd (US 24) and 8-mile road			○	
Miami Beach			○		
New York	New Jersey				
	Route 4–17				○
	First Ave at UN	○			
	Grand Concourse at Fordham Rd	○			
Seattle	Aurora	○			
Washington	Massachusetts Ave	○			
	Connecticut Ave, Dupont Circle	○			

The fly-overs and underpasses on roadways normally add about 30–40 ft of width to a roadway. Ideally the number of lanes should be the same as those on the parallel roadway to avoid bottlenecks the start and end points.

17.2.3 Street Connectivity, Continuity, and Spacing

Street patterns in cities and suburbs reflect topography, settlement densities, and policies regarding street spacing and subdivision requirements. City street grids are usually closely spaced, while many suburban streets are circuitous, discontinuous, and usually spaced far apart.

Systems of streets with circuitous and discontinuous routing patterns increase trip lengths and discourage walking trips and bus transit use.

Compared to cul-de sac neighborhoods, traditional neighborhoods built on grid systems combined with higher development (mixed use) densities experience a lower VMT per capita and higher utilization rates of non-motorized modes [13]. On traditional grid networks, local streets provide an alternative to arterials for short trips and lessen the traffic demand on arterials (e.g., see Fig. 11.2).

17.2.3.1 Connectivity

Many suburban residential streets were designed to limit through traffic. To accomplish this goal they were laid out to be disconnected. Lack of connectivity, however, made it difficult for police and emergency vehicles to provide a quick response. This condition generates excess travel that could be eliminated by making the necessary connections of routes within residential areas as shown in Fig. 17.9. Connecting the missing links within residential areas can reduce the VMT (and congestion) on connectors and arterial streets.

17.2.3.2 Continuity

Many collector streets and some arterial streets are discontinuous. The discontinuity results in double loading of available streets by the displaced traffic. The VMT increases, as do turning movements and peak period congestion. This condition is illustrated in Fig. 17.10.

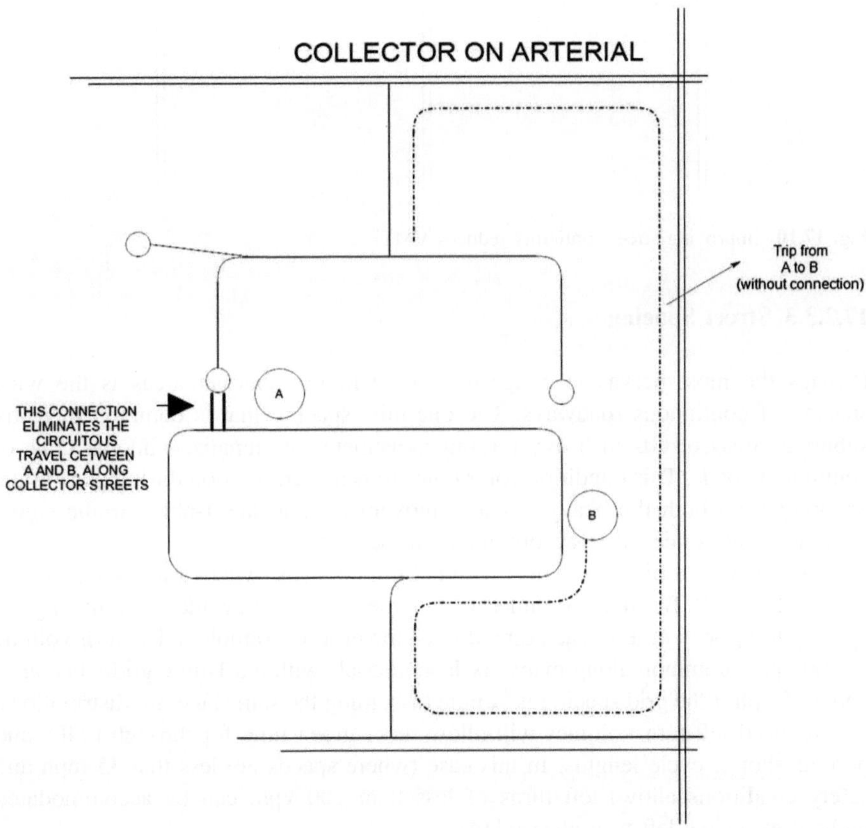

Fig. 17.9 Improving local street connectivity reduces excess travel

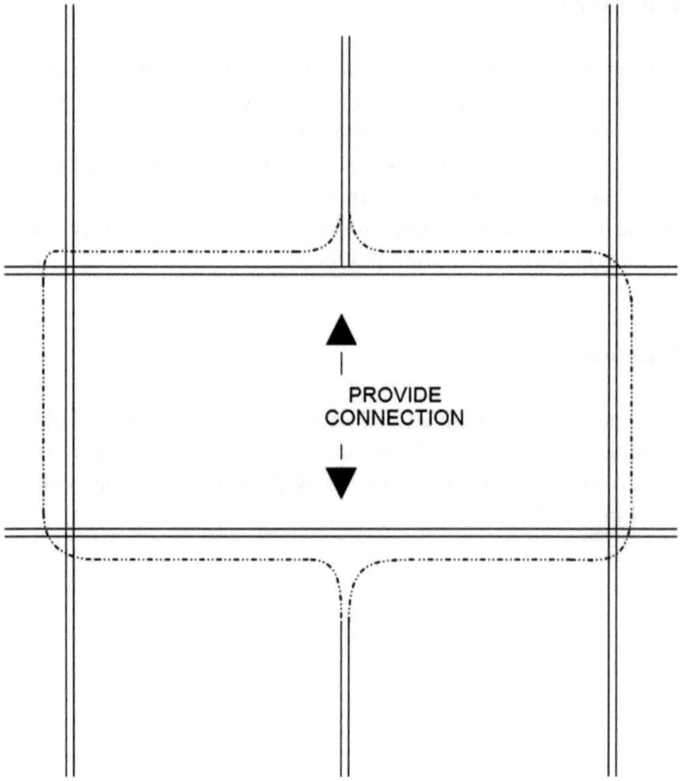

Fig. 17.10 Improving street continuity reduces VMT

17.2.3.3 Street Spacing

Perhaps the most pervasive congestion problem in suburban areas is the wide spacing of continuous roadways. The one-mile spacing that is common in many suburban areas results in heavy turning movements at signalized intersections of continuous roads. This condition contributes to heavy congestion during peak hours resulting from both the heavy left turn movements and the 4-phase traffic signal operation that is needed to accommodate these turns.

This situation is difficult to deal with in built up areas (where grade separation is difficult to build). But in areas under development it can be avoided by reducing the spacing to ½ or ¼ mile in each direction of travel. For example, a left-turn volume of 300 vph—common along many six-lane arterials within a 1-mile grid—becomes about 75 vph if the grid spacing is ¼ mile (assuming the same lane use distribution). The reduced left-turn volumes will allow more green time for through traffic and permits shorter cycle lengths. In this case (where speeds are less that 35 mph and safety conditions allow) left turns of less than 100 vph, can be accommodated without exclusive left-turn phases [14].

Table 17.4 Illustrative guidelines for suburban street access and spacing

Facility						
Item	A Freeway	B Strategic arterial[a] "Boulevard"	C Arterial	D Continuous access collector	E Local access collector	F Local streets
Access						
Access control	Full	Partial	Partial	Safety	Safety	Safety
Direct property access	None	Right in/out[b]	Full[b, c]	Full	Full	Full
Interchange with	B, C	A, B, C	A, B, D	B, C, E	D, F	E
Continuity	Continuous	Continuous	Continuous	Continuous	1/2 MI	300′–1,000′
Spacing						
Street spacing	4–6 MI	6[d]	1[e]	1[e]	1/4	1/8–1/16
Cross street spacing	1–2	½	1/2	1/4	1/8–1/16	100′
Traffic signal spacing based on through band width (%)	–	50–60	40–50	30–40	–	–

Source Reference [14], © National Academy of Sciences, Washington, DC, 2000. Reproduced with permission of the Transportation Research Board

[a] "Expressways" would have similar features except that there would be no direct property access. Direct property access could be prohibited where reasonable alternative access is available

[b] Direct property access may be prohibited where reasonable alternative access is available. Residential access would be prohibited

[c] Left turn exits from developments may be prohibited

[d] Locate midway between freeway whenever freeway spacing exceeds 5 miles

[e] Combined spacing of arterials and continuous collectors would be 0.5 miles

Illustrative suburban street access and spacing guidelines for suburban areas are shown in Table 17.4.

These guidelines would apply where population densities range from 2,500 to 7,500 people per square mile. For higher population densities, closer, continuous street spacing should be provided.

17.2.4 New Roads and Roadway Widening

Adding capacity through road construction and widening has been the traditional way to address traffic congestion problems. New roadways, particularly Interstate freeways, played a major role in expanding capacity to keep pace with population and automobile growth and in decentralizing urbanization in the decade following World War II.

17.2.4.1 Applications

Since the 1970s the pace of new general purpose roadway construction has slowed down. Some freeway reconstruction and expansion are still needed to reduce recurring congestion, to improve safety through the removal of design deficiencies on existing facilities, and to better integrate major roadway with their surrounding neighborhoods.

General-purpose roadway construction and widening is most frequently implemented in areas experiencing rapid population growth. It usually consists of building new freeways and arterials or adding lanes or shoulders to existing facilities. This strategy is often implemented in the fast growing areas in the south and west of the US (e.g., Dallas, Houston, Phoenix, Atlanta) where land for highway construction is more available. However, in the built up metropolitan areas of the northeast (e.g., New York, Boston, Philadelphia) physical and environmental constraints on highway expansion usually limit the opportunities for major new road construction.

There are several notable examples of new road construction that address both congestion and quality of life concerns. These include Boston's Central Artery Tunnel (the "Big Dig") and a new cable stay bridge across the Charles River that have significantly reduced congestion and created a linear park over the tunnel; Seattle is fitting its long standing Alaskan Way Viaduct with a new tunnel that will result in faster traffic movement and improves access between the city and Elliot Bay.

Most freeway construction, however, is now in the form of tolled facilities—both as a financing mechanism and to control demand through pricing. Compared to un-tolled new roads, tolled facilities have the potential of maintaining desirable operating speeds in the long term if tolls can be increased in response to increased demand.

17.2.4.2 Guidelines

New freeway and arterial roadway designs should reflect the design standards and guidelines set forth in the AASHTO "Green Book" [15].

Some specific guidelines from a planning, design, and mobility perspectives include the following:

- Complement rather than compete with rapid transit lines
- For right hand driving, locate all entrances and exits to the right of the main travel lanes
- Limit freeway interchanges to four legs—especially where two freeways intersect
- Avoid short or complex weaving sections
- Maintain "lane balance" to avoid bottlenecks (when 2 lanes merge with 3 lanes, 5 lanes should be provided past the merge point)
- Do not converge freeways in areas of high trip density
- Connect—not split—areas or neighborhoods.

17.2.4.3 Effects

The congestion effects of major new road construction are multifaceted. New freeways and arterial roads open new areas for land development, improve regional accessibility, and relieve traffic loads from local streets. But they also can increase travel miles (VMT) and create new points of congestion around areas where they intersect. New commercial/retail settlements typically locate near freeway interchanges and can create congested conditions especially where access management and interchange designs are inadequate.

A related concern of freeway construction is the transfer of economic activities from built-up center cities to suburban interchanges. Even a simple bypass around a small community can produce a shift in accessibility between the small community and the new area opened by the bypass. For example, this is apparent from Fig. 17.11. As the new bypass reduces congestion in the community center "O" by diverting though traffic away from the center, new areas served by the bypass (A, B, C) will become more attractive to developers as their accessibility increases relative

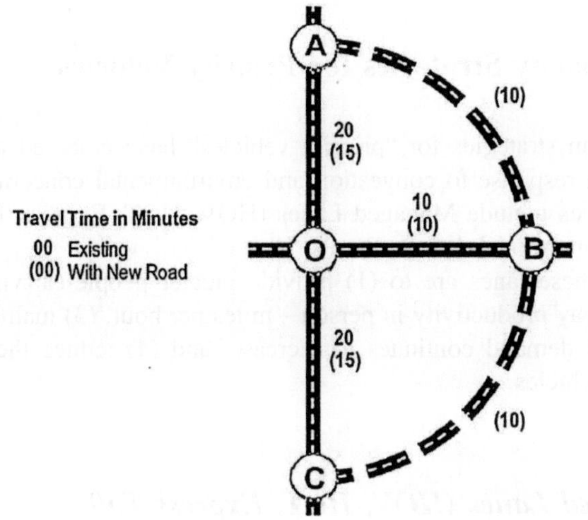

	Travel Times (Min.)	
	Before	After
O-A	20	15
O-B	10	10
O-C	20	15
A-B	30	10
A-C	40	30
B-C	30	10

	100 Divided By Sum of Travel Times From All Points	
	Before	After
A	1.1	1.8
B	1.4	3.3
C	1.1	1.8
O	2.0	2.5

Fig. 17.11 Illustrating the effect of perimeter road on area accessibility. *Source* Reference [16]

to that of the existing community center. While the new perimeter road increases accessibility for all four centers, the largest increase is shown for the new interchanges (A, B, C) and the lowest for the existing center "O" [16].

In larger metropolitan areas, freeway systems account for about 50 % of all vehicle miles of travel—even though they provide only a small fraction of the total lane miles.

One key question regarding the congestion relief benefits provided by added new freeway capacity is the is the issue of "induced demand"—both in the short term, as travelers make longer or more frequent trips; and in the long term, as people and businesses make location decisions that may lead to additional travel. This issue is discussed in Sect. 17.4.

In his 1979 book Urban Transportation System—Politics and Policy Innovation, Alan Altshuler summarized the congestion reduction benefits of road capacity enhancement: "What seem clear is that even over the long run, areas that invest heavily in road capacity seem to maintain higher speeds than areas that do not. Their main arterials may in time become heavily congested during peak hours, but they will normally continue to operate at higher speeds than older, unimproved roads" [17].

17.3 New Capacity Strategies for Priority Vehicles

Capacity expansion strategies for "priority vehicles" have emerged over the last quarter century in response to congestion and environmental concerns. The "priority vehicles" lanes include Managed Lanes (HOV, HOT, Express Toll), Truck-only Lanes, and Intermodal Access Roads.

The goals of these lanes are to (1) provide greater people-carrying capacity, (2) increase roadway productivity in person—miles per hour, (3) maintain corridor mobility as travel demand continues to increase, and (4) reduce the number of single occupant vehicles.

17.3.1 Managed Lanes (HOV, HOT, Express Tolls)

Managed lanes on freeways and arterial roads work best where there is extensive traffic congestion, adequate roadway geometry that can adapt to needed modifications, frequent bus service, and suitable ways to enter and leave the managed lanes. Strong traffic generators such as the city center and other outlying centers along the corridor are essential.

Managed lane treatments vary by type of facility (freeway or arterial streets), substantial volumes of eligible vehicles (buses, car pools, and or trucks), methods of operations (concurrent or contra-flow), hours of operation, and availability of ancillary facilities (park-and-ride lots or garages), toll collection infrastructure, and pricing.

17.3.1.1 High Occupancy Vehicles Lanes (HOV)

HOV lanes reserved for the exclusive use of multi occupancy vehicles, including transit vehicles. The objective of HOV lanes is to reduce single occupant vehicles use. This is achieved where the trip times of HOV users is competitive with the trip time of single occupant vehicles. In fact, improved travel time is often cited as the major reason for using an HOV lane in surveys of HOV lane users [18]. By controlling the minimum vehicle occupancy, HOV demand can be limited to demand-to-capacity (D/C) ratios that can sustain desirable speeds.

HOV lanes are desirable where the general purpose freeway lanes are congested (operating at Service Level "E"); the lanes are long enough to save users at least 10 min, and there is sufficient HOV demand in the corridor. However, they should not be provided by taking lanes from the general purpose lanes in the peak hours of travel.

Urban areas in the US that have HOV systems of at least 50 miles include: Los Angeles, Seattle, Denver, Salt Lake City, and Washington DC

Types of HOV Lanes

As shown in Fig. 17.12, HOV lanes can be created in a number of ways [18]:

- Two-Way Concurrent Flow—This is the most common application. Additional lanes are provided adjacent to the freeway median in each direction of travel. They are normally separated by painted lines from the general purpose lanes and sometime include additional space for vehicles entering and leaving the lane.
- Contra-flow—this typically entails use of the inside lane of a freeway (located in the opposite direction of peak flow) by high occupancy vehicles traveling in the peak direction of flow. The lane is typically separated from the off-peak direction by plastic pylons or movable concrete barriers.
- Contra-flow lanes are used in New York City on some expressways leading to Manhattan from Queens and Brooklyn.
- Physically Segregated Median Lanes—A two-way roadway located within the freeway median physically separated from the general purpose freeway lanes. The rebuilt Katy Freeway in Houston has a segregated median HOV roadway.
- Queue Bypass—HOV lanes provided to bypass recurring congestion points can reduce travel times and increase travel time reliability of high occupancy vehicles.

17.3.1.2 High Occupancy Toll (HOT) Lanes

These lanes are typically available without charge to HOVs, but they charge a toll to other vehicles at a price set to vary with traffic demand to maintain free-flowing traffic.

Where HOV lanes do not generate sufficient demand, while adjacent general use lanes are congested, HOV lanes can be converted to High Occupancy Toll (HOT) Lanes.

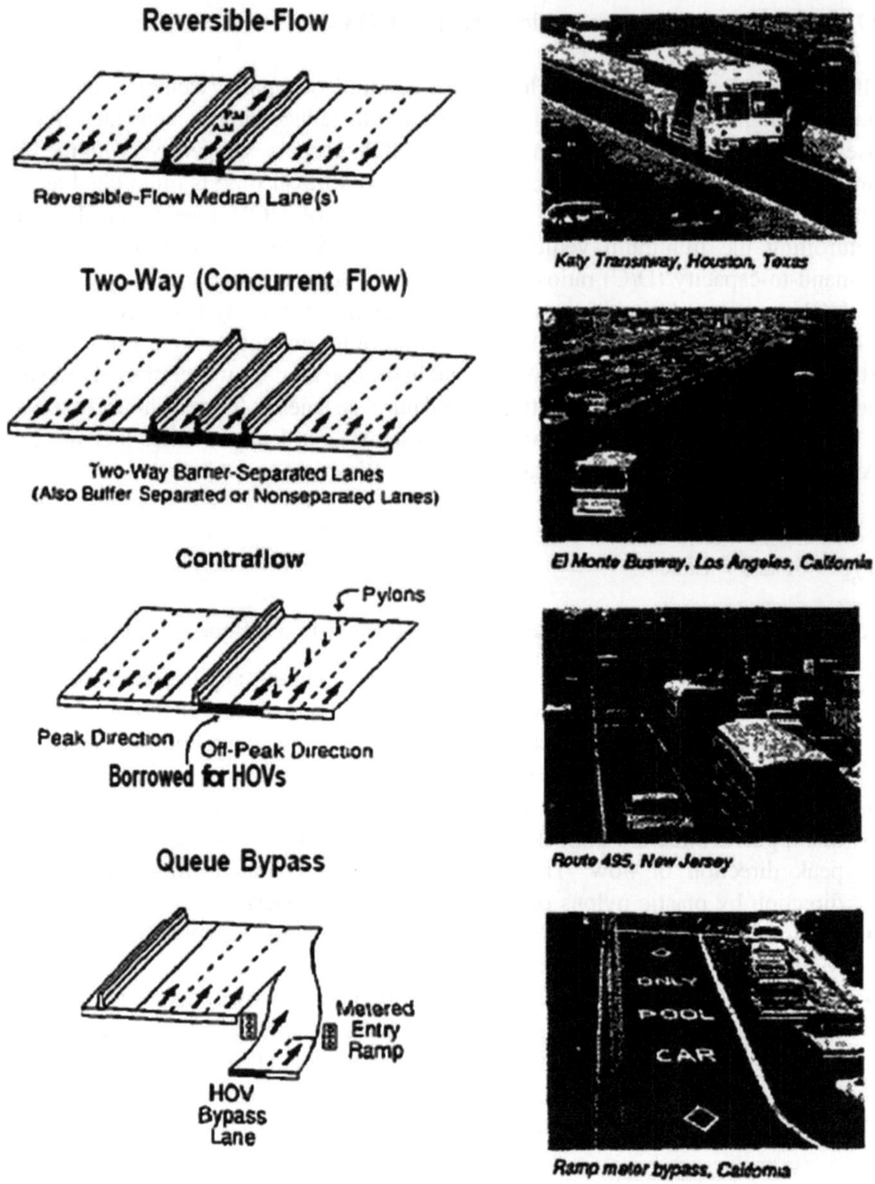

Fig. 17.12 Freeway HOV lane application. *Source* Reference [18], Fig. 2.8

For example, to use the HOV lane on the Katy Freeway in Houston, vehicles with one or two person are tolled while free access is limited to vehicles with three or more persons per vehicle.

17.3.1.3 Value Pricing/Express Toll Lanes

This strategy is typically applied in corridors with high traffic congestion. It consists of creating additional lane (s) capacity only for those vehicles that pay a fee. The fee is set to vary to maintain a desired traffic speed.

17.3.1.4 Effects

HOV, HOT, and Express Toll lanes can be effective strategies in highly congested corridors where other means of congestion relief are not available. While they might slightly reduce congestion on the general use lanes, they are effective in reducing the negative impacts of congestion for time-sensitive trips and for those travelers who value travel time reliability.

The study by the Institute of Transportation Engineers [18] reported travel time savings from HOV applications ranging from less than 5 to 24 min, with savings rates ranging from 0.4 to 3.5 min per mile. As shown in Fig. 17.13, these values vary by location and length of the HOV lane.

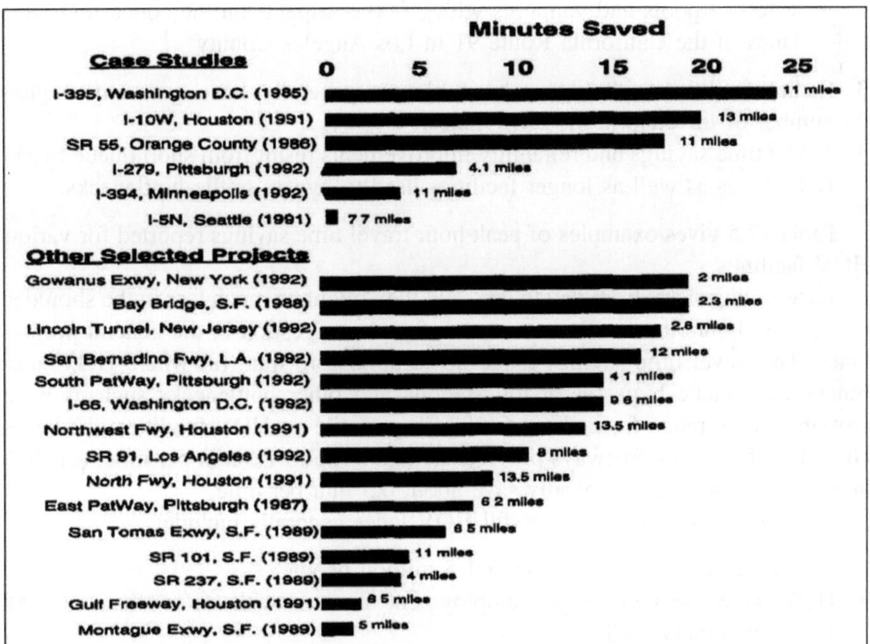

Fig. 17.13 Minutes saved using the HOV lane. *Source* Reference [18], Fig. 2.10

A comprehensive analysis of traveler response to HOV facilities is provided in Chap. 2 of TRCP Report 95 [19]. Some of the salient findings from this chapter are summarized as follows:

1. The attractiveness of HOV facilities depends upon the amount of time that users save, the trip time reliability afforded, the type and frequency of bus service, facility location and orientation, HOV eligibility requirements, years in service, availability of park-and-ride facilities. Corridor congestion levels (freeway lanes and parallel arterials), the size and configuration of urban area population and major attraction centers are critical determinants of use.
2. Most HOV facilities carry more people per lane than do the adjacent general purpose lanes in the peak hour—and sometimes in the entire peak period. Illustrative examples of AM peak-hour vehicle and person volumes on HOV facilities include:

    - About 500–600 buses carrying 23,000 passengers on the NJ Route 495 bus-only contra-flow lane approaching the Lincoln Tunnel to New York City
    - 1,200 vehicles (including 22 buses) carrying 3,600 people (including 1,100) bus passengers) on the exclusive Northwest HOV lane in Houston
    - 1,200 vehicles (including 64 buses) and 5,600 people (including 2,600 bus passengers) on the I-5 North concurrent flow lanes in Seattle, and
    - 1,300 carpools and vanpools with 3,000 occupants on the concurrent HOV lanes of the California Route 91 in Los Angeles County

3. Radial facilities with higher bus volumes generally have serve the highest number of travelers in the HOV lanes.
4. Travel time savings and reliability improvements result from short queue bypass HOV lanes as well as longer facilities used to bypass traffic bottlenecks.

Table 17.5 gives examples of peak hour travel time savings reported for various HOV facilities.

Time savings vary from day to day, and they might be much less in the shoulders of the peak hours than in the time span of peak congestion in the general purpose lanes. The travel time savings range up to almost 40 min: (a) where HOV lanes function as queue bypasses at toll stations and other bottlenecks such as water crossings, they range from about 6–20 min per mile on HOV facilities; (b) longer HOV facilities along freeways provide savings of up to 1.6 min per mile; (c) HOV lanes on arterial streets typically save about 0.5 min per mile.

Contributing factors to successful HOV lanes generally include:

- Metro area population of at least 1.5 million people
- HOV lanes serving major employment centers with more than 100,000 jobs—preferably a CBD
- Geographic barriers that concentrate development and constrict travel
- Potential for at least 25 buses per hour using the NOV facility

Table 17.5 Examples of reported AM peak-hour travel time savings associated with HOV facilities and bus lanes

Facility	Length (miles)	Year[a]	Travel time savings[b]	
			Total (min)	Minutes per mile
Exclusive freeway HOV lanes				
Houston, Texas				
I-45N (North)	13.5	1996	14	1.0
I-45S (Gulf)	12.1	1996	4	0.3
I-10W (Katy)	13	1996	17	1.3
US 290 (Northwest)	13.5	1996	22	1.6
US 59 (Southwest)	12.2	1996	2	0.2
Los Angeles, California				
San Bernardino transit way	12	1992	17	1.4
Minneapolis, Minnesota				
I-394 (exclusive and concurrent flow)	11	1992	5	0.5
Washington, DC				
I-95/I-395 (I-95 and Shirley Hwy)	27	1997	39	1.4
I-66 (exclusive and concurrent flow)	27	1997	28	1.0
Concurrent flow freeway HOV lanes				
California				
SR 55, Orange County	11	1986	18	1.6
SR 91, Los Angeles	8	1992	10	1.2
SR 101, San Francisco Bay Area	11	1989	5	0.5
SR 237, San Francisco Bay Area	4	1989	4	1.0
Bay Bridge, San Francisco Bay Area[c]	2	1998	20	10.0
Massachusetts				
I-93 (N) Boston[d]	2.5	1999	10 (max)	4.0 (max)
Maryland				
I-270	8	1997	5–6 (AM peak)	0.6–0.8
			9–12 (PM peak)	1.1–1.5
Miami–Ft. Lauderdale–Palm Beach				
I-95	45	1998	6 (AM/northbound)	0.1
			7 (PM/northbound)	0.2
			16 (AM/southbound)	0.4

Source Reference [19]

[a] Year travel time savings documented

[b] Comparison of travel time in the HOV lanes over the general-purpose lanes (in known cases, unless otherwise noted) for commuters traveling the full length of HOV facility

[c] Queue bypass on approach to toll plaza

[d] Queue bypass on approach to merge and lane drop

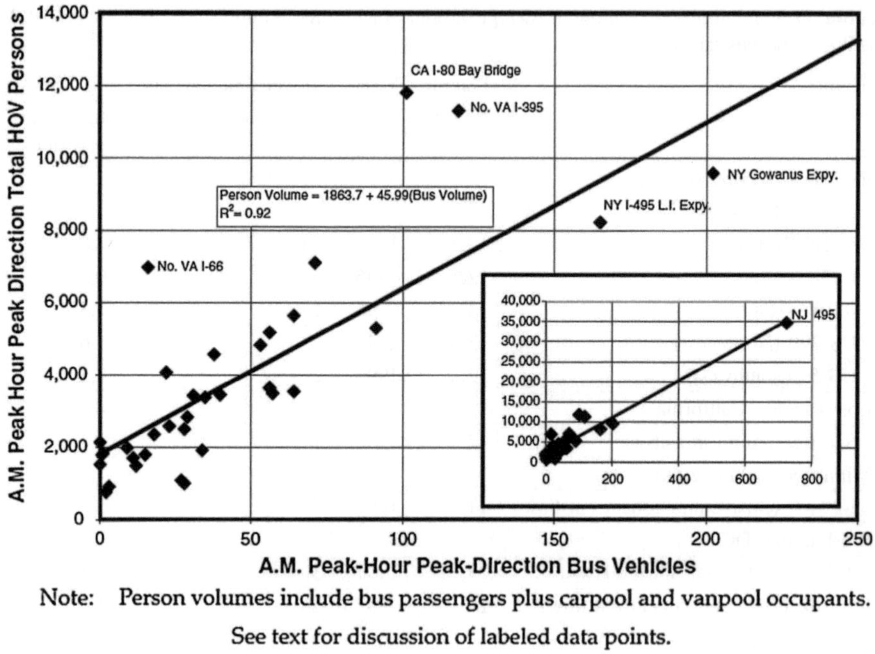

Fig. 17.14 Total peak hour person volumes in the peak direction on 35 HOV facilities related to bus vehicle volumes. *Source* Reference [19]

- Peak-hour freeway congestion in the general purpose lanes
- HOV time savings of at least 1.0 min per mile or 7.5 min per trip
- Availability of park-and-ride facilities and HOV price discounts or free passage on tolled facilities
- Willingness to accept several years of initial operation at marginal utilization, while usage develops, is critical and may be essential for HOV success.

Many HOV facilities—especially those oriented to the downtown business districts—have relatively high bus volumes. This relationship is apparent from the scatter diagram shown in Fig. 17.14.

17.3.2 Truck-Only Lanes

When first designed, many modern highways were expected to carry moderately low ADT with a traffic mix of up to 10 % trucks, for up to 30 years.

"Today, more than 50 years later, the Interstate system's VMT has more than doubled (its original estimates), and trucks weighing more than 80,000 lb account for 30–40 % of the daily VMT [20]."

Adding new mixed use lanes to existing highways to serve additional personal travel demand *and* truck freight traffic, might not solve the twin safety and congestion problems associated with the mixing of truck and car traffic. Crashes involving large trucks affect congestion dramatically, and approximately 12 % of all highway—related fatalities involve large trucks [20].

Truck-only facilities provide safety benefits by separating cars from trucks in heavy truck corridors (e.g., sections of the NJ Turnpike). They also serve to provide the flexibility/redundancy when a crash or other disruption in one roadway requires the rerouting of traffic.

In some situations, segregated truck-only roads might be desirable. Similar to bus and high-occupancy vehicle lanes, truck lanes/roads will need special access. Truck ways reduce congestion on parallel heavily traveled freeways and arterial roads. However, constructing additional truck lane capacity that is not fully used in peak hours, while the adjacent mixed use lanes are congested, could lead to political problems from users of congested lanes.

17.3.3 Freight Intermodal Access Roads

This strategy involves the construction of new roads or improving existing roads to improve travel time and reliability of truck traffic serving major freight/intermodal terminals. Linking intermodal facilities, warehouses, and highway interchanges can provide increased efficiency in goods movement for specific locations in a metropolitan area. In center cities intermodal access improvements are of a multimodal character (consisting of local streets connecting to the intermodal facility and of arterial highways that connect to the freeway network).

17.4 The Issue of Induced Traffic

17.4.1 What Is Induced Traffic?

"Induced traffic" is a widely used term among professionals that describes the observed increase in traffic volume after the capacity of an existing road is increased by operational improvements, or when new capacity is added by roadway widening or the construction of a new road. It is used by transportation planners in sizing new or improved roads, and also by advocacy groups that are usually opposed to roadway expansion.

As stated in Ref. [21], "The term often appears in the popular press, and has been used by some advocacy groups to support their argument that 'we can't build our way out of congestion,' because any increase in highway capacity is quickly filled up with additional traffic. …the term is often is often misused to imply that increases in highway capacity are directly responsible for increases in traffic. In fact, the relationship between increases in highway capacity and traffic is very complex, involving various travel behavior responses, residential and business location decisions, and changes in regional population and economic growth."

Travel demand on new or expanded roadways consists of several major components. These include (1) the traffic diverted from other roadways, other times, or other modes, and (2) the new, or "induced" traffic demand resulting from the improved service level made possible by the added capacity.

The components of induced traffic include longer and/or more frequent trips, as well as the additional traffic generated from new land developments "induced" by the improved access provided by the new or expanded road.

17.4.2 What Is the Source of Induced Traffic?

The conceptual analysis below illustrates the effect of capacity expansion on trip costs and traffic volumes for both existing and future travel demand. The results are shown graphically in Fig. 17.15 through Fig. 17.17.

Two time frames are discussed below to clearly differentiate (1) the proportion of traffic growth in metropolitan areas that is induced by highway capacity expansion and (2) the proportion that is the product of population and employment growth: one for existing demand, the other for future years demand.

17.4.2.1 Existing Demand

Increasing roadway capacity reduces the generalized cost of travel (including both travel time and out-of-pocket costs). Lowering the cost of travel increases the volume of traffic on the roadway (Fig. 17.15)—an outcome consistent with microeconomic theory of consumer demand.

Travelers who are attracted to the improved route include those who, to avoid the congested route, had (1) diverted to other less congested routes; (2) switched to different modes or times; (3) traveled to other destinations; or (4) decided not to take a particular trip. In addition, because of higher speed, trip distance tends to be longer on the improved facility.

The volume of traffic diverted and induced by the improved route [($V_{1,0}$)–($V_{0,0}$)] is shown in (Fig. 17.15). This additional traffic experiences a lower congestion on the improved route ($P_{1,0}$) and it creates lower congestion on the routes which lost trips diverted to the improved route.

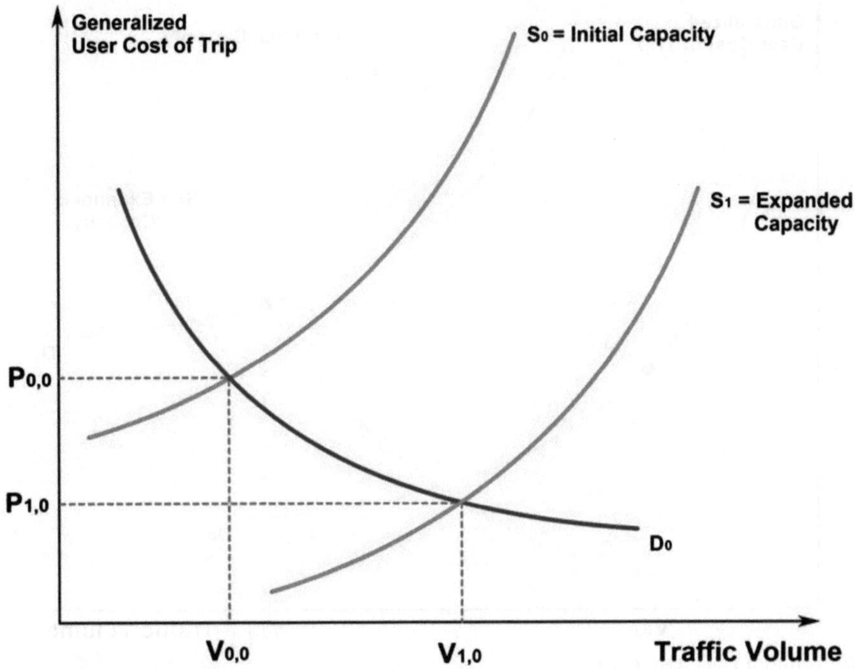

D_0 = Existing Demand
S_0 = Initial Capacity
D_1 = Expanded Capacity
$P_{0,0}$ = Initial User Cost; $V_{0,0}$ = Traffic Volume for Initial Capacity and Demand
$P_{1,0}$ = User Cost with Expanded Capacity; $V_{1,0}$ = Traffic Volume with Expanded Capacity

Fig. 17.15 Effect of capacity expansion on generalized trip cost and traffic volume

17.4.2.2 Future Years Demand

Diverted and induced travel is also a factor in assessing the congestion impacts for future travel years. Some observations are given for two cases (a) if road capacity is increased, and (b) if roadway capacity is not increased.

If the Road's Capacity Increases

Depending on the amount of added capacity, the route will make the area more attractive to land development[1] that, in time, will increase the travel demand of the

[1] It should be noted, however, that while improving transportation accessibility in a particular area may make land more attractive for development, other factors play an important role. These include: land acquisition and development costs; availability of water, sewer, and electric power;

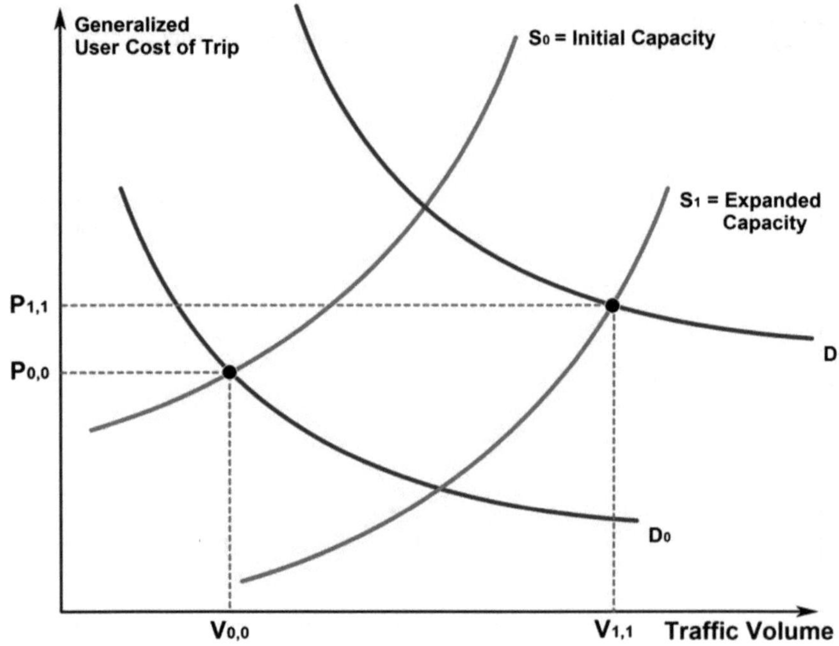

$P_{0,0}$ = Initial User Trip Cost

$P_{1,1}$ = Future Year User Cost with Expanded Capacity (S_1) and Future Year Demand (D_1)

Fig. 17.16 Effect of capacity expansion on land development growth and related new trips

improved route to (D1) and will add new trips [($V_{1,1}$)–($V_{1,0}$)] to the improved route (Fig. 17.16).

Because the growth in land use activities attributable to the improved route represents a share of future development growth (that probably would have located in other parts of the metro area if the route's capacity remained unchanged), the new trips induced by the improved route might not represent a net increase in trips for the metro area, unless it attracts development that would have otherwise located in areas accessible by transit.

As shown in Fig. 17.16, the demand function (D1) generated by the growth of land development induced by the improved roadway, will add traffic volume to the improved road ($V_{1,1}$) which, in time, will reach the same congestion level as it existed when the road was at its original capacity ($P_{1,1}$) = ($P_{0,0}$). This induced traffic volume can have both positive and negative consequences:

(Footnote 1 continued)
quality of schools and other public services; and zoning ordinances that determine the type, density, and location of development.

Positive Consequences: The capacity increase will increases personal mobility and accessibility as more travelers are able to travel to destinations which best meet their needs to maximize their wellbeing. The added capacity furthers the growth in economic development and land values in the corridor—bringing significant social and economic benefits by accommodating more activity. However, the rate of congestion growth on other routes is likely to decrease because they will attract less development growth and generated fewer trips.

Negative Consequences: The increase in traffic volume could increase air and noise pollution along the improved roadway. Other areas in the region become less competitive in attracting growth—as their accessibility relative to that provided by the improved corridor declines.

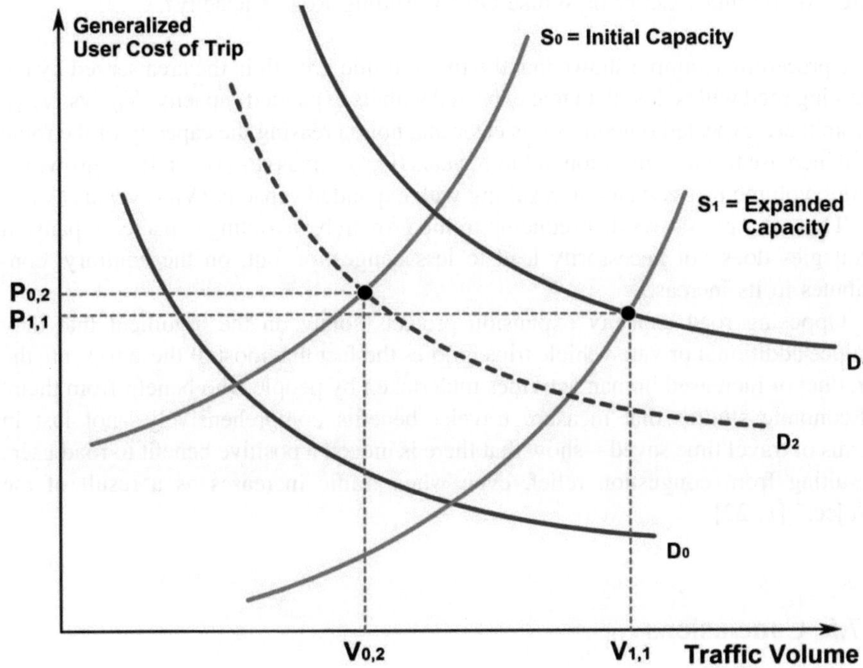

$P_{1,1}$ = Future Year User Cost with Expanded Capacity (S_1) and Future Year Demand (D_1)
$P_{0,2}$ = Future Year User Cost with Existing Capacity (S_0) and Future Year Traffic Demand (D_2)
$V_{0,2}$ = Future Year Traffic Volume for Existing Capacity (S_0) and Future Year Traffic Demand (D_2)
$V_{1,1}$ = Future Year Traffic Volume with Expanded Capacity (S_1) and Future Year Traffic Demand (D_1)

Fig. 17.17 Comparison of future year generalized travel cost and traffic volume with and without expanded road capacity

If Capacity Is Not Increased

In a growing region with increasing traffic demand, nearly every road will experience traffic growth (from development activities with direct access to the road, or from traffic passing through) in the area.

In the example cited, if the congested road in question does not increase its capacity (S_0), the corridor will receive less future years development with a smaller traffic demand limited by natural growth of the metro area (D_3) than it would experience (D_1) with added capacity This condition is shown in Fig. 17.17 where the added volume from "natural" growth [($V_{0,3}$)–($V_{0,0}$)] increases the generalized cost of travel from $P_{1,1}$ to $P_{0,3}$.

Are Communities Better or Worse Off by Adding Road Capacity?

The preceding example shows that the future traffic growth in the area served by the existing road will be less than that expected with its expanded capacity ($V_{0,2}$ vs. $V_{1,1}$). From these expected outcomes it is clear that not increasing the capacity of the route will increase traffic congestion in future years ($P_{0,2}$) versus ($P_{1,1}$) even if the growth in traffic volume is less than that resulting with expanded capacity ($V_{0,2}$) versus ($V_{1,1}$).

This example shows that reducing traffic growth by avoiding capacity expansion strategies does not necessarily lead to less congestion but, on the contrary, contributes to its increase.

Opposing road capacity expansion projects solely on the argument that they induce additional private vehicle trips ignores the fact that most of these trips are the product of increased human activities undertaken by people who benefit from them. "Economic studies that measure traveler benefits comprehensively—not just in terms of travel time saved—show that there is indeed a positive benefit to road users resulting from congestion relief, even when traffic increases as a result of the project" [1, 22].

17.5 Conclusions

This chapter has shown various ways to expand the capacity of the existing roadway systems to provide congestion relief. These include the removal of bottlenecks caused by lane imbalance, reducing conflicting flows at intersections and interchanges as well as adding new capacity to meet increasing travel demands.

However, adding new capacity is a two edged strategy: it reduces congestion in the short term, but it also generates additional vehicle traffic in the long term. This effect promotes endless public debate about the merit of this strategy—especially in built up areas where it creates major disruptions to communities and neighborhoods.

In his 1979 book on The Urban Transportation System—Policies and Policy Innovation [17], Alan Altshuler indicates that "Congestion tends to be greatest where development is most intensive; and it is precisely in such areas that highway construction entails the most severe community disruption, the most intensive public controversy, and the highest dollar cost. Thus the highways with the greatest congestion relief potential are also the least feasible to construct."

However, where added capacity is provided, its lasting effect on congestion relief (especially in metropolitan areas exceeding 2 million people) can only be realized by combining it with strategies that reduce the need to travel by car—while maintaining acceptable levels of mobility and accessibility. These strategies will be discussed in the next six chapters.

References

1. Cambridge Systematics, Inc. (2002) The positive impacts of transportation investment, NCHRP Project 8–36, Task 22, Final Report, Working paper 4—compilation of working papers. Transportation Research Board, Washington, DC
2. Prepared by: Cambridge Systematics, Inc., Resource Systems Group, Inc. (2008) NCHRP 20–24A Task 63: Effective management practices for congestion management: Final Report. Requested by the American Association of State Highway and Transportation Officials (AASHTO)
3. Schrank D, Lomax T, Eisele B (2011) TTI's 2011 urban mobility report. Powered by INRIX traffic data. s.l. Texas Transportation Institute, The Texas A&M University System, College Station
4. American Highway Users Alliance and Cambridge Systematics, Inc. (2004) Unclogging America's arteries—effective relief for highway bottlenecks, 1999−2004
5. Nevers B, Steyn H, Mereszczak Y, Clark Z, Rouphail N, Hummer J, Schroeder B, Bugg Z, Bonneson J, Rhodes D (2011) NCHRP report 707: guidelines on the use of auxiliary through lanes at signalized intersections. Transportation Research Board, Washington, DC
6. Brewer MA (2012) NCHRP synthesis 432: recent roadway geometric design research for improved safety and operations—a synthesis of highway practice. Transportation Research Board, Washington, DC
7. Hughes W, Jagannathan R, Sengupta D, Hummer J (2010) Alternative Intersections/Interchanges: Informational Report (AIR), Report No. FHWA-HRT-09-060, FHWA, Washington, DC
8. Dorothy PW, Maleck TL, Nolf SE (1997) Operational aspects of Michigan design for divided highways. Transportation research Record 1579−1997 geometric design and its effects on operations, pp 18–26
9. Koepke FJ, Levinson HS (1993) Case studies in access management. Transportation Research Board, Washington, DC
10. American Association of State Highway and Transportation Officials (2001) A policy of geometric design for highways and streets. s.n. Transportation Research Board, Washington, DC
11. Rodegerdts L, Bansen J, Tiesler C, Knudsen J, Myers E, Johnson M, Moule M, Persaud B, Lyon C, Hallmark S, Isebrands H, Crown RB, Guichet B, O'Brien A (2010) NCHRP report 672—roundabouts: an informational guide, 2nd edn. Transportation Research Board, Washington, DC
12. Bared JG, Edara P, Jagannathan R (2005) Transportation research record 1912: design and operation performance of the double crossing intersection and diverging diamond interchange. Transportation Research Board, Washington, DC
13. Outwater M et al (2014) Effect of smart growth policies on travel demand. Strategic highway research program (SHRP2). Transportation Research Board, Washington, DC

14. Levinson H (2000) Street spacing and scale. In Transportation Research Circular E-C019: Urban Street Symposium: Conference Proceedings, pp.B-7/1–B-7/16. Dec. 2000
15. American Association of State Highway and Transportation Officials (2011) A policy of geometric design for highways and streets. s.n. Transportation Research Board, Washington, DC
16. Levinson HS, Falcocchio JC (2011) Urban development and traffic congestion. s.l. American Society of Civil Engineers, Reston, pp 948–956
17. Altshuler A, Pucher JR, Womuck JP (1981) The urban transportation system—politics and policy innovation. MIT Press, Cambridge
18. Meyer Michael D (1997) A toolbox for alleviating traffic congestion and enhancing mobility. Institute of Transportation Engineers, Washingto, DC
19. Turnbull KF, Levinson HS, Pratt RH (2006) TCRP report 95, traveler response to transportation system changes, Chapter 2, HOV Facilities. Transportation Research Board, Washington, DC
20. Bucklew KJ (2012) Dedicated truck lanes—an innovative way forward. TR News 279: history in the making— 91st annual meeting highlights, March–April 2012
21. Induced Travel: Frequently Asked Questions (2012) Federal highway administration. http://www.fhwa.dot.gov/planning/itfaq.cfm. Accessed 12/03/2012
22. DeCorla-Souza P, Cohen H (1997) Accounting for induced travel in evaluation of metropolitan highway expansion, presented at the 77th Annual Meeting of the Transportation Research Board, January 1998, Paper No. 980132

Chapter 18
Overview of Mitigation Strategies that Reduce Traffic Demand

18.1 Introduction

Managing travel demand is increasingly recognized as a means of addressing urban traffic congestion—especially in large metropolitan areas. Commonly called "transportation demand management" (TDM), the strategy focuses on reducing the demand for single occupant vehicles. Emphasis is typically placed on reducing vehicle-miles of travel (VMT).

The travel time and travel time variability benefits of new capacity (or of existing capacity restored) cannot be *sustained* without mechanisms that preserve these gains in future years. To sustain the life of these benefits traffic demand reduction strategies are needed because when there is pent-up demand the capacity added is soon fully utilized.

Therefore reducing automobile travel demand becomes a necessary strategy to keep congestion at manageable levels and to maintain mobility in future years.

The major benefit from demand reduction strategies on roadways congested for several hours during each peak period, is from reducing the duration rather than the intensity of congestion.

Traffic demand reduction strategies can be aimed at specific routes, areas or zones. They can have region-wide applications, and they can be applied during specific time periods. They are intended to modify person travel and goods movement behavior by encouraging a mode shift away from private vehicles or a time shift in trip making; by diverting trips from congested locations, and/or by reducing the need to travel.

The synergistic effects of combining strategies can further help to relief congestion. Examples include (1) coordinating transit investments with land use planning, (2) coupling bottleneck reductions with congestion pricing, and (3) coordinating traffic operations improvements with pricing policies.

J.C. Falcocchio and H.S. Levinson, *Road Traffic Congestion: A Concise Guide*,
Springer Tracts on Transportation and Traffic 7,
DOI 10.1007/978-3-319-15165-6_18

Traffic demand reduction strategies can be categorized into two groups: those that *directly* aim at changing travel behavior (e.g., congestion pricing), and those that are intended to change behavior *indirectly* (e.g., through transit service expansion). Illustrative strategies in each group along with their expected effects and implication challenges are shown in Table 18.1.

A brief description of these direct demand reduction and indirect demand reduction strategies follows.

18.2 *Direct* Demand Strategies

Direct demand strategies focus on changing traveler behavior through policies that rely on various pricing or regulatory mandates.

These include:

- Pricing strategies for roadways (Chap. 19)
- Regulatory Restrictions on Car Use (Chap. 20)
- Freight Demand Management (Chap. 20).

18.3 *Indirect* Demand Strategies

Indirect demand strategies include actions that encourage a reduction in private vehicle use.

These strategies focus on reducing private vehicle use through land use planning and design, the enhancing of alternative modes of travel, and reducing the need to travel (e.g., telecommuting).

They include:

- Employer and Institutional Participation in the Work Commute (Chap. 21)
- Reducing the Need to Travel (Chap. 21)
- Parking Supply and Pricing (Chap. 22)
- Land Use Changes (Chap. 23)
- Transit and Pedestrian/Bicycle Improvements (Chap. 23).

18.4 Implications

Strategies that can effectively reduce the demand for roads (and parking spaces) will require urban areas to adopt a common vision of how they should develop. Achieving this vision at the regional level is a challenging task as it requires coordinating land use decisions at the local level with transportation decisions at the regional level.

Table 18.1 Congestion management strategies that reduce the demand for motor vehicle travel

			Effectiveness		Extent of application		Implementation issues		
	Strategy	Substrategy	Local	Area wide	Current	Potential future	Cost	Non cost barriers	Time frame
1	Land use		Low	Medium	Limited	Moderate (all new/redevelop areas)	Low or cost savings	High	Long-term
2	Road pricing		High	High	None	Extensive	High/revenue generating	High	Mid-term
3	Freight demand management		Varies	?	Limited (two major ports)	Limited (ports, CBDs?)	Low	Medium to high	Short-term
4	Nonmotorized improvements		Low	Low	Moderate	Moderate	Low to medium	Low to medium	Long-term
5	Parking policy		High	Low	Limited	Moderate	Low to medium	Low to medium	Short-term
6	Park-and-ride		High	Medium					Mid-term
7	Regulatory restrictions		High	Medium	Limited	Limited	Low to medium	High	Mid-term
8a	Transit enhancements	Bus service and operations improvements	Low	Low to medium	Limited	Extensive (bus systems)	Low to high	Low to medium	Mid-term
8b		Rail transit construction	Low	Low to medium	Moderate (new rail in ~30 metro areas)	Moderate (larger metro areas)	High	Medium to high	Long-term
9a	Commuter choice/workplace TDM	Transit and ridesharing programs/incentives	Low to medium	Low	Limited to moderate	Moderate (larger metro areas/activity centers)	Low to medium	Medium	Short-term
9b		Telecommuting and alternative work schedules	Low to medium	Low	Moderate	Moderate (30–40 % workforce)	Low	Medium	Short-term

Source Adapted from reference [1]

Reference

1. NCHRP 20-24A Task 63: Effective Management Practices for Congestion Management: Final Report. Requested by: American Association of State Highway and Transportation Officials (AASHTO). Prepared by: Cambridge Systematics and Resource Systems Group. November 2008

Chapter 19
Direct Demand Strategies—Pricing

19.1 Introduction

Pricing strategies have emerged since the 1970s as a means of better allocating road space to reduce congestion and to more equitably cover the costs that road users contribute to the congestion. Congestion (or value) pricing is supported by many economists, planners, and public officials. Within the United States, it has the support of the US Federal Highway Administration.

The sections that follow describe the evolution of road pricing in the US, define the concept, give examples of application world-wide, and suggest possible future directions of road pricing in reducing congestion.

19.2 Evolution of Road Pricing

Road pricing has a long history in the United States. Tolled bridges, tunnels, and turnpikes predate the automobile. They were designed to generate revenues to help pay for the construction, operation, and maintenance of these facilities. However, the main methods of road finance since the early 20th century were the state and federal motor vehicle fuel taxes.

By about 1930, tolls were collected on parkways in Westchester and Nassau Counties in New York. The Merritt Parkway between New York City and New Haven that opened about 1938 had two toll stations.

The Pennsylvania Turnpike—the first major interstate toll road—opened just before World War II (WWII). Tolls were also established on major water crossings such as the Holland and Lincoln Tunnels, NY, and the San Francisco—Oakland and Golden Gate bridges in California.

After WWII, major toll roads were built throughout the United States—starting with the New Jersey Turnpike. Many of these facilities placed tolls on entering and

J.C. Falcocchio and H.S. Levinson, *Road Traffic Congestion: A Concise Guide*,
Springer Tracts on Transportation and Traffic 7,
DOI 10.1007/978-3-319-15165-6_19

exiting ramps to avoid delays to main line traffic. However, most capacity additions to urban systems resulted from building the untolled Interstate Highway System that dramatically reduced urban traffic congestion and improved mobility. Increasing suburbanization of the population and employment in the ensuing decades soon produced traffic demands that approached or exceeded the capacity of many urban freeways. With freeway congestion increasing and as environmental and social concerns about adding new capacity emerged, many cities looked for other ways of improving mobility and reducing congestion—albeit for selected users.

Since the 1970s US transportation agencies have made considerable efforts in managing freeway congestion. Bus lanes and high-occupancy lanes were added on many freeways. The Federal Highway Administration permitted tolls to be charged on managed High Occupancy Toll Lanes (HOT Lanes) that were added to the Interstate system and other freeways. The growing congestion on many urban freeways has increased the need for reducing traffic demand. With the emergent advanced technologies (sensors, computing, communication), pricing strategies in the form of "congestion pricing" or "value pricing" became very attractive strategies to reduce congestion.

However, broad application of congestion pricing in the United States has been limited because of continued political, institutional, and public resistance to the concept. Therefore, most congestion pricing and other travel demand reduction strategies are primarily found in Europe, South America, and Singapore.

19.3 Congestion Pricing

19.3.1 Definition and Concerns

Congestion pricing is a way to reduce traffic demand during busy traffic periods by charging a fee (or toll) to road users. In economic terms it is the charging of higher prices to reduce the consumption of roadway capacity—especially when and where congestion occurs. Congestion pricing is a new "usage" charge or tax that is applied where demand exceeds capacity [1].

19.3.1.1 Basic Concept

To avoid adding new capacity that is only used part of the time, and to reduce operating costs, variable charges have been used successfully to manage peak demand in many sectors of the economy. It has been widely used by the telephone and utility industries, airlines, railways, hotels, and even restaurants. The Washington DC metro system charges higher fares for peak hour trips and lower fares during off peak periods.

Road user congestion pricing has been proposed by economists for more than four decades. Many economists view congestion pricing as the best way of relating user charges to the total cost of using, building, operating and maintaining the facility. They see congestion pricing as the most viable approach to reducing congestion.

Early studies by Vickery and Walters- among others- provided the conceptual and theoretical framework. These studies attempted to quantify the levels of congestion charges through econometric analysis [2–4].

The economic solution to congestion, according to Lyle Fitch and Associates, is "to impose charges for driving in congested hours high enough to keep out temporarily those whose driving is least important as measured by their willingness to pay for it. Prices should vary according to the level of demand. Thus they should be higher for peak than off-peak hours. If revenues exceed costs, the excess should be used to provide additional road space up to the point where supply meets demand" [3].

19.3.1.2 Purpose and Objectives

The two main reasons for congestion pricing are (1) reducing congestion and (2) generating revenues to finance transportation investments. Figure 19.1 shows how these relate to specific highway and transit treatments and their effect on enhancing regional competitiveness, economic opportunity, quality of life and sustainability [5].

More specifically, the purpose of congestion pricing is to reduce road traffic demand in congested networks of fixed capacity. Typical objectives of congestion pricing programs include:

- Achieving economic efficiency by balancing user costs with external costs that users impose on each other
- Maximizing the person or vehicle throughput of roadways and networks
- Providing reliable travel times during peak travel hours
- Encouraging shorter trip lengths and increasing transit use
- Increasing the use of underutilized HOT lanes.

19.3.1.3 Types of Applications

The terms "congestion" pricing and "value" pricing are used interchangeably in the literature. However, because drivers face different conditions involving different choices (e.g., choosing to pay or not the charge on congested networks vs. choosing to use a priced facility offering improved service) a more descriptive definition for each term is suggested below:

Congestion pricing is the price charged drivers for using congested roads. This condition requires all drivers to pay the charge for using the roads during peak periods. Those who do not pay the charge are forced to use alternative times, modes, or destinations for their trip.

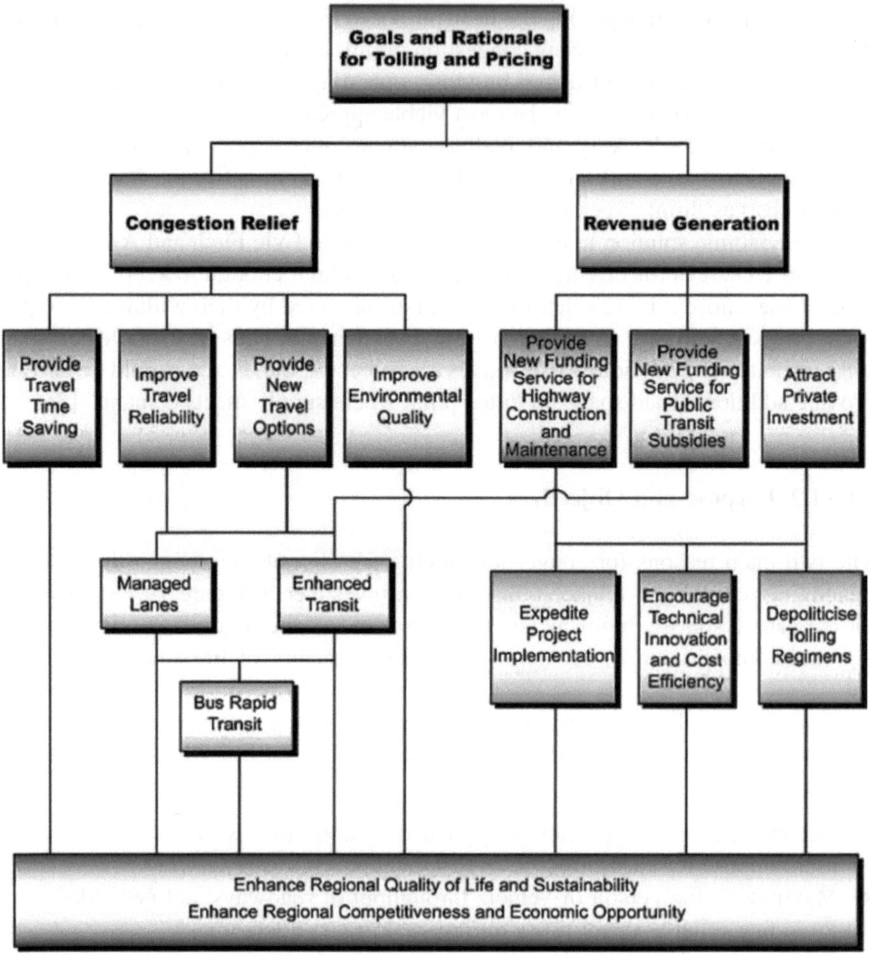

Fig. 19.1 Goals and objectives of tolling and pricing. *Source* Reference [5], p 9. Figure 1.2

Value pricing is the price charged drivers for using special lanes that guarantee higher travel speeds. This charge applies only to drivers who choose to travel at a higher speed during peak hours because they value the travel time savings benefits provided by the priced lane. Those travelers not using special lanes can continue using the general purpose lanes.

Congestion pricing (or value pricing) can be provided in several ways. It can be implemented at a single facility such as a tunnel or bridge, along a single roadway by converting high occupancy lanes to a single or multiple high occupancy toll lanes, at a cordon around a central business district, or along expressways and major roads throughout the region.

Table 19.1 Classification of pricing options

Differentiation eligibility/exemption/discounts	
By time-of—day/congestion level	Flat/fixed
	Variable/preset by TOD and direction
	Variable/real-time dynamic
By vehicle characteristics/type	Auto
	Low emission auto
	Motorcycle
	Single-unit truck combination truck transit bus
By vehicle occupancy	SOV
	HOV-2
	Registered HOV-2
	HOV-3
	HOV-4+
By place of residence	Resident of a certain area
	Visitor
By method of payment	Cash
	Transponder/ETC
By day of week	Weekday
	Weekend
By season	Summer/spring
	Winter/fall

Source Reference [5], p 13. Table 2.1

Facility pricing is common in the United States and Canada. However, the concept of facility use pricing based on fixed fees is being reexamined to favor methods of user charges based on variable usage—mileage based charges.

Area-wide (cordon) pricing is found in several European cities and in Singapore. The various methods of pricing include: time of day, type of vehicles, vehicle occupancy, place of residence, day of week, season of year, and type of user. These are shown in Table 19.1.

19.3.1.4 Setting Prices

Economists perceive traffic congestion as a pricing problem—without some form of marginal cost pricing, motorists have no opportunity or incentive to save money by avoiding traveling in heavy traffic. Fitch and Associates [3] have reported that the cost of peak hour auto travel is about 2.5–3.0 times the average cost—on a per mile basis.

Congestion prices for a roadway can be set in one of two ways: (1) they can be keyed to the marginal cost of providing the additional capacity needed to accommodate peak-period travel, or (2) they can be keyed to the marginal social cost imposed by the peak period traveler on existing users—the marginal cost the extra driver imposes on all drivers in the highway. Using only the travel time cost component of a trip, the relationship of average user cost and marginal social cost (aggregated for all users) is illustrated in Fig. 19.2.

The marginal social cost (MSC) can be calculated as:

$$\mathbf{MSC}\,@V_{n+1} = \mathbf{AC}@V_n + \mathbf{V_n}[(\mathbf{AC}@V_{n+1}) - (\mathbf{AC}@V_n)] \tag{19.1}$$

where:

GTC Generalized trip cost = [(time cost) + (vehicle operating cost) + (road maintenance cost)]
AC Average driver trip cost
MSC Marginal social cost = (the trip cost to the average driver) + (the cost imposed on all drivers by an additional driver to the traffic stream)
$\mathbf{V_n}$ Vehicles per hour per lane
$\mathbf{V_{n+1}}$ V_n + one additional vehicle
$\mathbf{V_{max}}$ The maximum number of vehicles per hour per lane

Illustrative Example

The congestion price may be defined as an incremental charge (congestion charge) paid by a motorist using a congested road to ensure that she or he pays a

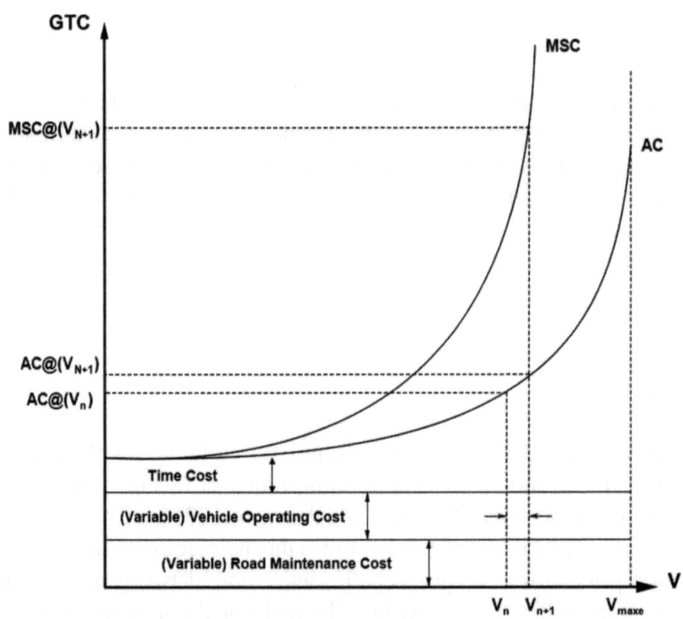

Fig. 19.2 Average cost and marginal (social) cost curves

price that matches the MSC imposed by the driver on the rest of the drivers. This definition is illustrated Fig. 19.3, using only travel time as the generalized trip cost.

In this example a congestion charge (CC) of $0.80/mile is calculated as shown below.

Assumptions

$$\text{ATTR (demand)} = \text{average travel time rate} = (12 - V/250)\ \text{min/mile} \quad (19.2)$$

$$\text{AART (supply)} = \text{average travel time rate} = \left[1 + (V/C)^6\right]\ \text{min/mile} \quad (19.3)$$

$$\text{MTTR} = \text{marginal travel time rate} = \left[1 + 7(V/C)^6\right]\ \text{min/mile} \quad (19.4)$$

$$\text{Value of travel time} = \$15.00/\text{h} \quad (19.5)$$

Findings

Setting Eq. 19.2 equal to Eq. 19.3, finds that at a volume (V) of 2,500 vehicle/lane/h, the TTR = 3.00 min/mile; and setting Eq. 19.2 equal to Eq. 19.4, finds that at a volume (V) of 1,800 veh/lane/h, the TTR = 4.71 min/mile. But when the TTR = 4.71, the equilibrium volume (V) corresponds to 1,800 veh/h/lane. And when (V) = 1,800 veh/h/lane it is possible to travel at a rate of 1.53 min/mile.

Therefore the optimum congestion toll (b–c) is achieved at the price equivalent of $0.80/mile or [(4.71–1.53) min/mile] * [($15.00/h)*(1 h/60 min)].

In this example the congestion charge of $0.80/mile will reduce the traffic volume from 2,250 to 1,800 veh/h/lane, and the TTR of an average driver from 3.0 min/mile to 1.53 min/mile.

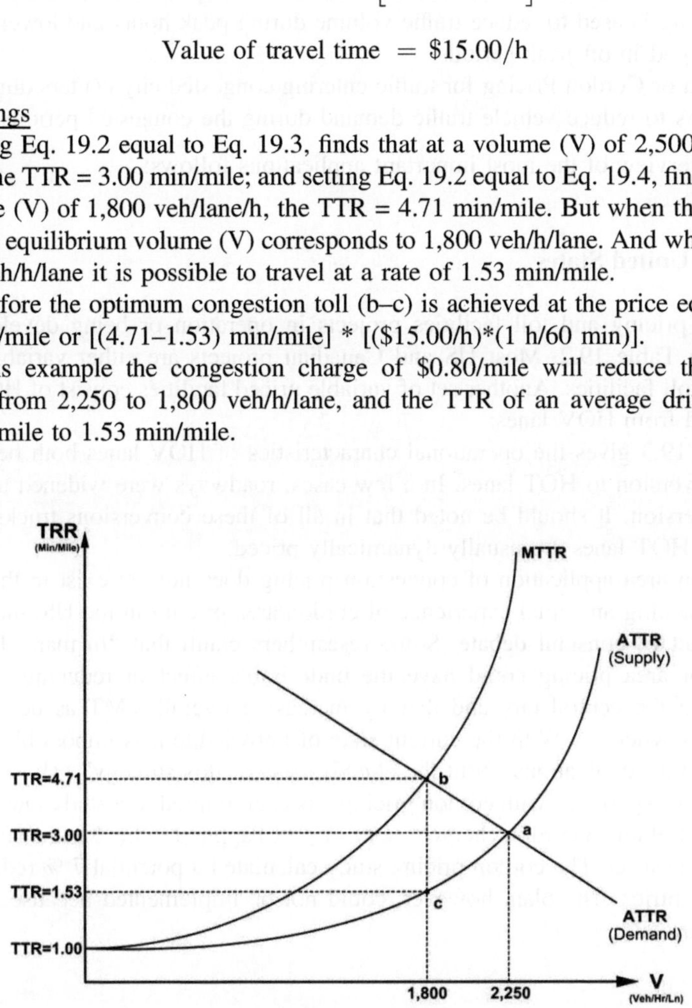

Fig. 19.3 Determining the congestion price for a facility

19.3.2 Overview of Applications

Congestion pricing applications are found in large urban areas—both within and outside the United States. They generally consist of three types of application:

(1) Variably priced Managed Lanes—these projects consist of cases where High Occupancy Vehicle (HOV) lanes are converted into High Occupancy Vehicle (HOT) lanes that allow low- occupancy passenger cars by charging a variable cost that guarantees uncongested flow, while permitting HOVs to use the lane free of charge; or the construction of new HOT lanes.
(2) Variably priced toll facilities (Tunnels, bridges, single roadways)—this pricing creates conditions that maintain desirable traffic speeds. In these cases higher pricing is used to reduce traffic volume during peak hours and lower tolls are charged in off peak periods.
(3) Area or Cordon Pricing for traffic entering congested city centers during peak hours to reduce vehicle traffic demand during the congested periods.

An overview of the most important applications follows:

19.3.2.1 United States

Variable pricing and toll facilities projects in operation or being developed are shown in Table 19.2. Most US and Canadian projects are either variable priced lanes or toll facilities. Another set of variable priced facilities consist of HOT lanes converted from HOV lanes.

Table 19.3 gives the operational characteristics of HOV lanes both before and after conversion to HOT lanes. In a few cases, roadways were widened as part of the conversion. It should be noted that in all of these conversions trucks are not allowed. HOT lanes are usually dynamically priced.

Cordon area application of congestion pricing does not yet exist in the United States. Lacking an actual experience of cordon/area pricing in the US, the issue is the subject of constant debate. Some researchers claim that "In many US cities cordon or area pricing could have the undesirable effect of reducing economic activity in the central city and thereby increasing overall VMT as development moves elsewhere.... With the current state of knowledge it is impossible to make any broad generalizations about the effectiveness of this strategy" [7].

The US experience with cordon pricing has been limited to a study that modeled the impact of an $8 cordon charge (7 a.m.–7 p.m.) applied to the Manhattan Central Business District. The cordon pricing study calculated a potential 7 % reduction in vehicles entries. The plan, however, could not be implemented because it lacked State approval.

Table 19.2 Variable priced-lanes projects (value pricing) and toll facilities projects in the united states—in operation and in the pipeline

Operating	Pipeline	
Variably priced managed lanes	Variably priced managed lanes	
Alameda County, CA I-680	Austin loop 1	Los Angeles I-10
Denver I-25	Baltimore I-95	Los Angeles I-110
Houston I-10 katy freeway	Bay area, CA I-580	Orange County, CA I-405
Houston northwest freeway	Bay area, CA I-80	Provo I-15
Miami I-95	Bay area, CA US 101	San Antonio loop 1,604
Minneapolis I-394	Charlotte I-77	San Bemardino/riverside counties, CA I-10
Minneapolis I-35 W	Dallas DFW connector	San Bemardino/riverside counties, CA M5
Orange County, CA SR 91	Dallas I-30 Tom Landry	San Bemardino/riverside counties, CA SR-91
San Diego I-15	Dallas I-35 Thornton	San Diego I-15
Seattle SR 167	Dallas I-35E stemmons	San Diego I-5
Salt lake city I-15	Dallas I-635/LBJ	San Diego I-805
	Dallas NTE (I-820/SH 121)	San Diego SR 52
	Denver US 36	San Jose SR 237/I-880
	Fort Lauderdale I-595	San Jose SR 85
	Georgia GA 400	San Jose US 101
	Georgia I-75/I-575	Seattle I-405
	Georgia I-85	St. Paul I-35E
	Houston area reversible lanes except I-10 Katy	Virginia I-395/I-95
	Las Vegas I-15	Virginia I-495 capital beltway
Toll facilities with variable pricing	Toll facilities with variable pricing	
Lee County, Florida bridges	Maryland Intercounty connector	
New Jersey Turnpike	Seattle Alaskan way	
Orange County, California San Joaquin Hills (73) and Foothill/Eastern (241, 261,133) toll roads delaware route 1	Seattle SR-520	
San Francisco-Oakland Bay Bridge		
The Port Authority of New York and New Jersey Bridges and Tunnels		
Virginia Dulles Greenway		
Cordon and area pricing	Cordon and area pricing	
None	San Francisco	

Source Reference [6], p 16. Table 2.2

Table 19.3 Characteristics of HOV operations before and after HOT conversion

Project	HOV operations before conversion	HOV operations after conversion
I-25 Express lanes Denver, CO	• Two-lane reversible facility • 2+ HOVs and registered hybrids allowed access • Motorcycles allowed access • Under 10 % violation rate • 6-min bus headways from park-and-ride lots	• No capacity added—conversion required operational changes only • No occupancy requirement changes • HOVs and hybrids not required to carry transponder but must use a "declaration" lane at the toll gantry mid-way down the project • Free motorcycle access continued • SOVs pay toll for access • No trucks allowed (same as before conversion)
I-95 Express lanes Miami, FL	• One-lane directional facility • 2+ HOVs and hybrids allowed access • As high as 80 % violation rate • Limited transit service	• Added one new lane of capacity and converted existing HOV lane to comprise the two-lane directional priced facility (4 lanes total) • Only 3+ HOV with prior registration may use priced lanes at no charge • SOV hybrid users must have a FL State Decal and an I-95 Express decal to use lane at no charge • SOVs, non-registered 3+ HOVs, non-registered hybrids, and HOV2 pay toll • No trucks allowed (same as before conversion)
I-10 "Katy freeway" managed lanes Houston, TX	• Previous single-reversible HOV lane operated with 3+ restriction in peak hours and 2+ outside the peak for most of the daytime hours	• Built two new managed lanes in each direction • 2+ HOV and motorcycles travel for free 5–11 am and 2–8 pm. Required to pay at all other times • HOVs not required to carry a transponder but must enter the facility through "declaration" lane • SOVs, hybrids, and small commercial vehicles allowed access for toll
Minnesota "MnPass lanes" I-394 I-35 W	• 2+ HOV and motorcycles allowed access • I-394: Two-lane reversible and single-lane directional facility • I-35 W: limited single directional lanes	• I-394: No capacity added—conversion required operational changes only

(continued)

Table 19.3 (continued)

Project	HOV operations before conversion	HOV operations after conversion
Minneapolis, MN	• Significant transit service	• I-35 W: Freeway modified and reconstructed with new capacity designated as priced lanes • 2+ HOV travel at no charge • HOVs not required to carry transponder • Free motorcycle access continued • Hybrids and SOVs allowed access for toll • No trucks allowed
SR-91 Express lanes Orange County, CA	• Opened in 1995 as first privately funded tollroad built in US in 1940 s. Project did not exist as an HOV lane as it opened as a priced lane under private ownership • Purchased by Orange County Transp. Authority in 2003 • Generally allowed 3+HOVs with transponders free use • No trucks	• Two-lane directional facility (4 lanes total) • Limited ingress and egress points only on each end • HOV3 motorists are typically allowed to use the facility free of charge, with the exception of the p.m. peak period from 4:00 to 6:00 p.m. eastbound, when they are required to carry a transponder and pay 50 % of the established toll • All other users pay toll via transponder • Limited transit service • No trucks allowed
I-15 Express lanes San Diego, CA	• 2+ HOV, hybrids with HOV Access Clean Air decal and motorcycles allowed access • 8 mile 2-lane reversible facility • Limited access on each end • Limited transit service	• No capacity added initially—conversion required operational changes only • No occupancy requirement changes • All HOVs and hybrids with HOV Access Clean Air decals are not required to carry transponders • Free motorcycle access continued • SOVs pay toll for access • No trucks allowed • Project has since been expanded and lengthened to a facility that can operate as 3–1, 2–2 or 1–3 directional configuration

(continued)

Table 19.3 (continued)

Project	HOV operations before conversion	HOV operations after conversion
SR-167 Hot lanes Seattle, WA	• 2+ HOVs and motorcycles allowed access • 11 miles single-lane directional facility • Only two adjacent general-purpose lanes in each direction • Unlimited access locations to HOV lane • Limited transit service	• No capacity added—conversion required operational changes only • No occupancy requirement changes • HOVs not required to carry transponders • Free motorcycle access continued • SOVs and hybrids pay toll for access • Access to HOT lane at designated locations only • No trucks allowed
I-15 Express lanes Salt lake city, UT	• 2+ HOV, hybrids with decals • and motorcycles allowed access • Single directional lanes in both directions • Unlimited access • Limited transit service	• Started with decal program to registered SOVs willing to pay $50/month for unlimited use, transitioning to toll for SOVs with transponders • No capacity added—conversion required operational changes only • No occupancy requirement changes • All HOVs and hybrids are not required to carry transponders • Free motorcycle access continued • SOVs pay toll for access • No trucks allowed • No transit service changes

Source Reference [6], P 68. Table 4.1

Table 19.4 Examples of congestion pricing applications outside the united states

	Countries visited	Purpose/ objective	Type of pricing	Milestone dates	Technology	Measured impacts	Annual revenues and cost (in USD)[a]	Distribution of net revenues
Demand management	**Stockholm, Sweden:** Congestion lax	Manage congestion (primary)	Cordon pricing in city center by time of day at SEK10 to SEK20 (about US$1.50 to US$3) per crossing of cordon line into and out of city center	Trial: January–Jul-06	Automated number plate recognition (ANPR) to assess tax to vehicle owner	20 % reduction in traffic congestion in the city center	Gross revenues (2019) SEK850 million (US$118.5 million)	Collected by national government and transferred to the city of Stockholm
		Promote transit and protect environment (secondary)		Referendum: September 2006		10–14 % decrease in emissions	Net revenues (2009): SEK530 million (US$74 million)	Net revenues used to invest in transit and new roads
				Permanently reinstated: August 2007			Overhead costs: SEK320 million (US$44.5 million), about 37 % of revenues	Net revenues used to invest in transit and new roads
	London, United Kingdom: Congestion charge	Manage congestion (primary)	Area pricing in central London and its western extension	Started in central London: February 2003	ANPR to track compulsory payment compliance and identify violators	Initial traffic reductions of 25 % and 19 % (central London and western extension, respectively)	Gross revenues (2008): £268 million (US$435 million)	Net revenues used for transit (80 %) and other transport (20 %) improvements within greater London
		Promote transit and protect environment (secondary)	Flat daily rate of £8 (US$13)	Price increased from £5 to £8 (60 % increase) in July 2005			Net revenues: £137 million (US $222 million)	
				Western extension: February 2007			Overhead costs: £131 million (US $212 million), about 50% of revenues	
				Repeal of western extension: planned in 2010			Capital costs for central London zone: £130 million (US$211 million)	

(continued)

Table 19.4 (continued)

	Countries visited	Purpose/ objective	Type of pricing	Milestone dates	Technology	Measured impacts	Annual revenues and cost (in USD)[a]	Distribution of net revenues
Revenue generation	**Singapore:** Electronic road pricing (ERP)	Manage congestion (primary)	Cordon and express way pricing by time of day and vehicle class	Cordon pricing via manually enforced paper permit system in 1975	Dedicated short-range communications (DSRC) in-vehicle units with removable stored-value smart card for payment	Achieves free flow road speed targets of 45–65 km/h on expressways and 20–30 km/h on arterials.	Gross revenues (2008): SGS125 million (US$90 million)	Net revenues returned to vehicle owners through tax rebates-heavy investment from general fund in transit and highway systems
		Promote transit (secondary)		Transition to ERP in 1998, followed by expressway pricing			Net revenues: SGS100 million (US$72 million)	
					ANPR for enforcement		Overhead costs: SG$25 million (US$18 million), 20 % of gross revenues	
	Germany: Heavy goods vehicle (HGV) charging on highways	Generate revenue and promote user-pays principle (primary)	Truck tolls for HGVs greater than 12 metric tons on the autobahn and limited portions of other national highways based on distance traveled, number of axles and emissions class	Opened in January 2005	Global positioning System (GPS) for vehicle location	Violations less than 2 %	Gross revenue (2008): €3.5 billion (US$5 billion)	Net revenues for roads (50 %), rail (38 %), and waterways (12 %)
						Empty track trips declined by 7 %		
		Protect environment and encourage mode shift to rail and water (secondary)			Global System for Mobile Communications (GSM) for data transmission	58 % shift from dirtier track models (Euro class 1, 2, 3) to cleaner trucks (Euro class 4, 5)	Overhead costs: 15–20 % of gross revenues	
					DSRC and ANPR for enforcement		Average toll rate: £0.163 per km (US$0,378 per mi)	€560 million (US $815 million) per year for trucker "harmonization" program
					Manual booking system via kiosk terminals and Internet for those without onboard units			

(continued)

Table 19.4 (continued)

	Countries visited	Purpose/ objective	Type of pricing	Milestone dates	Technology	Measured impacts	Annual revenues and cost (in USD)[a]	Distribution of net revenues
	Czech Republic: Truck Charging on Highways	Generate revenue and promote user-pays principle (primary)	Truck charges on selected national highways based on distance traveled, number of axles, and emissions class	Opening January 2007	Transponder-based DSRC system with gantries on mainline highways	Average toll rate of US $0.35 per mi on freeways	Gross revenue (2008): CZK6 billion (US$340 million)	Net revenues for roads and highways, railway lines, and inland transport mutes
				Originally for HGVs >12 metric tons				
		Advance environmental objectives (secondary)		Expansion to include trucks >3.5 metric tons in January 2010	ANPR for enforcement		Overhead costs: 30 % of gross revenues	
							Average toll rate: CZK4.05 per km (US $0.36 per mi) for highways; CZK1.90 per km (US$0.17 per mi) for fiist-dass roads	
Planned	**The Netherlands:** National distance-based lax	Planned to manage congestion, replace vehicle tax revenue, and promote user-pays principle (primary)	National distance-based road pricing of all vehicles (commercial trucks and private cars) on all roadways	Phased implementation originally planned to begin in 2011, with all trucks cove red by 2012 and all vehicles by 2018	Underdevelopment, likely GPS for vehicle location, GSM-based data communication, and DSRC interrogation with ANPR for enforcement	2020 forecasted results: 10–5 % reduction in vehicle-miles traveled	Gross revenues (2019 forecasted): €9 billion (US $13.1 billion)	Revenues intended to replace existing vehicle ownership taxes
						40–60 % reduction in delays	Overhead costs: to be determined (capped in law at 5 % of gross revenues)	
		Promote transit and protect environment (secondary)		Implementation on hold because of parliamentary elections in June 2010		10 % reduction in $C0_2$		
						6 % increase in public transit use	Capital costs (estimated): €5.7 billion (US$8.3 billion)	

Source Reference [8], p 9. Table 1
[a]Prevailing exchange rates of 2010.

19.3.2.2 International Applications

Examples of congestion pricing in cities outside the United States are shown in Table 19.4.

The general benefits produced from congestion priced facilities include:

- Faster travel and more predictable travel times
- Increased vehicle throughput
- Reduced fuel consumption and emissions
- Increased revenues for transportation improvements.

The Stockholm, London, and Singapore examples apply pricing on a cordon surrounding the central area to reduce congestion.

Key findings include:

Purpose

- The primary purpose for implementing congestion pricing in each city was to "manage congestion"
- The secondary purpose was to promote transit and protect the environment (Stockholm and London).

Type of Pricing

- Stockholm applies a congestion tax (that varies by time of day) per crossing of cordon line into and out of city center
- Singapore applies cordon and expressway pricing by time of day and vehicle class
- London applies area pricing in central London at a daily flat rate of 8 pounds from 7 a.m. to 7 p.m.

Measured Impacts

- Stockholm reported a 20 % reduction in traffic congestion in the city center, and up to 14 % reduction in emissions (10)
- London reported a 26 % reduction in vehicle trips in 4 years following implementation in 2002(9)
- Singapore's variable price achieves traffic speed targets of 45–65 km/h on expressways and 20–25 km/h on arterials.

Distribution of Net Revenues

- In Stockholm net revenues are used to invest in transit and new roads
- In London 80 % of net revenues are used for transit and 20 % are used to improve other forms of transportation
- In Singapore the net revenues are returned to vehicle owners through tax rebates —heavy investments from general funds are made in transit and highway investments.

These results have been achieved in large part by the high cost of motor vehicle use and by the presence of strong transit systems in these cities that provide a viable alternative to driving.

19.3.3 Effects of Congestion Pricing on Travelers

The various congestion pricing applications vary in their benefits and impacts. Key factors include the type and location of the application, the intensity and extent of peak period congestion, and the amount of the congestion charge.

19.3.3.1 Probable Impact on Drivers

Probable impacts on drivers are as follows:

- Drivers who value the time savings more than the congestion charge are better off when the charge is high enough to reduce trip time and/or trip time variability.
- Drivers who chain trips and find that the congestion charge exceeds the value of the travel time saved and would experience an increase in door-to-door trip times if transit service is not direct or not easily accessible, are worse off.
- Motorists who value the travel time savings of a particular trip less than the congestion charge for that trip will forego making that trip in the peak period and may reschedule it at other times when no charge is imposed, or by using other modes, or by traveling to different destinations.
- Those who have no modal alternatives available are worse off if they are unable to afford the congestion charge, and are forced to travel outside the peak hours, or need to find new destinations that may be of inferior value to them.

19.3.3.2 Equity Issue—Low Income Travelers

Equity issues involving the mobility of low-income drivers often arise. Usually they can be minimized where net revenues from congestion tolls can be allocated to assist low-income travelers in satisfying their mobility needs. In the case of HOT or Toll lanes, low income travelers, who cannot afford paying the charge, can use the general purpose lanes.

19.3.4 Implementation Considerations

Implementing area-wide congestion pricing in US cities requires a number of conditions that are not always present: (1) transit systems must be able to serve those who will be unable to afford the charge, (2) the level of congestion must become unbearable, and (3) the cost of driving and parking should be very high.

In effect, most US large urban areas in the US do not have an extensive and effective transit system—especially in low density areas where the most of the population lives. As a result, improving the transit system (as well as other modes) will require extensive investments before congestion pricing becomes acceptable to the public.

In many US cities, although peak hour congestion has grown considerably in the last two decades, congestion levels (intensity, extent, and duration) are below those experienced in Stockholm, London, or Singapore before these implemented congestion pricing. Therefore, the pressure to reduce traffic congestion by reducing traffic demand may not be as great in US cities.

Finally, the cost of driving (e.g., the price of motor vehicle ownership, gasoline and parking) in London, Stockholm, and Singapore is considerably higher than in the US. This higher cost of driving increases the willingness to accept the need to shift from the automobile to improved transit by those who are unwilling/unable to pay the congestion change.

Moreover, road pricing strategies aimed at reducing peak period congestion do not necessarily reduce daily VMT because auto trips foregone during the peak hours could be scheduled outside peak hours or may be made to different destinations where the charge is not applied.

19.4 Other Road User Charges

Other road pricing strategies aimed at reducing daily vehicle travel and/or increase revenues from road users include: truck tolls mileage-based fees, and pay-as-you-drive insurance [7].

Examples that apply pricing to generate revenue from goods movement vehicles —in Germany (heavy toll vehicles), Czech Republic (truck charging on highways), and the Netherlands (a planned National Distance—based tax) are discussed next.

19.4.1 Truck Tolls

Country-wide truck tolls have been established in Germany and the Czech Republic.

19.4.1.1 Germany Heavy Goods Vehicles

The tolling program began in January 2005. It is the world's first satellite-based countrywide electronic tolling system and applies only trucks weighing more than

12 tons on autobahns, and a few other national roads. Truck tolls are based on the number of axles, vehicle emission ratings, and distance traveled. Fifty percent of the revenues are applied to roads, 38 % to rail, and 12 % to waterways. The system covers about 7,400 road miles including about 30 miles of local roads. Since its inception, empty trucks declined by 7 and 58 % of the trucks shifted from dirtier to cleaner emission models.

19.4.1.2 Czech Republic Tolling

The goal of the Czech Republic's truck tolling system is to generate revenues fro the foreign-based trucks that account for 40 % of trucks on Czech highways. Tolls are based on distance traveled, vehicle classification, and emission rating.

The tolling system was implemented in January 2007, originally for trucks carrying more than 12 metric tons and expanded in January 2010 to include trucks greater than 3.5 metric tons. The average charge is $0.36 per mile for highways, and $0.17 for first class roads. This charging scheme is viewed as the first phase of a more comprehensive road pricing policy.

19.4.2 Mileage-Based Fees

Mileage-based fees are a possible alternative to the motor fuel tax. They provide a mechanism for offsetting lower revenues from reduced motor fuel consumption (due to higher efficiency from newer motor vehicle models). A mileage-based tax also could be applied for congestion mitigation purposes because it could vary by type of facility and/or time of day (degree of congestion). The technology has been applied elsewhere (e.g., Singapore) and also could be applied to ramp metering and congestion pricing.

Mileage charges are being considered by several states, and they have been suggested as a long range alternative for transportation funding in a Transportation Research Board policy study [9].

The State of Oregon has studied mileage-based pricing [10]. The study focused on mileage-based fees and peak period driving charges. Its primary goal was to reduce traffic demand during the most congested periods, and the secondary goal was to raise revenues and replace the existing fuel-based taxes.

Washington State applied estimated VMT elasticities to mileage-based fees. Results yielded estimated reductions in the morning peak VMT of 4 %, and in afternoon peak close to 11 % [7].

A Transportation Research Board Policy Study [9] recommends that "the states and federal government should explore the potential of road user metering and mileage charging." It suggested a program with a national focus, including federal aid for research and testing. This policy study indicated that road user metering and mileage charging appear to be the most promising long term approach to pricing reform. Charges could vary by type of road, time of day, traffic conditions, and miles traveled.

Mileage-based charges would offer a more stable long term revenue base. Their implementation, however, would require public support and technical testing of the concept in a demonstration project.

19.4.3 Pay-as-You-Drive Insurance

Pilot studies of mileage-based user fees in Minnesota indicated that converting the statewide fixed-cost insurance to per-mile cost insurance would reduce the vehicle miles traveled by 6.6 % on weekdays and 8.1 % on weekends [7].

"A national elasticity-based estimate produced consistent results, suggesting that a pay-as-you-drive insurance charge of $0.06/mile could reduce total VMT by 10 %." [7].

However, that these results come from a sample of the population that participated in the pilot studies, and they might not represent the impact of on VMT if pay-as-you-drive insurance were implemented for the whole population of insured drivers.

19.5 Implications and Guidelines

Pricing has emerged as an important strategy for reducing travel demand or and/or generating revenues to finance transportation improvements. The demand reductions vary with the specific type and location of application. Electronic road pricing, as applied on an area wide or cordon area basis in Europe and Singapore, achieves substantial traffic demand reductions. But in the United States it is presently difficult to implement.

Dynamically priced toll facilities spread the peak demand to earlier or later times, thus reducing peak congestion.

Variably priced managed freeway lanes provide congestion-free travel for motorists willing to pay, but usually have had little effect on the freeway congestion in the general use lanes.

Variably priced VMT charges (replacing or as a complement to motor fuel taxes) provide a long term alternative for congestion relief. Such charges could apply to freeways and to major arterials, and they could vary by traffic conditions. But they will require political support in changing the motor fuel taxes as the primary revenue source.

The following guidelines are suggested for consideration when pricing is used as a strategy to reduce congestion:

- Congestion pricing should be focused on places that experience severe recurring congestion
- Implementing road pricing strategies requires the support of the impacted communities, and that of public agencies and legislative bodies. Effective public outreach and communications are key elements of successive pricing programs
- Pricing programs should be easy to use and understand and should avoid complex payment options
- Suitable/acceptable travel alternatives should be available for motorists who are unable or unwilling to pay the toll
- The collection of revenues from road charges should be efficient and enforceable
- The net revenues should be sufficient to at least cover operations and maintenance costs of the collection system, and the balance should be used for transportation improvements
- Congestion pricing should be high enough to reduce road congestion, and
- The congestion charge to users should be linked to the benefits received by them.

References

1. The Federal Highway Administration, US Department of Transportation (2008) Congestion pricing—a primer. Washington, DC
2. Vickery WS (1959) Statement on the pricing of urban streets use. s.l.: In: Hearings, US congress, joint committee on metropolitan Washington problems
3. Fitch LC, associates (1964) Urban transportation and public policy (Chapter 4, Appendix I). Chandler Publishing, New York
4. Walters A (1961) The theory and measurement of private and social cost of highway congestion. Econometrics 29(4):676–699
5. Perez BG, Batac T, Vovska P (2012) NCHRP Report 722: assessing highway tolling and pricing options, volume 1: decision-making framework. Transportation Research Board, Washington, DC
6. Perez BG, Giordano R, Stamm H (2011) NCHRP Report 694, evaluation and performance measurement of congestion pricing projects. Transportation Research Board, Washington, DC
7. Cambridge Systematics and Resource Systems Group (2008) Effective practices for congestion management. In: prepared as part of national cooperative highway research program (NCHRP) Project, Transportation Research Board, pp 20–24 (63)
8. Arnold R, Smith VC, Doan JQ, Barry RN, Blakesley JL, DeCorla-Souza PT, Muriello MF, Murthy GN, Rubstello PK, Thompson NA (2010) Reducing road congestion using road pricing in Europe and Singapore. In: Cooperation with american association of state highway and transportation officials, Federal highway administration, US department of transportation, Washington, DC
9. The fuel tax and alternatives for transportation funding, committee for the study of the long term viability of fuel taxes for transportation finance. Transportation Research Board, Washington, DC, 2006.
10. Oregon's mileage fee concept and road user fee pilot program, final report. s.l. Oregon Department of Transportation, 2007.

Chapter 20
Direct Demand Strategies—Regulatory Restrictions

20.1 Introduction

This congestion relief strategy entails reducing traffic demand through *regulatory road use strategies* that limit the number and type of vehicles from using roadways or from entering certain areas that become congested when vehicles access is not restricted. These restrictive strategies include:

- Traffic Free Streets and Zones—prohibiting private vehicles from using specific streets or from entering specific zones. This strategy is characterized by travel bans by time of day that prohibit private vehicle entry into certain areas (e.g., historic centers in many European cities) or streets (e.g., transit only streets in the US and Canada). These bans are suitable where narrow and irregular dense street networks are not able to serve modern vehicular traffic and high pedestrian volumes.
- Rationing Road Space—This strategy aims at reducing vehicle traffic demand in cities that experience severe congestion problems by rationing street use. The rationing is typically done by license plate restrictions that alternates the days when only odd or even numbered plates are allowed in traffic—either in specific areas or citywide.
- Restricting Truck Use—this strategy (1) bans through trucks from boulevards, parkways, and local streets; or (2) restricts access to residential streets unless trucks have specific destinations in these areas.

20.2 Traffic-Free Streets and Zones

Traffic streets and zones installed in cities throughout the world have three main objectives: (1) to reduce traffic congestion, (2) to improve the pedestrian environment, and (3) to make the area more attractive to customers and visitors and thereby

J.C. Falcocchio and H.S. Levinson, *Road Traffic Congestion: A Concise Guide*, Springer Tracts on Transportation and Traffic 7,
DOI 10.1007/978-3-319-15165-6_20

increase its economic activity. The emphasis placed on these objectives varies depending on area's location, street and development patterns, and community attitudes.

20.2.1 Europe

Pedestrian streets were first introduced in Europe as it reconstructed its cities during the late 1940s following WWII. They were reintroduced in the 1960s after attempts to create wider streets caused adverse environmental effects on historic city centers and did not alleviate congestion.

The traffic free streets and zones were created by eliminating vehicular traffic from existing streets in specific parts of cities usually for architectural, environmental, and commercial reasons [1].

European cities often adopted the traffic-free zoning approach because it suited their historical physical conditions, and eliminated the congestion that would otherwise occur by allowing motor vehicle access and avoided the need for wider streets. By 1975, nearly every major European city had banned cars from its retail district and retail core—a few examples are described below.

Today's Cologne (Germany) pedestrian street system was created as part of the reconstruction of the city from its World War II destruction. The system is centered on two perpendicular streets that connect with the Cathedral Square and the Railroad Station. Parking facilities are located on the periphery with easy pedestrian access to the automobile restricted area.

Essen's (Germany) pedestrian system links the center of the city with major transportation terminals and parking garages. Parking garages in the proximity (less than 1,200 feet) of shopping/entertainment activities on pedestrian areas are priced to serve short term demand. Parking garages outside the ring are intended for long term (all day) parking and form a part of a park-and-ride program.

The downtown of Copenhagen (Denmark) pedestrian system is the most extensive in Europe, after Venice (Italy). The narrow streets across the city connect plazas and open spaces offering visitors a pleasant experience.

20.2.2 United States and Canada

In the United States and Canadian cities, the initial goals of pedestrianization were quite different from those of European cities. The focus was more on improving the quality and attractiveness of downtown retail areas and, at the same time, improving the retail economy of downtown rather than reducing traffic congestion. When improvements in the retail environment were not realized (as in the Westminster Mall in Providence, R.I.) some malls were reopened to traffic.

The reported success of the Kalamazoo (MI) mall opened in 1959, and the Lincoln Road Mall in Miami Beach opened in 1960, spurred mall development elsewhere. By the mid 1970s there were more than 70 malls in the United States and Canada.

The malls and their surrounding commercial areas were designed similar to those of the regional shopping centers. Improved streets parallel to the malls with easy access to parking facilities located adjacent to or near the stores and office buildings. Traffic engineering treatments along parallel streets and at terminal points made these malls "congestion neutral."

Some cities had malls that permitted buses (or light rail) through them. Examples include the NIcollet Mall in Minneapolis, and the 16th Street Mall in Denver. In these cases bus travel times were reduced since there was less conflict with automobile traffic. Buses on the 16th Street Mall limit the time spent at bus stops to board passengers and enable buses to maintain the traffic signal progression set for them.

Most pedestrian (and pedestrian—transit streets) are linear. Traffic is diverted to adjacent parallel streets that sometimes operate one-way. Cross streets usually continue across the mall to maintain continuity of the street system.

To minimize congestion, it is essential to provide a suitable transition of displaced traffic to parallel streets. Sometimes the connections are provided via one-way operations on cross streets.

20.2.3 Guidelines

The size, configuration, and design of traffic-free areas depends on each city's geography, street system, and development patterns both within and in the environs of the area under consideration. Local history and geography, and congestion severity, play an important role in developing traffic free streets and zones. Typically pedestrian streets are less than one mile long, and sometimes they include pedestrian cross-streets. Some general guidelines [2] are shown in Table 20.1.

A discussion and extension of these guidelines include:

- Pedestrian streets (malls) should be clear of obstructions and continuous. Walking should be safe and pleasant
- Sometimes traffic—free streets are interrupted by cross streets that are needed to maintain local circulation patterns
- The streets should be wide enough to avoid overcrowded walking conditions. Where transit vehicles are allowed on the traffic—free streets, an additional 25 feet of width is needed
- There should be no parking and good-loading facilities directly along traffic-free streets. These activities should be located along parallel or cross streets

Table 20.1 Typical guidelines for the design of pedestrian and transit streets

• Develop as part of an overall central area plan
• Obtain positive support from business and civic interests
• Coordinate with ongoing development and transportation actions
• Connect major "magnets" such as department stores and office buildings
• Provide an adequate supply of conveniently located and readily accessible off-street parking (shopper parking should be located in the same or adjacent block to the mall; ideally within 400 feet, except in larger cities). Parking supply and price should complement, rather than compete, with transit; they will depend on the intensity of the city center and its reliance on public transportation
• Provide convenient transit service. Buses and rail vehicles may operate directly on the mall where routing patterns and physical conditions permit
• Develop suitable routes for vehicles currently using the street to be pedestrianized. Parallel one-way streets with suitable transition at each end of a pedestrian street provide one such solution
• Enable pedestrians to walk from one side of the street to another with little or no interference from vehicles
• Provide goods and service vehicle access from parallel streets, alleys, or cross streets. Where this is not practical, trucks can use mall before noon
• Design mall to accommodate police, fire, and other emergency vehicles and to allow efficient maintenance of public utilities

Source Reference [2], Page 27

- Suitable parallel access streets are needed to accommodate the displaced traffic. Their capacities should be adequate to avoid congestion. Sometimes one-way couplets are provided on adjacent streets
- Public transport service should be provided along parallel streets
- Suitable transition of displaced traffic to parallel streets is essential to avoid congestion
- Landscaping, chairs, benches, tables, fountains, and other amenities should be provided along traffic-free pedestrian streets

20.3 Road Space Rationing

Road space rationing is a travel demand strategy that attempts to reduce the traffic congestion and the negative externalities generated by travel demands that constantly exceed the available capacity. It is achieved by restricting access into the congested areas or city centers based upon the last digits of the vehicle license plate numbers on pre-established days and during certain time periods (e.g., peak hours or 7 a.m.–7 p.m.). It has been applied for several decades in large cities in Europe and Latin America. It also has been applied to manage traffic for special events such as the Beijing Summer Olympics in 2008, and the London Summer Olympics in 2012.

The earliest known application of limiting traffic access was in ancient Rome about 45 B.C. when Julius Caesar declared the center of Rome off-limits, between the hours of 6 a.m.–4 p.m., to all vehicles except those transporting priests, officials, and high-ranking citizens [3].

The basic concept is straightforward. Motorists are restricted from entering the designated area on a given day based on the last two digits of their license plate on a given day. There is about a 20 % reduction in weekday travel and congestion when based on the last two digits. Several cities use more digits and extend the prohibitions to more than 1 day per week [4].

Road space rationing based upon license numbers has been implemented in the following megacities [4]:

- 1982: Athens, Greece
- 1986: Santiago, Chile
- 1989: Mexico City, Mexico
- 1997: Sao Paulo, Brazil
- 1998: Bogota, Colombia
- 2003: La Paz, Bolivia
- 2005: San Jose, Costa Rica
- 2008: Countrywide, Honduras
- 2010: Quito, Ecuador

Application of road space rationing by license plate numbers in large US and Canadian cities is not likely in the foreseeable future because: (1) traffic congestion in even the largest metropolitan areas has not reached the levels found in Athens, Greece, Mexico City, or Central and South American cities, (2) street patterns are better suited to accommodate motor vehicle traffic, (3) public transportation service is not usually able to serve dispersed residential patterns, and (4) public acceptance would likely be low.

20.4 Truck Travel Restrictions

Truck travel restrictions afford an opportunity to reduce congestion along heavily traveled streets and roads, and in centers with a high concentration of traffic. They also improve pedestrian safety and enhance neighborhood quality of life by reducing noise and air pollutant emissions.

20.4.1 Current Status

Truck travel is restricted along parkways and boulevards in several US cities including Boston, Chicago, New York, Washington, and San Francisco. They also apply to suburban and intercity parkways such as the Westchester and Long Island parkways, in the New York City area, the Taconic Parkway in New York State, and the Baltimore-Washington Parkway.

Many cities and suburbs have adopted various truck travel restrictions to reduce congestion and neighborhood intrusion of trucks. Examples of common restrictions include:

- "no standing" restrictions along arterial streets that prohibit peak period on-street truck loading and unloading
- Tractor trailer trucks (usually more than 33 feet in length) are prohibited in downtown Atlanta, the Chicago "Loop", and Lower Midtown in Manhattan
- Trucks are restricted by size and weight from entering many residential areas (except for deliveries)

20.4.2 Suggested Guidelines

Implementing truck travel restrictions by time of day and/or size of vehicle to reduce congestion should reflect the different needs of carriers, shippers, receivers, and the affected community.

The need for restricting truck travel should be based on location, congestion levels, land use patterns, environmental requisites, and community values. Key considerations include: (1) amount of street congestion, including the proportion of delay caused by truck movement and loading/unloading activities, (2) ability of other streets to accommodate the displaced trucks, (3) actions already taken to improve passenger and goods flow (such as peak period parking bans), (4) potential effects on impacted activities, and (5) ease of implementation and enforcement.

The daytime or peak period restrictions of truck travel should be considered only after peak-period curb parking and loading activities have been prohibited and are effectively enforced. In addition, alleviating traffic congestion in the city center should have a higher priority than problem locations in the rest of the urban area. Finally, suitable means of goods delivery and pick up access to and from impacted streets or areas should be available. This commonly entails having access from adjacent or cross streets. Restrictions in central areas might require special truck tunnels such as those found in Dallas (Texas) and New Haven (Connecticut).

Sometimes all-day bus lanes are provided along commercial streets to improve the speed and reliability of bus service. In these cases time "windows" can be established that permit goods delivery and unloading during designated midday hours. Examples are the "windows" along Fordham Road in the Bronx, New York, that permit deliveries from about 10 a.m. to noon on one side of the street and from 1 p.m. to 3 p.m. on the other side. New York City Transit's "Select Bus" service uses the lanes from 7 a.m. to 7 p.m.

20.4.3 Effects

The travel time benefits and reduced congestion resulting from restricting truck trips should be estimated for the hours that the restrictions would apply. These benefits can be translated into monetary terms, and then should be compared to the costs resulting from rescheduling deliveries and pickups. Time-of-day restrictions can pose problems for time-sensitive deliveries such as bakeries and restaurants, and provisions for these deliveries should be provided where practical.

20.5 Implications of Regulatory Restrictions

Traffic and road use regulations are generally applied to reduce congestion by increasing the available roadway capacity. But they also can be used to reduce traffic demand and achieve a better demand-capacity balance. Some key implications of regulatory restrictions follow.

20.5.1 Traffic-Free Streets and Zones

Pedestrian/transit streets and zones have reduced central area traffic congestion in many European cities. They have also concurrently enhanced the historical/architectural character of their historic centers

Within US and Canadian cities these applications have generally improved the urban environment, but usually have remained congestion-neutral. But congestion was reduced where the use of diagonal streets was removed from the traffic network and converted from vehicular use to pedestrian use, or where transit-only streets were created.

The 2011 experience in the Manhattan CBD of converting street segments carrying vehicular traffic to only pedestrians use simplified complex intersections, and reduced delays to motorists, and bus passengers. Pedestrian zones were created along sections of Broadway in Manhattan. Converting sections of Broadway to exclusive pedestrian use simplified the traffic conflicts at intersections in Times Square and Herald Square, substantially reduced traffic backups and delays on the Avenue of the Americas, and has created open spaces for visitors, employees, and shoppers. Figure 20.1 shows a photo of the Macy's and Herald Square pedestrian plaza at Broadway and 33rd Street in Manhattan with tables, chairs, umbrellas and plantings.

Fig. 20.1 Pedestrian plaza at Macy's and Herald Square, Manhattan. *Source* Reference [5]

20.5.2 Rationing of Road Space

Road space rationing has succeeded in reducing severe congestion in many of the world's megacities in Europe and Latin America. But they have not been applied/adopted in US and Canadian cities where congestion levels are not as severe, and where residential land use is more spread out and cannot support frequent transit service.

20.5.3 Truck Restrictions

The congestion reduction benefits from truck use restrictions are generally limited to the restricted zones/routes of their application. They can be only achieved if their negative residual effects on other activities can be mitigated. For example:

- prohibiting trucks from using congested streets in city centers reduces congestion—but it may constrain essential commercial vehicle access to building
- restricting peak-period truck use of freeways could reduce freeway congestion—but it could add trucks to local streets

20.5.4 Conclusions

Regulatory restrictions pertaining to road use can reduce both travel demand and traffic congestion. They can effectively reduce traffic congestion, improve pedestrian environments, and improve the overall quality of the affected areas. The application of regulatory restrictions on traffic flow requires community consensus—especially in the United States and Canada.

References

1. Brambilla R, Longo G (1977) For pedestrians only—planning, design, and management of traffic-free zones. Whitney Library of Design, New York
2. Transportation Planning for Pedestrian and Transit Streets (1981) Compendium of Technical Papers, ITE Technical Council (Herbert S. Levinson, Chairman), 51st Annual Meeting, August 16-20, Boston, MA
3. Black WR (2010) Sustainable transportation: problems and solutions. s.l., The Guilford Press, New York. p 10
4. Road Space Rationing. s.l. Wikipedia, the Free Encyclopedia
5. Keating, Added by Brigid. s.l. on http://WWW.OPENGREENMAP.ORG/GREENMAP/NYC-ENERGY/HERALD-SQUARE-26243

Chapter 21
Indirect Demand Strategies—For Employers, Institutions, and Public Agencies

21.1 Introduction

Travel demand strategies and actions can reduce vehicle traffic during peak periods and thereby help reduce recurring congestion. The strategies focus on reducing the use of single-occupant automobiles for commuter trips, and in spreading the peak travel periods through work schedule changes. They are generally applied by large employers and large institutions such as medical centers and universities.

Specific strategies include (1) providing incentives for group riding (or carpooling) such as guaranteed rides home, (2) contract transit and financial incentives, (3) "cash outs" of employee parking, and (4) alternative work arrangements such as flexible or staggered hours and telecommuting.

The strategies and actions are drawn from information contained in TCRP Report 95 [1]. This information is complemented by reported work-schedule changes in various cities and by demand management strategies established by the state of Washington's transportation agencies.

21.2 The 82-Program Sample

Chapter 19 of TCRP Report 95 contains detailed information on the effects of support actions documented for a sample of 85 agencies. The performance metrics used to measure the effectiveness of the various actions consist of "vehicle trip reductions (VTR)" percentages that are associated with specific strategies.

J.C. Falcocchio and H.S. Levinson, *Road Traffic Congestion: A Concise Guide*,
Springer Tracts on Transportation and Traffic 7,
DOI 10.1007/978-3-319-15165-6_21

21.2.1 Overview

Table 21.1 shows six types of strategies supported by employers for reducing vehicle trips to work: transit enhancements, restricted parking, parking fees, transportation services, modal subsidies, telecommuting, and compressed work weeks.

Trip reduction comparisons are provided for programs that incorporate high, medium, and low levels of employer support. The numbers in parenthesis represent the sample sizes used in calculating the average vehicle-trip reductions. Overall, all programs reported a 16.9 % reduction in vehicle trips. The "all" row in the table shows that, employers who provided a high level of support achieved a 19 % vehicle trip reduction (VTR) as compared with about 15–16 % for lower levels of support.

Table 21.1 Vehicle trip reduction percentages related to employer support actions and levels of support

Other condition	VTR by level of overall employer support (sample size)			
	High	Medium	Low	All
All	19.0 % (32)	15.9 % (33)	15.0 % (17)	16.9 % (16.9)
Transit availability				
High	28.4 % (10)	28.2 % (6)	24.3 % (8)	26.0 % (24)
Medium	10.1 % (5)	15.3 % (10)	3.2 % (3)	11.9 % (18)
Low	15.9 % (17)	13.6 % (17)	8.6 % (6)	12.3 % (40)
Restricted parking				
Yes	29.9 % (12)	23.8 % (11)	18.0 % (11)	24.1 % (34)
No	12.5 % (20)	12.0 % (22)	9.6 % (6)	11.9 % (48)
Parking fees				
Yes	24.4 % (14)	27.3 % (7)	22.8 % (9)	24.1 % (30)
No	12.5 % (18)	12.0 % (26)	9.6 % (8)	11.9 % (52)
Transportation services				
Yes	26.5 % (15)	15.5 % (14)	24.2 % (5)	21.6 % (34)
No	12.4 % (17)	16.2 % (19)	11.2 % (12)	13.6 % (48)
Modal subsidies				
Yes	20.5 % (26)	19.8 % (25)	16.9 % (13)	19.5 % (64)
No	12.7 % (6)	3.7 % (8)	9.1 % (4)	7.9 % (18)
Telecommuting				
Yes	16.6 % (10)	14.6 % (6)	28.2 % (1)	16.5 % (17)
No	20.1 % (22)	16.2 % (27)	14.2 % (16)	17.1 % (65)
Compressed work week				
Yes	19.7 % (15)	21.5 % (10)	8.3 % (3)	19.5 % (28)
No	18.4 % (17)	13.5 % (23)	16.5 % (14)	15.8 % (54)

Source References [1], pp 19–20, Table 19.1 and [16] © National Academy of Sciences, Washington, DC. Reproduced with permission of the Transportation Research Board.

21.2.2 Specific Actions

Table 21.2 shows the vehicle trip reduction results obtained for various types of alternative transportation services provided by employers under different travel conditions.

Although the sample size generally is too small to be statistically significant, the table provides anecdotal data that shows reductions in vehicle trips of varying magnitudes. For example,

- The highest impact on trip reduction (36.6 %) was achieved where parking fees were implemented jointly with "company vehicles"
- And where they were implemented jointly with Vanpool services, the trip reduction increased to 23.6 %

Table 21.2 Vehicle trip reduction (VTR) percentages related to transportation services provided by employers

Other condition	VTR by level of overall employer support (sample size)					
	Transit	Vanpool	Transit and vanpool	Company vehicles	All with services	No services
All	18.9 % (4)	21.3 % (11)	18.8 % (11)	24.6 % (11)	21.6 % (34)	13.6 % (48)
Transit availability						
High	35.3 % (2)	24.4 % (4)	27.2 % (3)	33.5 % (2)	28.8 % (11)	23.6 % (13)
Medium	14.1 % (1)	25.6 % (1)	12.6 % (1)	15.0 % (3)	21.0 % (6)	8.0 % (13)
Low	<0 (1)	18.6 % (6)	16.1 % (7)	21.8 % (6)	15.5 % (17)	11.1 % (22)
Restricted parking						
Yes	35.3 % (2)	28.4 % (6)	23.1 % (5)	20.7 % (5)	27.9 % (17)	20.3 % (19)
No	2.6 % (2)	12.8 % (5)	15.2 % (6)	17.2 % (6)	15.4 % (17)	9.3 % (29)
Parking fees						
Yes	35.3 % (2)	34.1 % (5)	23.6 % (4)	36.6 % (5)	31.4 % (15)	18.2 % (16)
No	2.6 % (2)	10.7 % (6)	16.1 % (7)	14.6 % (6)	13.9 % (19)	11.3 % (32)
Level of support						
High	n/a (0)	21.7 % (6)	25.6 % (5)	33.5 % (6)*	26.5 % (15)	12.4 % (17)
Medium	15.8 % (3)	12.8 % (2)	12.9 % (5)	14.0 % (5)*	15.5 % (14)	16.2 % (19)
Low	28.2 % (1)	26.3 % (3)	14.1 % (1)	n/a (0)	24.2 % (5)	11.2 % (12)
Modal subsidies						
Yes	28.2 % (3)	25.4 % (9)	25.4 % (7)	30.8 % (8)*	26.7 % (24)	14.7 % (40)
No	0.0 % (1)	13.4 % (3)	7.3 % (4)	8.3 % (3)*	9.4 % (10)	6.0 % (8)
Telecommuting						
Yes	28.2 % (1)	9.6 % (2)	14.5 % (3)	21.9 % (4)	18.3 % (9)	14.6 % (8)
No	15.8 % (3)	23.9 % (9)	20.4 % (8)	26.2 % (7)*	22.9 % (25)	13.5 % (40)
Compressed work week						
Yes	20.5 % (3)	22.2 % (3)	15.4 % (2)	24.2 % (7)*	23.0 % (12)	16.2 % (16)
No	14.1 % (1)	21.0 % (8)	19.5 % (9)	25.4 % (4)*	23.2 % (22)	12.0 % (32)

Note Asterisked sample (*) include cases combined with other transportation services
Source Reference [1], pp 19–28, Table 19.4

- But where employee parking fees were implemented jointly with transit enhancements, vehicle trip reductions increased to 35.3 %
- Where employee parking fees were implemented without transportation services enhancements, vehicle trips were reduced by 18.2 %

A number of employers from the 82-program sample achieved substantial vehicle trip reductions. A sample of sixteen transportation service programs is summarized in Table 21.3. The programs are listed according to type of transportation services provided (transit, vanpool, company vehicles); type of parking policies (restricted, priced, HOV discounts); financial incentives (transit subsidy, carpool subsidy, travel allowance, other monetary); and level of support (high, medium, low). They show vehicle use reductions ranging from 13 to 62 %.

21.2.3 Effect of Financial Incentives on Single Occupancy Vehicles (SOV) Commuters

Some employers provided financial incentives (e.g. parking charges or alternative mode travel allowances) to their employees to reduce single occupancy vehicle (SOV) use and encourage greater use of higher occupancy vehicles in commuting, with increasing financial incentives (Fig. 21.1).

21.3 Alternative Work Arrangements

21.3.1 Introduction

Alternative work arrangements modify the times when travel takes place and/or the frequency of travel. They have their origin in the staggered hour programs operated at large defense plants during World War II. They began with the need to reduce congestion on public transit systems by spreading peak hour travel demands; they have since also addressed automobile travel demand. Alternative work hours are generally found in large cities and involve the participation of large employers.

Work schedule changes to reduce the amount of work-related peak period travel can be accomplished in several ways.

- Staggered Hours—staggering up or rearranging employee starting and quitting times to distribute arrivals and departures on the shoulders of the peak hour
- Flexible Work Hours—starting and quitting times can be adjusted at the discretion of employees provided that they work the required number of hours each day. Usually employees are required to be at work during specific periods each day
- Compressed Work Week—the work week is reduced to four days and the number of hours each day is increased

Table 21.3 Examples of employer transportation services programs

Employer	Type	Size	Setting	Transit availability	Support level	Type of service[a]	Parking[b]	Financial incentives[c]	VTR[d] (%)
Puget sound blood center	Medical	200	CBD fringe	High	Medium	TR	R, P, D	T, O	42.4
Swedish hospital	Medical	2.250	CBD fringe	High	Low	TR	R, P, D	T, A	28.2
Univ. of Washington	University	17,400	CBD fringe	High	High	TR, VP	R, P, D	T	62.5
Sears (hoffman estates)	Comm./Svc.	5,400	Exurban	Low	High	TR, VP		T	42.4
P.L. Porter	Ind./Manuf.	230	Campus	Low	Medium	TR, VP		T, V	23.0
Pacofic Bell (San Ramon)	Utility	6,900	Exurban	Low	High	TR, VP	R		21.5
So. California gas	Utility	1,800	Exurban	Low	High	VP, CV	R, P, D	T, A	47.4
Travelers insurance	Prof./Office	10,000	CBD	High	Low	VP	R, P, D	T, V	42.4
Atlantic Richfield	Prof./Office	2,000	CBD	High	High	VP	R, P, D	V, A	34.5
Bonneville power	Utility	100	CBD Fringe	Medium	Medium	VP	R, P, D	T	25.6
Rockbestos	Ind./Manuf.	400	Exurban	Low	Low	VP			29.0
Johnson and Higgins	Prof./Office	180	CBD	High	High	CV	R, P	T	44.2
City of Simi Valley, CA	Gov't	150	Suburban	Low	High	CV		T, A, O	43.5
CH2 M Hill (Bellevue)	Prof./Office	400	Sub CBD	Medium	Medium	CV	R, P, D	T, C, A	38.9
Bellevue City Hall	Gov't	650	Office park	Medium	Medium	CV	R, P, D	T, V, A	30.0
Wm. H. Meecer	Prof./Office	120	CBD	High	High	CV	P, D	T	22.7

Notes
Codes TR = Transit, VP = Vanpool, CV = Company Vehicles
Codes R = Restricted, P = Priced, D = HOV Discounts
Codes T = Transit Subsidy, V = Vanpool Subsidy, C = Carpool Subsidy, A = Travel Allowance, O = Other Monetary
VTR = Vehicle Trip Reduction, defined as the percentage by which the vehicle trip rate (vehicle round trips per 100 employees commuting) for the program is less than the vehicle trip rate for the control population
See accompanying text and also case study "University of Washington's U-PASS Program—Seattle, Washington" Footnote 21
Source Reference [1], pp 19–33, Table 19.5

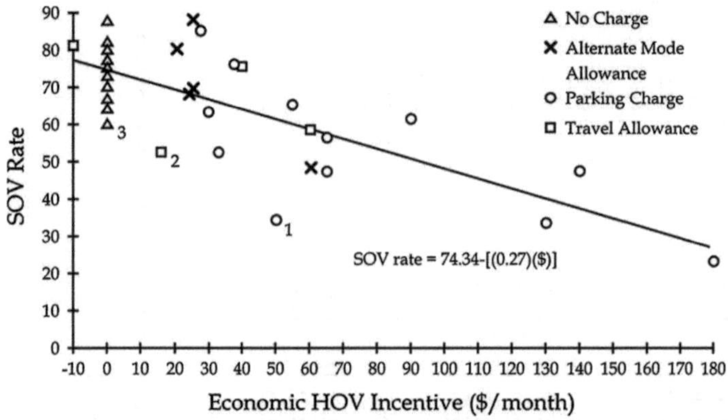

Fig. 21.1 Effect of economic incentives on SOV use rate. Key *1* Reduced parking supply. *2* Superior transit service. *3* Alternative work hour program (two cases). *Source* Reference [1], pp 19–60, Fig. 19.1

- Telecommuting—working at home some number of days per week or month. Advanced communication technology make telecommuting a feasible alternative to commuting to work for some professions

21.3.2 Staggered and Flexible Hours

Work rescheduling emerged as a congestion-reducing and fuel conservation strategy in the mid-1970s. Table 21.4 contains examples of companies that instituted staggered and flexible work hours programs and the number of employees involved, before 1976 [2].

21.3.2.1 Effects

The results achieved by these early programs in Manhattan (New York City), Toronto, Ottawa, St Paul, Bishop's Ranch (California), San Francisco, and Hawaii, are described in chronological order below.

a. Lower Manhattan (New York City)

Staggered work hours were introduced in Lower Manhattan on April 1, 1970, by the Port Authority of NY and NJ, and by PATH (Port Authority—Trans Hudson Transit), in cooperation with the Downtown—Lower Manhattan Association. The area had a worker population of 486,000 of whom about 85 % commuted by rail transit.

The program, in which about 100,000 employees participated, was one of the most ambitious staggered hours programs enacted. It was designed to relieve

Table 21.4 Examples of staggered and flexible work hour programs in the United States—mid 1970s

Location	Number of employees	
	Staggered work hours	Flexible work hours
Chicago, IL		
Montgomery Ward		500
East Hanover, NJ		
Sandoz Inc.		1,300
East Meadow, NY		
Lufthansa Airlines (executive office)		300
Madison, WI		
City/county/state employees	17,000	
Minneapolis, MN		
Control data corporation (and most of its US offices)		20,000
New York, NY		
400 Manhattan firms	220,000	
Palo Alto, CA		
Hewlett-Packard (15 facilities)		16,000
Philadelphia, PA		
Center city employees	33,500	
St. Paul, MN		
3M company	12,000	
St. Paul company (and several of its regional offices)		5,000
Washington, DC		
Six federal departments	50,000	
White plains, NY		
Nestle company		700

Source Reference [3]

congestion on subways, commuter rail lines, buses, and building elevators. It successfully achieved these goals:

- 6 % fewer passengers were carried on the Lower Manhattan subway lines during the busiest 10 min period
- At the PATH terminal, the volume of passengers during the busiest 15 min evening period dropped 13 % (from 7,500 to 6,500), while the passenger volume during the lightest 15 min period rose by 48 % (from 3,100 to 4,600).

In September 1974, the Port Authority of NY and NJ began a flexible hours experiment that lasted 8 months and involved 850 headquarters staff. The experiment consisted of employees who were on a conventional work schedule and employees who were on staggered hours.

It was found that employees who changed their work hours from a fixed schedule to flextime reduced their peak arrival and departure volumes by significant amounts; while those who changed from staggered work schedules to flextime reported an insignificant change.

As shown in Table 21.5, where most employees changed from a conventional fixed schedule to flextime (on floor A), fifteen-minute work-floor arrival and departure peaks decreased by 13 % points (from 31 to 18 %), and 10 % points (from 35 to 25 %), respectively. In contrast on floors "B" and "C" where the change to flextime involved workers who previously were on staggered hours, there was little or no change in peak hour volumes [3].

b. Toronto (Canada)—1973

A staggered work hour demonstration program was initiated October 1973 in Toronto. The program involved 11,000 public employees of the city's Queen's Park complex. The peak arrival and departure times became flattened to hours adjacent to the peak hour. As a result, one third of the employees experienced a decrease in travel time; and about one third indicated a more comfortable and convenient

Table 21.5 Port authority of New York and New Jersey flextime experiment—15 min peaking before and after flexible work hours

Work floor/work hours programs	Peak 15 min arrivals	Peak 15 min departures		
	Percent of 7:30–10:00 AM Arrivals (%)	Peak AM time period	Percent of 3:30–6:00 PM departures (%)	Peak PM time period
Floor A				
Before (conventional hours)	31	8:45–9:00	35	4:30–4:45
After (flexible hours)	18	8:45–9:00	25	4:30–4:45
Floor B				
Before (staggered hours)	20	8:15–8:30	26	4:00–4:15
After (flexible hours)	20	8:15–8:30	27	4:00–4:15
Floor C				
Before (staggered hours)	28	8:15–8:30	25	4:00–4:15
After (flexible hours)	24	8:30–8:45	26	4:15–4:30
Floor D (control)				
Before (floating day)	24	8:15–8:30	30	4:00–4:15
After (floating day)	29	8:15–8:30	25	4:00–4:15
Floor E (control)				
Before (conventional hours)	27	8:30–8:45	30	4:15–4:30
After (conventional hours)	27	8:15–8:30	28	4:15–4:30

Note Arrivals and departures were surveyed on the individual work floors. The surveys included some employees not participating in the flexible work hours experiment
Source References [1], pp 19–71. Table 19.18 and [3]

trip. Over 90 % of the employees involved expressed a favorable reaction to the demonstration program.

c. Ottawa, Ont. (1970s)

The Canadian government is the city's dominant employer. At the time of the experiment, it accounted for about 50 % of the central area's workforce. The effect of staggered hours for government employees reduced the peak hour to peak period ratios for all transit passengers by 8–19 % [4].

d. St. Paul, Minnesota (mid-1970s)

The 3M Company implemented a staggered work hour program to reduce peak period auto congestion on roads in the vicinity of the plant. After the program was in effect for some time the highway department took a traffic count on these roads that showed a traffic reduction during the peak period.

e. San Francisco Flextime (1979)

Results of the Flextime Demonstration Project in San Francisco are shown in Table 21.6.

Employees from companies participating in the demonstration project had a more uniform distribution of arrival times than the rest of downtown employees. While 61 % of all downtown employees arrived to work between 8:00 and 8:30 am, only 14–40 % of employees from companies adopting variable work hours arrived in the same am period.

f. Hawaii (1988)

A staggered demonstration project was conducted between February 22 and March 18, 1988 to determine whether spreading the arrival times of public employees would relieve peak period congestion. The State of Hawaii changed official office hours for public employees from 7:45 a.m.–4:30 p.m. to 8:30 a.m.–5:15 p.m. About 4,000 workers, or 6–7 % of the downtown workforce participated in the project.

Table 21.6 Employee arrival times at three san francisco employers adopting variable work times

Arrival time (AM)	Fireman's fund (self-staggered start) (%)	CSAA (flextime) (%)	Metropolitan life (flextime) (%)	All downtown employees (%)
7:00–7:30	31	16	53	8
7:30–8:00	34	31	24	13
8:00–8:30	20	40	14	61
8:30–9:00	10	7	6	1
After 9:00	5	6	3	17
Total	100	100	100	100

Note Earliest sanctioned arrival time at CSAA was 7:30 AM
Source References [1], pp 19–70, Table 19.17 and [17]

The shift in start times reduced travel times in peak periods—average travel time of 45 min was reduced by 3–4 min. Saving in travel time, however, differed by route and time of day. The project spread out the peak travel hour which improved conditions for those traveling during the most congested time periods. But it made conditions slightly worse for those already traveling during less congested time periods [5].

g. Bishop Ranch (exurban San Francisco)—1990

Flextime policies were a major part of commute assistance for employees relocating from downtown San Francisco to Bishop Ranch—a major business park.

Two years after opening, a survey of 14,800 employees showed that the percentage of employees starting before 7 am increased from 8 to 17 %, and the percentage staring after 9 am, increased from 1 to 9 %. The percentage of workers leaving before 4 pm increased from 12 to 17 % [6].

21.3.3 Compressed Work Week

The compressed work week (CWW) usually involves four 10 h days in a 40 h week (4/40 plan), or 80 h in 9 days (9/80). Those employees whose employers provide this flexibility reduced their commute trips by 20 %. A 1976 study [2] found that the 4/40 plan was in effect for more than one million workers at an estimated 3,000 companies in the United States. Thus the 4/40 plan could reduce the number of commuter trips by 20 % if it was universally adopted. In addition, it could redistribute the peak hour commute trips outside the peak hour at least once per day.

However, the effectiveness of the CWW in reducing peak hour traffic demand was constrained by its lack of universal adoption. This is understandable as it would require every household to reorganize their daily schedules around the CWW—a complicated and complex undertaking.

Some businesses adapted readily to the four day work week while others found it difficult—if not impossible. As with staggered and flexible work hour programs, businesses that are primarily administrative adapted well to the CWW, while businesses that cater to customers' hours or are linked to delivery schedules, or inter-industry requirements, would not be able to adapt.

21.3.3.1 Examples and Effects

Several applications of the CWW were cited in Chap. 19 of TCRP Report 95 [1]. The applications of Colorado, California, and Washington are briefly described as follows.

a. Denver, CO

In the early 1980s, a carefully controlled CWW experiment was implemented in Denver. About 9,000 federal employees—65 % of all federal employees in Denver-were involved in the program. The most popular schedules were 80 h in 9 days, and 40 h in 4 days. The maximum percentage of total arrivals to work in the peak half hour dropped from 47 to 34 % [7].

The Denver Federal employees compressed work week schedule found a net household VMT reduction for both work and non-work travel averaged about 15 % for participating employees. These percentages are for total household travel during the 7 day week.

b. Ventura County, CA

As part of its effort to meet trip reduction requirements of the Air Pollution Control District's Rule 210 (predecessor to Regulation XV), Ventura County tested a variable work hours program consisting of a combination of flextime and both 9/80 and 4/40 work weeks. Commuter Transportation Services, Inc. (CTS) conducted a 6 month pilot project to determine the impact on ridesharing and organizational effectiveness. A total of 367 employees were involved, with 172 adopting a 9/80 schedule, 33 adopting a 4/80 schedule, and 76 adopting a flextime schedule. The remaining 86 employees either did not opt for one of the variable work hour schedules or discontinued participation in the 6 month demonstration program.

Survey data from the 367 employees indicated that drive-alone rates declined from 82.2 to 76.6 % over the course of the project, while ridesharing rates increased from 8.0 to 12.8 %, and use of "other" modes of travel increased from 9.8 to 10.6 % [8].

c. California Air Resources Board

This 1995 study of CWW on employee travel by the California Air Resources Board found that 2,600 Southern California employees on CWW schedules reduced their net number of trips by an average of 0.5 person trips, per week. Those working a 9/80 or a 4/40 schedule drove an average of 20 fewer miles per week [9].

d. Washington State Commuter Trip Reduction Program

Employee participation in compressed work week schedules among participating employers grew steadily from 14.5 % in 1993, to 20.0 % in 2005 [10]. An update on how this program has grown is presented in Sect. 21.4 of this chapter.

21.3.3.2 Implementation Considerations [2]

Implementation of the compressed work week generally involves little, if any, costs. The 4 day work week (4–10) is the more commonly applied pattern. As with staggered and flexible work hour programs, businesses that are primarily administrative adapt well to a 4 day work week, while businesses that cater to customers' hours or are linked to delivery schedules or inter-industry requirements may not be able to change from their 5 day schedules.

The benefits of the CWW include reduced traffic and transit congestion, improved employee and management morale, reduced absenteeism, and increased leisure time. However, state and federal work rules, employee fatigue, or personal obligations outside work (e.g., child care schedules), were reported to constrain participation in the CWW program.

21.3.4 Telecommuting

A review of available literature indicates that the following conditions are conducive to telecommuting [1]:

(a) long commute distances and travel times
(b) pay parking at the work place
(c) high income and a college education
(d) professional occupations

Information on the trip reduction effects of telecommuting is limited. Two examples follow:

21.3.4.1 1992 National Telecommuting Survey

A national telecommuting survey obtained information from 16 organizations representing almost 5,000 telecommuters. Most organizations were government agencies and communications companies. The number of telecommuters ranged from seven to 2,600 with a mean of 110, and a median of 82. Most of telecommuters were in professional or managerial positions.

The most common telecommuting schedule was 1 day per week, which accounted for about 55 % of all telecommuters. About 18 % telecommuted 2 days per week, 15 % 3 or 4 days per week, and 12 % 5 days per week [11].

21.3.4.2 Southern California Association of Governments Survey of 2002

This home-based survey of 5,000 residents reported that out of 2,766 workers, 24.6 % telecommuted at least 1 day per week; 7 % were home-based business workers; and 68.4 % commuted [12].

21.3.4.3 Teleworking Trends in the United States

Telecommuting participation is likely to increase as advanced communication technology lowers barriers to its implementation in the workplace. As shown in Table 21.7, the number of people who work from home at least 1 day per month, has been increasing—from 20.7 million in 2002 to 26.2 million in 2010, (reaching a 33.7 million in 2008)—at an average of close to 7 % per year [13].

Table 21.7 US teleworking trends (2002–2010)

Number of teleworkers, 2002–2010	
Year	Millions of workers
2002	20.7
2003	23.5
2004	24.1
2005	26.1
2006	28.7
2008	33.7
2010	26.2

Source Reference [14]

21.4 Washington State Commuter Trip Reduction (CTR) Program

The Washington State CTR is a good example of state actions to manage/reduce commuter traffic demand and contain congestion. Started in 1993, the program employs a number of alternative strategies to reduce auto use for commuting. These include CWW, flextimes, staggered hours, vanpools/carpools, etc.

A 2005 study of the program's effectiveness reported that employee participation in compressed work week schedule among CTR—covered employees grew steadily from 14.5 % in 1993, to 20.0 % in 2005. In 2005 about 2/3 of CTR program's employees were eligible to choose the compressed work week (CWW). Participation in 9/80 schedules (working 80 h in 9 days) doubled between 1993 and 2005 to 5.8 %. Participation in 4/40 schedules (working 40 h in 4 days) grew more slowly and remained at 7.3 % [15].

Twenty years of investment in the CTR program have built a foundation of partnerships between local governments and employers in managing transportation demand. The partners in the CTR program include:

- Major employers implement programs based on locally adopted goals for reducing vehicle trips and VMT. They include the Boeing Company, the Microsoft Corporation and the Starbucks Corporation. Employers also may form TMAs
- Local Governments provide technical assistance and services to help employers achieve the goals and they may initiate outreach and service programs for commuters. Local governments must develop a CTR plan that is consistent with the local comprehensive plan.
- Transit agencies operate bus and vanpool services and coordinate program implementation with local governments, employers, Transportation Management Associations (TMAs), and others.
- Six regional transportation planning organizations provide planning support and coordination across jurisdictions, ensuring consistency in transportation and economic plans.

- Washington State provides $3.9 million in grants to local governments every 2 years to support employers programs. Washington State DOT administers the funding, guides the program with policies and procedures, measures performance, and evaluates the program.
- The CTR Board is made up of representative of the various partners, sets policy direction, allocates the funding appropriated by the legislature, and reports to the legislature on the effectiveness of the program every 2 years. This innovative government structure is one of the program's strengths.
  - The CTR partnership begins with the state investment—a total of $5.7 million every 2 years.
  - 70 % of the state CTR funds are distributed to local governments, which also invest their own resources to assist employers in the development and implementation of CTR programs.
  - Washington State DOT applies the balance of state funds for technical support and program tools and for measuring, evaluating and reporting the program's performance.
  - Employers contribute far more to the program than they receive from the state and local investment: in 2006, employers invested $45 million in their CTR programs- more than $16 for every $1 invested by the state.

21.4.1 Program Performance

The program's goals were to reduce the drive-alone rate by 10 % and the vehicle miles of travel (VMT) per employee by 13 % at CTR worksites between 2007 and 2012. These targets were set to mitigate the congestion impacts from additional commuter trips on the transportation system generated by the expected job growth.

The 2009 CTR survey data indicated that if CTR Program participants in the Central Puget Sound region returned to driving alone to work at the same rate as they did before the program started, the freeway and arterial system would need to accommodate an additional 22,500 additional drive-alone vehicle trips during the morning peak commute period. These additional trips would increase freeway and arterial system delays in the morning peak by approximately 12,900 h, an increase of almost 10 % in freeway delays and an increase of almost 6 % in arterial delays—or 7.6 % increase in combined freeway and arterial delays.

- In 2009, the Puget Sound Regional Council estimated that the choices made by commuters in the CTR program since its start in 1993 avoided an increase of nearly 8 % in congestion for the central Puget Sound region. CTR worksites did better at reducing vehicle trips than the rest of the state and the nation (Fig. 21.2).
- During the economic recession, while Washington State lost approximately 140,500 workers, CTR employers added 14,393 workers. Because of the CTR program, daily vehicle trips to CTR sites increased by only 1,225, placing far

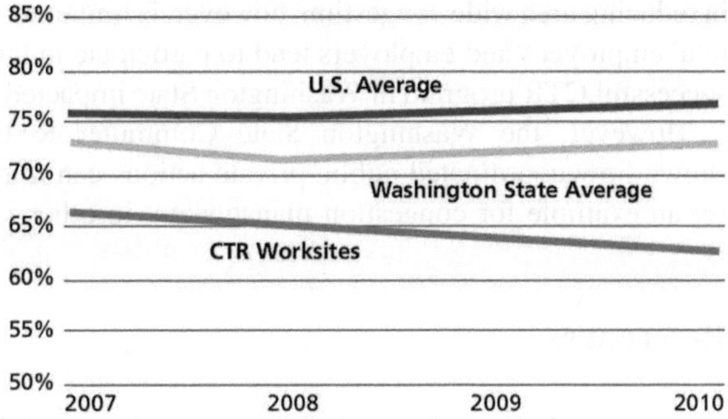

Fig. 21.2 Comparison of drive-alone rates for commute trips [15]. *Source* Reference [15], p 31, Fig. 2 © National Academy of Sciences, Washington, DC. Reproduced with permission of the Transportation Research Board.

Table 21.8 Comparison of employment at CTR worksites and Washington state [15]

	Employment at CTR worksites	Percent change	Daily vehicle trips to CTR worksites	Percent change	Employment in Washington state	Percent change
2007	513,720		356,861		3,154,787	
2010	528,113	+2.8	358,086	+0.3	3,014,335	−4.7

Source Reference [15], p 31, Table 2. © National Academy of Sciences, Washington, DC. Reproduced with permission of the Transportation Research Board.

less demand on the transportation system than expected and reducing the need for additional state investments in highway capacity (Table 21.8).

By averting vehicle trip growth associated with increased employment, CTR participants reduced VMT by 160 million each year between 2007 and 2010. This saved approximately 71,500 metric tons of GHG emissions annually—an equivalent of burning 389 rail cars of coal—and approximately 8 million gallons of fuel each year. CTR commuters saved more than $22 million in fuel expenses in 2010.

Washington State DOT has elevated demand management to one of the state's 3 primary investments in transportation solutions—joining efficient operations and strategic capacity additions.

Although CTR has demonstrated strong performance within its target markets, the program includes only 6 % of the VMT in the state.

21.5 Conclusions

This chapter has demonstrated how various employer and institutional strategies can reduce highway travel demand. It describes and assesses the various actions taken to reduce the number of single occupant cars used for commuting to and from work. It also shows how work schedule changes can reduce peak hour traffic and transit volumes.

Some applications can be effective at the local level as was the case in Lower Manhattan in the 1970s when peak hour transit crush loads were reduced. Their effect in reducing area wide congestion, however, is limited if only a small percentage of the total employees and employers tend to participate in their implementation. Even the successful CTR program in Washington State impacted only 6 % of the state's VMT.

However, the Washington State Commuter Reduction Program (CTR) has shown how coordinated public-private actions can mitigate congestion, and should set an example for congestion management in other large metropolitan areas.

References

1. Kuzmyac JR, Evans JE, Pratt R (2001) Transit Cooperative Research Program (TCRP) Report 95: traveler response to transportation system changes, chapter 19—employer and institutional TDM strategies. Transportation Research Board, Washington, DC
2. Keyani BI, Putnam ES (1976) Transportation system management: state of the art. Prepared by INTERPLAN corporation for urban mass transportation administration. US Department of Transportation, s.1. September 1976
3. Port Authority of New York and New Jersey, Planning and Development Department (1975) Flexible work hours experiment at the port authority of New York and New Jersey, 1974–1975. Urban mass transportation administration, US Department of Transportation
4. Safavian R, McLean KG (1975) Variable work hours—who benefits. Traffic Eng 45(3):17
5. Giuliano G (1992) Transportation demand management: promise or panacea. APA J 58 (3):327–335
6. Berald S (1990) Bishop ranch 1990 transportation survey. Bay Area Rides, Oakland (December 1990)
7. Cambridge Systematics, Inc (1981) Denver federal employee work week experiment—evaluation of transportation related impacts. Prepared for denver regional council of government, sl September 1981
8. Freas AM, Anderson SM (1991) Effect of variable hours work programs on ridesharing and organizational effectiveness: a case study, Ventura county. Transportation Research Record (TRR) 1321
9. Association for Commuter Transportation, Urbantrans Consultants, PB Parsons Brinckerhoff, and ESTE (2004) Mitigating congestion: the role of demand-side strategies. In partnership with the federal highway administration. US Department of Transportation, sl
10. Zhou L, Winter OL (2008) Empirical analysis of compressed workweek choices. Transportation Research Record (TRR) 2046
11. Rathbone NB (1992) Telecommuting in the United States. ITE J 62(12), December 1992
12. Walls M, Safirova E, Jiang Y (2000) What drives telecommuting's relative impacts of worker demographics, employer characteristics, and job types. Transportation Reseach Record (TRR) 2010
13. Vehicle Technologies Program: Fact #486: September 10, 2007 Teleworking Trends. sl http://www1.eere.energy.gov/vehiclesandfuels/facts/2007_fcvt_fotw486.html
14. Telework (2011) A worldatwork special report. Worldatwork, sl June 2011
15. Cotton K, Johnston K, Leotta K, Stark S (2012) Washington state's commute trip reduction program: reducing emissions and growing the economy by managing transportation demand. TR news 281. July–August 2012
16. Rutherford GS, Bacigett SI, Ishirnaru JM, Maclachlan S (1994) Transportation demand management: case studies of medium-sized employers. Transportation Research Record (TRR) 1459
17. Harrison F, Jones D, and Jovanis P (1979) Flextime and comntuting behcrior in San Francisco: some preliminary findings. Summary Report, Institute of Transportation Studies, University of California, Berkeley, CA

Chapter 22
Indirect Demand Strategies—Parking Supply and Price

22.1 Introduction

Parking supply, location, and price play an important role in managing travel demand and in relieving traffic congestion. Too few parking spaces can result in extra travel while searching for a parking space, increase street congestion, inhibit access to activities, and discourage new development. On the other hand, too many parking spaces (especially in the city center) can attract motorists and undercut transit use.

Similarly when parking space is underpriced, commuter parking is encouraged and parking revenues may not cover debt service costs. When parking is overpriced motorists might go elsewhere to shop, work or conduct business.

This chapter shows how parking supply and price can be managed to improve accessibility and help reduce congestion. Its findings and analysis are drawn from a broad range of sources. Key documents include:

- Chapters 3, 13, and 18 of TRCP Report 95 [1–3]
- Parking published by the ENO Foundation [4]
- An article describing changing perspectives related to Parking and Traffic Congestion [5]
- Parking and Pricing Implementation Guidelines, prepared by the Victoria Transport Policy Institute [6]
- The High Cost of Free Parking [7]

J.C. Falcocchio and H.S. Levinson, *Road Traffic Congestion: A Concise Guide*,
Springer Tracts on Transportation and Traffic 7,
DOI 10.1007/978-3-319-15165-6_22

22.2 Parking Supply Management

22.2.1 Changing Perspectives

The "parking problem" has persisted since the beginning of the Twentieth Century. It has changed in character, location, and scale. As urban areas expanded, motor vehicle use increased, environmental concerns emerged, and public attitudes changed. Figure 22.1 summarizes these various changes.

Once closely related to downtown access and congestion, parking concerns have spread throughout the urban region. Adequate parking supply, location, and price have become tools to foster economic development, influence mode choice, reduce traffic congestion, and improve air quality.

Eliminating and controlling curb parking in conjunction with increasing the supply of off-street parking helped city centers adapt to major increases in traffic demand and congestion during the first 40 years of automobile use. Beginning in the late 1940s, the provision and control of parking was used to address new issues of urban economic preservation and revitalization, as well as traffic congestion. Interstate highways and urban expressways offered new opportunities for the joint development of parking and highways.

In the 1960s, public opposition to urban expressway construction grew, and strong political support emerged for improving transit access to downtowns. Expanded park-and-ride facilities at outlying transit stations were used to encourage

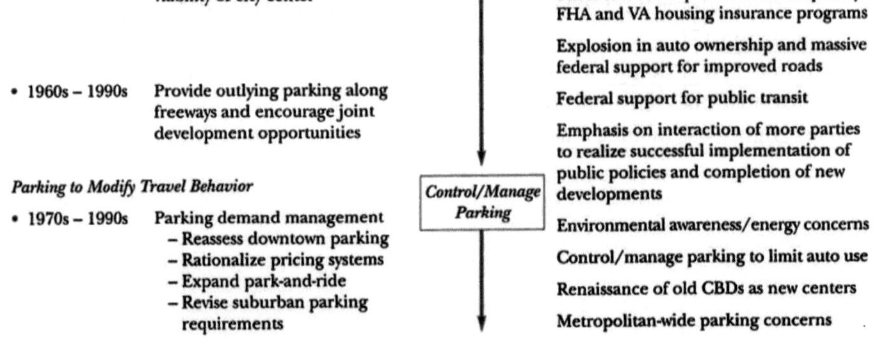

Fig. 22.1 Summary of changes in parking perspectives relating to traffic congestion. *Source* Reference [5]

commuters who drove downtown to switch to transit. By the early 1980s, the effective management of parking was recognized an effective strategy to achieve travelers and community objectives as well as manage congestion [5].

22.2.2 Objectives of Parking Management

Parking management, as defined in this chapter, is the comprehensive control of the quantity, location, price, and availability of parking. Parking supply and demand strategies include (1) limiting and controlling parking supply, (2) expanding park and ride facilities for transit and carpool riders, (3) providing preferential parking spaces for high occupancy vehicles, and (4) establishing pricing incentives. An acceptable mix of these strategies can help reduce congestion by encouraging peak period commuters to use transit or ride share.

22.2.3 Limiting Parking Supply

Historically, municipal parking codes established minimum parking ratios (number of spaces per unit of development) to ensure sufficient off-street parking, accommodate a site's needs and avoid spill back onto public streets or neighboring sites. Typical values were codified in various "parking generation" guides published by public agencies and the Institute of Transportation Engineers. More recently, parking requirements are sometimes framed as the maximum allowable parking rates, rather than minimum rates.

Transit oriented cities have increasingly recognized that a large supply of commuter parking in central areas has a negative impact on transit ridership and contributes to downtown congestion.

Analysis of modal choices in eight major Canadian cities, for example, found a strong inverse relationship between downtown transit use and employee parking supply [8]. Table 22.1 shows that A.M. peak hour transit share increases as the number of parking spaces per employee decreases.

A non-linear regression model found an r-square value of 0.92 for the equation explaining transit use to US and Canadian CBDs as a function of downtown parking supply per employee. But when US cities were included the relationship was much weaker yielding an r-square was 0.59 [8].

Accordingly, since the 1970s some larger cities in the United States and Canada have limited their downtown parking supply as a means of reinforcing transit ridership and reducing automobile use. The share of automobile trips was reduced by establishing ceilings on the parking supply and/or by modifying zoning requirements. These parking supply restraints work in a compact and densely developed CBD that is well served by public transit, and where there is the need to limit traffic congestion.

Table 22.1 Relationship between downtown parking supply and transit use in Canadian cities

City	CBD share of area employment (%)	CBD office space (1,000 ft^2)	Parking spaces per 1,000 ft^2	Parking spaces per CBD employee	Park and ride spaces per CBD employee	AM peak hour CBD transit share (%)
Saskatoon	20.7	3,600	3.5	0.79	–	14.6
Edmonton	20.2	15,133	2.1	0.51	0.029	32.0
Calgary	23.4	31,493	1.3	0.46	0.084	38.8
Montreal	14.9	87,996	1.0	0.38	0.270	48.7
Winnipeg	26.1	17,478	1.4	0.36	–	39.7
Vancouver	16.3	n/a	n/a	0.29	0.034	46.0
Toronto	25.37	61,570	1.5	0.29	0.122	64.1
Ottawa	31.7	21,024	1.1	0.28	0.008	48.8

Listed in order of decreasing ratios of long-term parking spaces per CBD employee
Source References [3], pp 17–18, Table 18.3 and [8]

Some congested suburban activity centers also have limited their parking supply as part of measures that contain traffic congestion and improve air quality. Downtown Bellevue, Washington, for example, with a 24,000 employees established a maximum of 2.4 spaces per 1,000 ft^2 of office building space. This has resulted in a reported 18 % reduction in vehicle trips [3].

Although a common practice is to set a ceiling (or limit) on the parking space supply in the city center, an alternative policy is to establish minimum and maximum zoning requirements for various land uses. Zoning requirements would reflect development density, level of transit service available in the area, and any restrictions on automobile access. An illustrative example of this mini-max strategy is shown in Table 22.2 [9]. This strategy can be modified to reflect the needs of any urban area.

Employees and institutions usually set their own parking supply requirements, including price and space incentives and disincentives. Sometimes priority parking is provided for high-occupancy vehicles. And some local jurisdictions restrict parking for non-residents in their neighborhoods.

22.2.3.1 Examples and Effects

Central business districts parking supply management is found in several large US and Canadian cities.

Boston instituted a downtown parking freeze in November, 1972 when additional expressway construction was stopped and a decision was made to improve transit access; New York City discouraged development of additional off-street parking in the Manhattan business district; Seattle limited the amount of downtown parking where there is good transit access; Portland set a ceiling of 40,000 spaces for its downtown parking supply; and San Francisco limits free-standing parking facilities in the city center.

Table 22.2 Suggested parking guidelines for major transit corridors

			Number of parking spaces per unit by distance from transit stop					
			0–500 ft		500–1,000 ft		1,000–1,500 ft	
Landuse	Activity	Unit	Minimum required	Maximum allowable	Minimum required	Maximum allowable	Minimum required	Maximum allowable
Residential	Single family	Housing unit	0.5	1.0	0.7	1.0	0.8	1.3
	Multi-family	Housing unit	0.4	1.0	0.6	1.0	0.8	1.3
Commercial	General office	Gross floor area, 1,000 ft^2	–	2.0	1.0	2.0	1.7	2.9
	Medical-dental office	Gross floor area, 1,000 ft^2	–	3.3	1.7	3.3	2.5	4.0
	Restaurant	Gross floor area, 1,000 ft^2	2.0	3.3	2.5	3.3	33	5.0
	Restaurant	Seats	–	0.17	0.17	0.25	0.17	0.25
	Hotel-Motel	Rental units	0.7	1.0	0.7	1.0	0.7	1.0
Industrial	Manufacturing, warehouse, wholesale	Employees	0.2	0.33	0.25	0.33	0.33	0.5
Institutional[a]	Auditorium	Seats	0.13	0.2	0.13	0.2	0.14	0.25
	Hospital	Beds	0.80	1.0	0.80	1.0	1.0	1.4
	Church	Seats	0.14	0.2	0.14	0.2	0.14	0.25
Educational	Elementary and Junior high school	Classroom and office	0.7	1.0	0.8	1.0	0.8	1.0
	Senior high school	Classroom and office	0.7[b]	1.0[d]	0.8[b]	1.0[d]	0.8[c]	1.0[e]
	College and university	Classroom and office	0.7[b]	1.0[d]	0.8[b]	1.0[d]	0.8[c]	1.0[e]

Source Reference [9], p 23

[a] Where public use of auditoria is likely, specific auditorium standards should apply

[b] Plus one space per 10–15 students, except where constrained by policy

[c] Plus one space per 8–10 students, except where constrained by policy

[d] Plus one space per 8–10 students, except where constrained by policy

[e] Plus one space per 5–8 students, except where constrained by policy

Examples of parking supply requirements in three cities are shown below.

- Portland had a limit of 0.7–1.0 spaces/1,000 ft^2 of office floor area. With this limit the transit share accounted for 48 % of all CBD arrivals. Before the limit this percentage was 45 %
- Seattle with a maximum rate of 1.0 space/1,000 ft^2 of office floor area; a minimum rate of 0.54 spaces/1,000 for areas with "good" transit service; and 0.75 spaces/1,000 for areas with moderate transit service.
- San Francisco has 45,000 parking spaces for 250,000 employees. No more than 7 % of buildings' gross floor area can be used for parking.

Table 22.3 compares the experiences of six US cities that have managed their downtown parking ratios.

22.2.3.2 Complementary Actions

Managing parking supply to reduce congestion in city centers and in other major activity centers requires several complementary actions. These actions include establishing time restrictions on the availability of on-street parking and implementing residential permit programs.

In addition, locating peripheral parking facilities adjacent to downtown areas can reduce the number of vehicles searching for parking spaces on downtown streets. Where these lots and garages are located within one mile of the CBD, most parkers can walk to their CBD destinations. A good example is Chicago's Grant Park lots and garages that are located just east of Chicago's Loop. These facilities have direct access to Lake Shore Drive and Michigan Avenue. But like other peripheral parking facilities they do not reduce Chicago's peak hour radial freeway congestion.

22.2.4 Managing Suburban Parking

Expanded commercial and residential developments in suburban areas have increased suburban traffic congestion. Abundant, relatively inexpensive land fostered low density spread out development for nearly 50 years, creating auto-oriented suburbs. Early suburban developments were characterized by large tracts of single family houses, planned regional shopping centers, and industrial parks. The rapid suburban growth since the 1980s included large office parks and mega centers. Suburban developments serve single purposes (e.g., residential, retail, offices, etc.) and are surrounded by large parking areas mandated by zoning requirements. Often streets are inadequate to serve the concentrated demand volumes; and transit service is of poor quality. These conditions result into severe peak-hour suburban traffic congestion.

The amount, location, and design of parking facilities have a strong influence on suburban traffic congestion. Traffic volumes are usually keyed to number of parking

Table 22.3 Experiences in six cities with managing parking ratios, Circa 1988

City	Downtown employment (population)	Parking supply	CBD parking rates	CBD parking policy	CBD traffic mitigation	Key findings
Los Angeles, CA	200,000	127,000 spaces off-street (projected)	up to \$0.50/h off-street	Minimum ratios (spaces per 1,000 ft^2): 2–3 citywide, reduced to 1.0 in CBD	Employer trip reduction plans required under Regulation XV (Southern CA air quality mandate)	Peripheral parking options not exercised by developers
	(SMSA 3.3 million)	(81,300 open to public, with 5,000 on-street)	up to \$l/h on-street	Developers in certain areas can provide up to 75 % of parking at remote locations	Trip reduction plan average vehicle employee ridership (AVR) target of 1.75 persons/vehicle	Many traffic mitigation plans not very good or effective
				If project >100,000 ft^2 in certain areas, *must* provide 25–40 % of parking off-site		CBD employee mode share 25 % transit, 60 % drive alone
Denver, CO	118,000	71,000 greater-CBD spaces, mostly private off-street (open to public)	\$60–80/mo. \$0.50/half h	Minimum ratio (spaces per 1,000 ft^2) for office 2.0 citywide, except no CBD maximum or minimum	Essentially none—note that parking requirements *not* set to encourage transit or ridesharing	Some developers provide as little as 0.25 spaces per 1,000 ft^2 in CBD
	(Denver 491,000)		off-street, public	Peripheral parking allowed as alternative	Price breaks for car and vanpools in certain city facilities	Transit mode share 28 % for core area employees; 13 % for greater downtown
	(SMSA 1.6 million)		\$0.20–\$1/h on-street	Density bonuses for CBD parking *above* 70 % of non-CBD ratio		Peripheral parking provided, but little used and hard to monitor

(continued)

Table 22.3 (continued)

City	Downtown employment (population)	Parking supply	CBD parking rates	CBD parking policy	CBD traffic mitigation	Key findings
Hartford, CT	90,000	21,000 spaces, 12,7000 open to public	\$120–180/ mo. garage	CBD Minimum ratio (spaces per 1,000 ft^2) for office 1.0, new CBD office parking must be underground	TMPs required for CBD developments, additional requirements for projects impacting state hwys	Parking ratio incentives and peripheral parking options not used
		(2,700 public on-street)	\$50–75/mo. Surface lot	30 % reduction for demand management	Rideshare Company encourages transit and ridesharing	City peripheral parking lot not well used
			\$0.50/h on-street	Peripheral parking allowed		SOV use to CBD 55 % and increasing
Los Angeles, CA	200,000	127,000 spaces off-street (projected)	up to \$0.50/h off-street	Minimum ratio (spaces per 1,000 ft^2) 2–3 city-wide, reduced to 1.0 in CBD	Employer trip reduction plans required under regulation XV [Southern CA air quality mandate]	Peripheral parking options not exercised by developers
	(SMSA 3.3 million)	(81,3000 open to public, with 5,000 on-street)	up to \$1 /h on-street	Developers in certain areas can provide up to 75 % of parking at remote locations	Trip reduction plan average vehicle employee ridership (AVR) target of 1.75 persons /vehicle	Many traffic mitigation plans not very good or effective
				If project >100,000 ft^2 in certain areas, *must* provide 25–40 % of parking off-site		CBD employee mode share 25 % transit, 60 % drive alone

(continued)

Table 22.3 (continued)

City	Downtown employment (population)	Parking supply	CBD parking rates	CBD parking policy	CBD traffic mitigation	Key findings
Denver CO	118,000	71,000 greater-CBD spaces, mostly private off-street (open to public)	\$60–80/mole \$0.50/half h	Minimum ratio (spaces per 1,000 ft^2) for office 2.0 citywide, except no CBD maximum or minimum	Essentially none—note that parking requirements *not* set to encourage transit or ridesharing	Some developers provide as little as 0.25 spaces per 1,000 ft^2 in CBD
	(Denver 491,000)		off-street, public	Peripheral parking allowed as alternative	Price breaks for car and vanpools in certain city facilities	Transit mode share 28 % for core area employees; 13 % for greater downtown
	(SMSA 1.6 million)		\$0.20–\$1/h on-street	Density bonuses for CBD parking *above* 70 % of non- CBD ratio		Peripheral parking provided, but little used and hard to monitor
Hartford, CT	90,000	21,000 spaces, 12,7000 open to public	\$120–180/ mo. garage	CBD Minimum ratio (spaces per 1,000 ft^2) for office 1.0, new CBD office parking must be underground	TMPs required for CBD developments, additional requirements for projects impacting state hwys	Parking ratio incentives and peripheral parking options not used
		(2,700 public on-street)	\$50–75/mo. Surface lot	30 % reduction for demand management	Rideshare Company encourages transit and ridesharing	City peripheral parking lot not well used
			\$0.50/h on-street	Peripheral parking allowed		SOV use to CBD 55 % and increasing

Population is that within the city limits of Central City unless otherwise indicated.
SMSA Standard Metropolitan Statistical Area
Source Reference [3], pp 18–23, Table 18.2

spaces provided to serve specific land use activities. Thus the provision of large amounts of parking at major suburban traffic generators concentrates traffic demand which creates traffic congestion. There is, however, a growing emphasis on adopting parking reforms and developing land use sites and design that foster shared parking, ridesharing and/or transit use. For example, up to two thirds of the parking space serving office buildings is empty in the evening peak hour.

Accordingly, attitudes toward providing suburban parking are beginning to change. Minimum parking requirements sometimes have produced an oversupply of parking spaces and are a disincentive for ridesharing, transit, or other commute alternatives work in the suburbs [10].

A few communities have reduced their parking requirements when they initiated as part of ridesharing or TSM programs [11]. Examples include Palo Alto, CA; Orlando and St Petersburg, FL; Montgomery County, MD; and Bellevue Washington. But these actions have been not been significantly replicated elsewhere.

22.3 Park-and-Ride

Limiting and managing downtown parking supply can effectively reduce traffic demands and traffic congestion when complemented with remote park-and-ride facilities along express transit lines. This strategy helps to keep downtown parking and traffic demands constant—even as the downtown employment grows.

- It reduces commuter traffic volumes on heavily traveled and often congested radial freeways and arterial roads
- It enables city center employment to grow without corresponding increases in downtown street congestion
- It builds transit ridership for downtown destinations from travelers that live in low density areas (where transit is not sustainable)
- It permits wider station spacing on express transit lines in outlying areas where there is usually little walk access to transit station

Remote parking facilities ("park- and-ride" and "park-and-pool") provide a means of accommodating future central area transit demands—especially where downtown parking space is stabilized. They take the form of park-and-ride facilities at express transit stops, and park-and-pool facilities where transit service is limited. They range in size from a few hundred spaces at outlying park and pool lots to large multistoried garages serving commuter rail and rapid transit riders. In effect the downtown commuter parking demands are transferred to outlying locations. This parking intercept strategy, shown in Fig. 22.2, has several advantages in reducing congestion [5].

The number of park-and-ride travelers multiplied by their trip lengths on the train, bus or carpool, is a measure of the VMT removed from the roadways and, in turn, of reduced traffic congestion on these same roads. These savings are balanced versus the additional traffic concentrated on roads used to reach the car-and-ride/pool facilities.

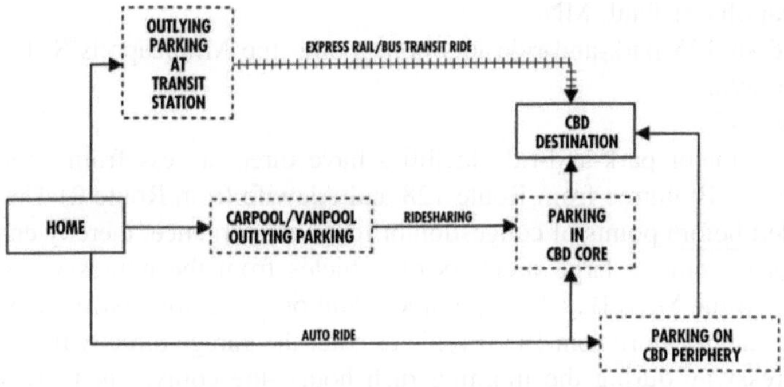

Fig. 22.2 Commuter parking public policy options. *Source* Reference [5]

In many metropolitan areas, the number of parking spaces at park-and-ride lots exceeds the parking supply in city centers. Parking facilities range from bus and carpool lots of less than 100 spaces, to garages of 3,000 spaces. And some of these facilities provide direct freeway access.

Park and ride lots for car poolers along express highways, while usually small in scale, enable group riding to employment centers.

Park-and-ride spaces can reduce radial freeway volumes as well as downtown street congestion by intercepting motorists in outlying areas. This results in fewer vehicle-miles of travel (VMT) and related air quality improvements. The potential VMT reduction of a perk-and-ride lot can be estimated by multiplying two times the park-and-ride spaces by the distance to the city center.

22.3.1 Examples

1. Connecticut Department of Transportation
   In addition to providing 17,000 spaces in 49 lots for Manhattan-bound rail commuters, the Connecticut Department of Transportation provides 17,600 spaces to 188 park-and-ride lots along freeways. The carpool lots reduce about 200,000 VMT daily—most in the peak hours. However, the effect of this reduction in VMT does not necessarily reduce the freeway's peak hour travel time as latent traffic demand will be attracted by the higher freeway speed made possible by the reduction in VMT. The park-and-ride lots benefit people accrues to (1) who switch to carpool and save on commuting costs, (2) employers who can reduce the number of parking spaces needed by their employees, and (3) traffic flow at intersections adjacent to parking facilities at the place of employment.
2. Houston TX
   Houston provides several large 1,000-space parking lots along freeways for people who carpool or ride buses. These park and ride lots have direct 'T' ramp connections to high occupancy vehicle (HOV) lanes that provide fast service to the downtown area.

3. Minneapolis-St Paul, MN
   More than 175 park-and-ride lots honeycomb the Minneapolis-St Paul metropolitan area.
4. Boston, MA
   Boston's major park-and-ride facilities have direct access from express highways (e.g., Braintree from Route 128 and Alewife from Route 2) The access is provided before points of congestion or route convergence, thereby enabling the garages to remove large numbers of vehicles from the expressways and city streets. At the Mass Bay Transportation Authority's 2,400-space Alewife transit station garage, more than 1,000 vehicles enter the garage directly from the Route 2 Expressway during the morning rush hour—the equivalent to two lanes of traffic on Cambridge streets.

22.3.2 Parking Planning Guidelines

Locating, selecting, planning and designing park-and-ride facilities calls for assessing and balancing many factors. Sites should be located where:

1. Land is available at affordable cost
2. Land is level and of suitable size and shape
3. Access is convenient from expressways and major roads
4. There is rapid and reliable transit service to the city center and other major activity concentrations
5. The access can be provide without increasing traffic congestion
6. Ridership potential is strong
7. Sites are secure
8. There is minimum negative environmental impact
9. The design is acceptable to the surrounding communities.

22.3.2.1 Location

Rapid transit park-and-ride facilities generally should be located at least 5–8 miles from the city center. They should be placed in *advance* of points of traffic congestion (bottlenecks)—especially where peak period freeway speeds are less than 30–35 mph. Their location should be perceived as safe and secure by parkers, pedestrians, and transit riders.

The park-and rode lots and garages should use land that is readily available, has good access for auto drivers and pedestrians, and minimizes environmental impacts.

Sites that are near junctions of radial transit lines and circumferential freeways can tap a wide catchment area. One such example is the large parking garage in suburban Boston that is located where Route I-95/128 interchanges with the MBTA-AMTRAK rail line.

The rapid transit service to the city center should be fast, frequent and reliable; headways should be 10–12 min or less during peak periods, and not more than 15 min midday. Headways of 20–30 min or less are desirable for commuter rail lines during rush hours.

22.3.2.2 Site Considerations

New facilities should be placed where they will not adversely affect the use of existing sites. Sites should be compatible with adjacent land uses. Site selection should give (in order of importance) priority to (1) land currently used for parking; (2) underdeveloped or unused land in public ownership; (3) undeveloped public land; and (4) developed private land.

Large park-and-ride facilities generally should not be located near town centers, or in densely developed areas.

Sites should be of suitable size and shape to serve anticipated demand, to permit efficient design of access, and of convenient passenger transfer facilities; and should allow easy parking maneuvers.

22.3.2.3 Facility Size

Very large or very small facilities should generally be avoided. Small lots do not have enough space to justify frequent transit service, and parkers would spill over to nearby areas. Very large facilities can result in long walking distances or possible under-utilization.

- Bus Rapid Transit (BRT) lots should contain at least 250 spaces. An optimum size range is 400–700 spaces—depending on demand
- Commuter rail and rail transit facilities usually range from 500 to more than 2,500 spaces. These facilities support frequent transit service and draw riders from a large catchment area. Larger garages and lots may create congestion on approach roads and require a grade-separated access road system.
- To accommodate daily fluctuations in park-and-ride demand, site occupancy should not exceed 95 % occupancy over a typical day. A design factor of 80 % occupancy is desirable to allow for ridership growth at the station

22.3.3 Travel Characteristics of Park-and-Ride Users

Examples of the types of transport used by park-and-ride patrons in their trips to their center city destinations are summarized below.

22.3.3.1 Bus Transit

As shown in Table 22.4, about half of all park-and-ride patrons previously drove alone to their ultimate destinations. Almost 75 % of park-and-ride arrivals drove alone, and most (97 %) traveled to work or business. 69 % of the lot users lived less than 6 miles away, and 63 % of the trips from the lot to destination were between 10 and 30 miles long [12].

Table 22.4 Overview of travel characteristics of park-and-ride users

Characteristics	Range	Number of lots[a]	Average[a]
Previous mode of travel			
Drove alone	11–65 %	305	49.2 %
Carpool/vanpool	5–28	303	23.2
Transit (bus or other)	5–49	304	10.4
Did not make trip	0–29	303	14.9
Arrival mode to facility			
Drove alone	38–91	146	72.6
Shared ride	3–36	146	11.0
Dropped off	0–31	117	11.1
Walked	0–21	132	4.4
Bus	0–10	132	1.3
Trip purpose			
Work or business	83–100	107	97.2
School	0–11	80	2.3
Other	0–17	80	0.5
Travel frequency (rd.-trlps/wk.)			
Three or less	2–15	101	6.6
Four	3–16	86	7.6
Five or more	71–93	86	86.8
Home-to-lot distances (miles)			
Three or less	6–74	163	46.4
Four to six	18–42	162	22.8
Six or more	8–69	162	29.2
Lot-to-destination distance (miles)			
Less than 10	0–100	190	6.9
10–30	0–100	190	63.2
30 or more	0–51	177	30.4

Source Reference [4]

[a] The “average” values shown are weighted by the number of park-and-ride lots surveyed. Partial or missing data from certain studies may cause the percentages not to total 100

22.3.3.2 Rail Transit

Studies of rail transit park-and-ride facilities show an average of 1.1 cars per space, and 1.2 transit boardings per parked car. Kiss-and-ride (passenger drop-off and pick-up) can represent 20–40 % of total peak hour transit station arrivals; median driving distances of 3–4 miles for park-and-ride patrons and 1–2 miles for kiss and ride are common. Occupancies frequently exceed 90 % of facility capacity.

22.3.3.3 Commuter Rail

Commuter rail lines usually rely on automobile access to stations [13]. As shown in Table 22.5, almost three quarters of all am peak period METRA (a commuter rail service in the Chicago metropolitan area) rail boarding bound for Chicago is by automobile—55 % drive-alone, 13 % dropped-off, and 6 % carpool. The parking space occupancy was approximately 86 %.

22.3.4 Supply and Use

Use of park-and-ride facilities varies by location and line-haul transit access mode. Usage is generally lowest for express and local bus lines, and highest for rail transit and commuter rail lines. Most facilities are well used and many are filled to

Table 22.5 METRA park-and-ride travel characteristics

System inbound (AM peak boarding)	
Station distance to CBD (miles)	Overall system
0–10	7 %
10–20	39 %
20–30	41 %
Over 30	13 %
Access modes to station	
Drive Alone	55 %
Walked	21 %
Dropped off	13 %
Carpool	6 %
Bus	4 %
Other	1 %
Station parking	
Capacity	68,301
Usage	58,882
Occupy	86 %

Source Derived from References [14], p 96. Exhibit 10.3 and [13]

capacity. Tables 22.6, 22.7 and 22.8 give the reported occupancies (utilization) for bus, HOV, and rail park-and-ride facilities.

- Occupancies (%capacity) of express and local bus lots range from 10 to 100 % (Table 22.6)
- Occupancies (%capacity) of HOV park-and-ride facilities range from 34 to 100 % (Table 22.7)
- Occupancies (%capacity) of rail park-and-ride facilities range from 26 to 100 % (Table 22.8)

Table 22.6 Examples of utilization of express and local bus park-and-ride facilities

System (year)	Number of lots	Number of spaces	Parked vehicles	Percent capacity (%)
Buffalo-Niagara Frontier (1995)	6	200	n/a	50
Calgary (1998)				
North	6	186	129	69
Southwest	2	103	103	100
Cincinnati Region (1998–1999)	25	2,089	1,296	62
Denver—Regional Transit District (1998)[a]	55	11,251	8,199	73
Des Moines Transit Authority (1995)	3	n/a	n/a	50–80
	1	n/a	n/a	10
Duluth Transit Authority (1995)	1	22	n/a	50
Houston-US 59 North/Eastex (1998)	3	2,418	1,130	47
Maryland—State Highway Admin. Lots (2003)	23	5,219	3,792	73
Sacramento (1989)	3	154	60	39
Santa Clara Valley Transportation Authority (1998)[a]	17	820	260	32
Seattle (1998)				
North District[a]	26	4,115	3,025	74
East District[a]	43	6,235	4,964	80
South District[a]	57	7,536	5,554	74
Miami Valley (Ohio) Regional Transit Authority (1995)	26	960	n/a	75
Metro-Dade (1993)	8	1,767	883	50
Norfolk-Tidewater Transport District (1995)	5	700	400	57
San Diego (1998)	31	2,125	850	40

n/a—Information not available except by inference based on the "Percent Capacity" values, which come from estimates or other derivations used by the reporting agencies

Source Reference [1], Table 3.5

[a] Some buses operating out of these lots use the various HOV lanes and other priority treatments

Table 22.7 Examples of utilization of HOV park-and-ride facilities

HOV system (year)	Number of lots	Number of spaces	Parked vehicles	Percent capacity (%)
Houston				
I-45 North (1998)	5	7,386	3,643	49
US 290/Northwest (1998)	3	3,852	2,069	54
I-10 West/Katy (1998)	3	4,525	2,764	61
US 59 South/Southwest (1998)	8	7,308	2,481	34
I-45 South/Gulf (1998)	3	3,018	1,694	56
Los Angeles-San Bernardino Transitway				
El Monte Station (1994)	1	2,100	2,100	100
Minneapolis-I-394 (1993)	7	936	558	60
Seattle-I-5 North/Community Transit (1998)	18	4,200	n/a	89[a] 91[b]

n/a—Information not available except by inference based on the "Percent Capacity" values, which come from estimates or other derivations used by the reporting agencies
Source Reference [1], pp 3–20, Table 3.4
[a] Major lots
[b] Minor lots

22.3.5 Relation of Park-and-Ride Size to Boarding Passengers

The number of boarding passengers per parking space (and spaces provided per boarding passenger) at rail transit and commuter rail stations in selected cities are shown in Tables 22.9, 22.10 and 22.11.

- Table 22.9 shows system spaces and boardings for rapid transit (heavy rail) and light rail
- Table 22.10 shows system spaces and boardings for commuter lines
- Table 22.11 shows parking spaces and boardings for a sampling of stations in seven cities

It is clear from these exhibits that park-and-ride spaces are a sizable number. The park-and ride spaces in several urban areas equal or exceed the downtown parking space supply.

The variability in these relationships comes from the different development densities around the stations service areas and the availability of alternative modes for station access.

- At rail rapid transit and light-rail transit stations that rely on auto access, there are about two to three boarding passengers per parking space

- Outlying commuter rail stations typically have about two passenger boarding per space
- Stations experiencing an excess of three boarding per space, are those typically located in areas of higher density with transit and walk access (Table 22.11).

Table 22.8 Examples of utilization of rail park-and-ride facilities

System (year)	Number of facilities	Number of spaces	Parked vehicles	Percent capacity (%)
Commuter rail				
Caltrain (1998)	34	4,125	3,210	78
Connecticut–New Haven Line(s) (1996)	35[a]	14,258	12,056	85
Go Transit–Toronto (1998)	8	32,052	30,139	94
MARC–Maryland/West Virginia (1995)	26	5,922	5,150	87
METROLINK–Los Angeles (1999)	46	14,500	n/a	75
Sound Transit–Puget Sound, Washington (2010)	10[b]	5,982	5,264	88
TriMet–Portland, Oregon (2010)	4[c]	699	280	40
Virginia Railway Express (1995)	13[d]	3,901	2,411	62
Heavy rail				
Chicago Transit Authority (1998)	15[a]	6,506	51–5,500	78–85
Metrorail–Miami (1993)	17	9,391	5,030	53
Metrorail–Washington, DC (1995)	39[a]	38,137	34,195	90
Southeastern PA Transp. Authority (1993)	3[a]	1,133	1,133	100
Light rail				
Buffalo (1995)	2	1,400	n/a	70
Calgary (1998)	11	7,354	7,126	97
Dallas Area Rapid Transit (1998)	8	4,190	n/a	86
Denver (2009)	20	11,739	8,517	73
Sacramento (1999)	9	4,120	n/a	55
San Diego Trolley (1999)	23	5,553	1,471	26
Santa Clara Valley Transp. Authority (2009)	21	6,471	1,700	26
TriMet–Portland, Oregon (2010)	23	9,606	5,261	55

n/a Information not available except by inference based on the "Percent Capacity" values, which come from estimates or other derivations used by the reporting agencies

Source Derived from [14], p 97, Exhibit 10.4)

[a] Parking fee charged at several or all facilities

[b] South Sounder line, includes adjacent and satellite lots

[c] Includes the parking facility operated by the City of Wilsonville

[d] Parking fee charged at several facilities in the survey year (fees since removed)

Table 22.9 Parking spaces and passenger boarding for rapid transit and light rail transit lines (stations with parking in selected cities)

City	Year	Parking spaces	Number of stations	Parking spaces per boarding passenger	Boardings per space
Heavy rail transit					
Atlanta	1990	17,700	9	0.1–0.4	2.3–13.6
Boston	2005/2006	17,500	15	0.1–0.5	1.8–8.3
Chicago	2000/2005	6,700	10	0.1–0.3	3.3–12.3
Cleveland	2005/2006	4,000	10	0.1–0.9	1.1–12.3
San Francisco	2003	47,100	29	0.1–1.1	0.8–10.2
Washington, DC	2000	58,200	33	0.1–0.7	1.5–16.9
Light rail transit					
Boston	2005/2006	2,000	6	0.1–0.7	1.5–15.0
Cleveland	2005	820	1	1.2	0.9
Portland	2006	7,000	17	0.1–0.8	1.2–6.7

Source Derived from Reference [14], p 98, Exhibit 10.5

Table 22.10 Parking spaces and passenger boardings at selected commuter rail systems

System	Spaces	Daily boardings	Spaces per boarding	Boardings per space
Boston (MBTA)—(2005/6)				
North Station	10,418	24,738	0.4	2.4
South Station	21,758	43,879	0.5	2.0
Chicago (Metra)—(2002)	85,563	149,187	0.6	1.7
Toronto (Go Transit)—2006[a]	27,180	46,670	0.6	1.7

Source Derived from Reference [14], p 98, Exhibit 10.6)

[a] Sample of system

22.3.5.1 Commuter Rail Ridership Effects of Additional Park-and-Ride Spaces

Adding park-and-ride spaces at commuter rail stations can increase rail ridership beyond that anticipated from normal growth in travel demand. These increases were quantified in a 2000 study [15] that looked at the ridership changes resulting from adding parking spaces to the New Haven, Bridgeport, and South Norwalk stations on Metro North's New Haven Line. The ridership results are shown in Table 22.12.

Table 22.11 Parking provisions at selected rail transit stations

Region	Location	Boarding passengers per weekday	Off-street Parking spaces	Boarding passengers/ parking space
Atlanta	Avondale	9,700	1,180	8.2
	Eastlake	2,800	610	4.6
	Hightower	10,300	1,400	7.4
	Chamblee	8,000	1,520	5.3
	Brookhaven	4,200	1,700	2.5
	Lenox	10,900	800	13.6
	Lindbergh	11,100	1,470	7.6
	Lakewood	4,300	1,900	2.3
	College Park	7,700	2,120	3.6
Boston	Wollaston	2,700	500	5.4
	North Quincy	2,400	800	3.0
	Quincy Center	7,500	930	8.1
	Commuter Rail—North[a]	11,000	3,360	3.3
	Commuter Rail—South[a]	3,800	2,640	1.4
Chicago	Ashland	4,750	264	18.0
	Cicero-Berwyn	2,700	360	7.5
	Cumberland	5,500	828	6.6
	Dempster	3,200	594	5.4
	Desplaines	4,750	596	8.0
	Howard	9,600	300	32.0
	Kimball	4,100	180	22.8
	Linden	3,500	456	5.5
	River Road	3,900	747	5.3
Cleveland	West Side (5 stations)	20,000	6,400	3.1
	East Side (3 stations)	10,000	900	11.1
Philadelphia	Bucks County[a]	4,000	1,800	2.2
	Chester County[a]	3,900	1,100	3.5
	Delaware County[a]	15,500	2,200	7.0
	Montgomery County[a]	19,500	4,300	4.5
	Lindenwold (New Jersey)	20,000	9,000	2.2

(continued)

Table 22.11 (continued)

Region	Location	Boarding passengers per weekday	Off-street Parking spaces	Boarding passengers/ parking space
San Francisco	Concord line (6 stations)	20,360	6,555	3.1
	Richmond line (5 stations)	9,130	3,381	2.7
	Alameda line (8 stations)	27,100	7,562	3.6
	Oakland line (3 stations)	7,300	1,087	6.7
	Daly City	8,860	1,877	4.7
Toronto	Islington	23,500	1,300	18.0
	Warden	24,600	1,500	16.4

Source Reference [4]
[a] Commuter railroad stations

Table 22.12 Changes in parking supply and demand at three Connecticut stations

	New Haven	South Norwalk	Bridgeport
Time period studied	1985–1999	1996–1999	1985–1999
Parking spaces added	+628	+325	+500
Additional rail ridership			
Gross ridership increase	+467	+250	+736
Ordinary growth (estimated 1.5 %/year)	+400	+55	+277
Ridership increase attributed to mode Shifts induced by parking ("New Riders")	+67	+195	+459
Additional rail ridership per parking space added			
Gross ridership increase/space added	0.74	0.77	1.47
"New Riders"/space added	0.11	0.60	0.92

External factors affecting Bridgeport included lowered train fares, free parking at State lot, and station area improvements
Source References [1], pp 3–12, Table 3.2 and [15]

The study estimated the number of new riders that would be attracted to the station by adding park-and-ride spaces. An estimate of ordinary ridership growth was subtracted from the total ridership growth, isolating the number of new riders then presumed to be attracted to commuter rail by the added parking. The Bridgeport station experienced a number of other changes during the study period, including a rail fare reduction, free parking, and a substantial improvement and cleanup at the station and its environs. The overall ridership increase per space was 0.74 in New Haven, 0.77 in South Norwalk, and 1.47 in Bridgeport. The "new riders" per parking space added were 0.11 in New Haven, 0.60 in South Norwalk, and 0.92 in Bridgeport.

22.4 Pricing Parking Space

22.4.1 Introduction

Pricing parking has been a long-established method for managing demand and generating revenue. It is common in central business districts and in major outlying employment concentrations such as medical centers and universities. Parking charges normally cover capital and operating costs.

Since the 1970s pricing also has become a major demand management action. Pricing can influence mode choice and traffic congestion.

22.4.2 Purpose

Parking facilities have been priced for many years to

- Increase turnover of the most convenient spaces
- Encourage the longer-duration parkers to use less convenient spaces
- Help reduce traffic volumes in congested areas
- Generate revenues by requiring parkers to pay their share of capital and operating costs

Pricing, along with parking supply management, can modify travel demands and reduce congestion by

- charging parkers who previously had free parking provided by employers
- charging parkers at major activity concentrations that are located in areas with good transit service
- increasing rates for all-day and monthly parkers, especially where rapid transit services are available
- increase availability of off-street parking space in congested parts of the city

22.4.3 Types of Parking Charges

The common types of pricing policies to manage parking demand include (1) high fees for commuters, (2) discounts for ridesharing, (3) "cash out" programs for employers, (4) parking taxes and transportation allowances. Some examples of various parking pricing policies (by type of facility) drawn from available literature are presented in Table 22.13. Their application could lead to reduced congestion.

Table 22.13 Examples of cost-related parking policy mechanisms

Type of facility	Applicable policy mechanism
Municipal parking facilities	
Off-street	• Increase cost per hour directly or in-directly via parking taxes (flat or selective rate based on location, day of week, or elapsed time) • Impose surcharges for all-day (over 4-h) use
On-street	• Increase meter cost per hour
Private commercial lots	• Impose parking taxes • Reduce or remove parking subsidies • Impose surcharges (flat or selective)
Free parking spaces provided by	
Employers for staff	• Impose a surcharge on employees for single-passenger vehicle parking; reduce to half for two-passenger car-pools and to zero for carpools with three or more members; subsidize transit-riding employees
Businesses for customers	• Imposed a surcharge on owner equivalent to estimated surcharge proceeds from commercial lots • Reduce or remove parking subsidies

Source Reference [16]

22.4.4 Price Elasticity

Parking demand price elasticity for individual employment sites and location generally ranges from −0.1 to −0.3. Thus there is up to 3 % reduction in parking use for a 10 % increase in parking fees [6].

A further discussion of pricing strategies to reduce VMT and their reported effectiveness follows.

22.4.5 Reducing Employer Subsidies

One way of reducing congestion is to reduce or eliminate free parking for commuter driver trips. This is sometimes done by "cashing out" the employee subsidies. The effectiveness of individual programs depends upon the availability of transportation alternatives and the enactment of local regulations to prevent drivers from parking elsewhere in the immediate area.

Results of a 1986 survey of free and subsidized parking in various parts of downtown Los Angeles are shown in Table 22.14. In all cases car use was reduced and transit use was increased when parking subsidies were eliminated [17].

California's "parking cash out" legislation, enacted in 1992, requires employers to offer employees the option to choose cash in lieu of any kind of parking subsidy. The legislation applied only to large employers located in air quality non-attainment areas who subsidized parking costs, and who could reduce the number of parking spaces leased for their employees without penalty.

Table 22.14 Sensitivity of mode share to parking subsidy policy, by Los Angeles subarea

	Location and average parking price	Mode share percentage					
		Financial core ($121)	Bunker hill ($100)	Civic center ($84)	Broadway —spring ($73)	South park ($59)	Entire study area ($85)
All employers	SOV (%)	62	70	60	39	67	61
	HOV	12	11	22	16	18	15
	Transit	25	16	17	40	15	22
	(# cases)	(870)	(1,314)	(2,225)	(448)	(155)	(5,012)
Free parking (subsidized)	SOV (%)	67	85	65	73	68	71
	HOV	10	5	18	27	21	13
	Transit	22	5	17	0	11	13
	(# cases)	(216)	(74)	(418)	(4)	(27)	(739)
No subsidies	SOV (%)	56	42	51	39	77	54
	HOV	7	14	28	0	11	8
	Transit	35	45	20	61	11	36
	(# cases)	(72)	(268)	(126)	(22)	(18)	(506)

The number of cases reported is for all modes in that subarea, unweighted. The mode share percentages use weighted survey responses. Parking costs are derived from a 1986 market survey. Parking subsidy characteristics were estimated from survey data
Source References [2], pp 13–20, Table 13.3 and [22]

The effects of "cash out" parking pricing on solo-drivers mode share are shown in Tables 22.15 and 22.16. Overall when the employer pays for parking more employees drive to work. But when the employer gives cash to employees who are free to use it or not for parking, fewer employees chose to drive and use the cash for other benefits. The net effect of the cash out policy resulted in fewer parking spaces needed to serve the same number of employees at the seven employment locations —from 0.72 spaces per employee to 0.53 spaces per employee [18].

The "before and after" impacts for eight selected employers who complied with this legislation are summarized in Table 22.16. After implementation of parking cash out the drive-alone mode share fell by about 15 % and the car pool share increased by about 55 % fell by about 12 and 7 % points, respectively; while the transit, walking and bicycle shares increased by 43, 48, and 27 % respectively. slightly. These changes reduced the annual vehicle trips per employee and the VMT per employee by 11 and 12 % respectively. As a result of the parking cash out program the average commuting subsidy per employee rose by only $2.00 per month [19].

Table 22.15 Case studies of parking pricing effects at seven employment locations

	Solo driver mode share			Cars per 100 employees			
Location, date (type of case study)	Employer pays for parking (%)	Driver pays for parking (%)	Difference (%)	Employer pays for parking	Driver pays for parking	Difference	Price elasticity of parking demand
1. Civic Center, Los Angeles, 1969 (with/without)	72	40	−32	78	50	−28	−0.22
2. Downtown Ottawa, Canada, 1978 (before/after)	35	28	−7	39	62	−7	−0.10
3. Century City, Los Angeles, 1980 (with/without)	92	75	−17	94	80	−14	−0.08
4. Mid-Wilshire, Los Angeles, 1984 (before/after)	42	8	−34	48	30	−18	−0.23
5. Warner Center, Los Angeles, 1989 (before/after)	90	46	−44	92	64	−28	−0.18
6. Washington, DC, 1991 (with/without)	72	50	−22	76	58	−18	−0.13
7. Downtown Los Angeles, 1991 (with/without)	69	48	−21	75	56	−19	−0.15
Average values	67	42	−25	72	53	−19	−0.15

"With/without" refers to a case study comparing the commuting behavior of employees with and without employer-paid parking. "Before/after" refers to a case study comparing the commuting behavior of employees before and after elimination of employer payment for parking. The estimated price elasticity of demand is the midpoint arc elasticity

Source Reference [2], pp 13–16, Table 13.10

Table 22.16 Average parking cash-out travel impacts for eight Southern California case studies

Measure	Average value		
	Before	After	Change
Drive alone mode share	76.8 %	65.3 %	−11.5 %-pts.
Carpool mode share	12.9 %	20.0 %	+7,1 %-pts.
Transit mode share	5.8 %	8.3 %	+2.5 %-pts.
Walk mode share	3.1 %	4.6 %	+1.5 %-pts.
Bicycle mode share	1.1 %	1.4 %	+0.2 %-pts.
Annual vehicle trips per employee	379	335	−11 %
Annual VMT per employee	5,348	4,697	−12 %

Analysis based on employee travel survey data—average of 8 case studies. Average one-way trip distance is 15 miles based on 1991 survey of commuters in the South Coast Air Basin
Source Reference [2], pp 13–18, Table 13.12

22.4.6 Pricing Downtown Parking for Employees and Shoppers

Pricing parking in central business districts should balance two differing objectives: (1) parking should be convenient and relatively inexpensive for shoppers and visitors—an essential requirement for city centers; (2) employee parking pricing should be high enough to discourage all day (and monthly) parking. This second objective is especially important where CBD employment exceeds 100,000 and where good public transport service is available (e.g., off-street rapid transit).

Thus the pricing of commuter parking will be different in Boise as compared with Boston; Dayton compared with Denver; and Cedar Rapids compared with Chicago.

From a congestion perspective, decreasing the parking supply and increasing the average parking price can reduce both SOV use and traffic congestion. But these actions require consistent and continuous enforcement of on-street parking and may not be acceptable to some communities.

To raise revenues and encourage commuter use of transit some agencies have considered (or established) a parking tax. However, experience in large US cities and in Melbourne, Australia, found that a significant proportion of drivers who contribute to the congestion problem had their parking costs paid by their employers. In addition, in the US for participating employers the IRS allows the amount their employees spend on commuter parking to be tax-free (as is the amount paid by those who commute by transit).

22.5 Congestion Relief Implications of Parking Policies

Providing adequate parking in center cities for workers and shoppers makes activity centers more competitive with surrounding locations. But as cities get larger, substantially increasing the amount of downtown floor area of office buildings could further increase traffic demand and congestion on radial highways and downtown streets.

Effective parking management can reduce highway travel demand and traffic congestion especially in large urban areas. Park-and-ride facilities located at outlying express (rapid) transit stations can intercept motorists, reduce VMT, increase transit ridership, and make it possible to place limits on downtown parking space. This strategy enables rapid transit and commuter rail lines to extend further into suburban and exurban areas.

Large suburban developments should be redesigned to encourage more transit and pedestrian access, and limit the number of parking spaces. Better management of parking and price is essential.

Congestion relief from parking management actions in both city and in suburban centers needs to be supportive of the commercial viability of these areas. Congestion relief strategies therefore will vary with city size, downtown employment, and parking location. Examples of parking guidelines follow.

- Effective management of on-street parking is essential in all cities. Some times and at some locations, parking should be prohibited to make curb lanes available to move traffic. In other situations, time limits and parking policies should be established to discourage all-day parkers
- A shortage of available curb parking spaces during busy periods especially in large central business districts has been reported to add VMT and congestion resulting from motorists who drive around searching for an available space [18]. By varying the cost of parking to levels that ensure that at least 10 % of the curb spaces are always available, this additional VMT can be practically eliminated and traffic congestion can be reduced [19–21].
- Downtown parking supply in large cities should be stabilized by setting a limit on the maximum number of spaces. This should entail revising zoning requirements, and should increase support for improving transit access
- Park-and-Ride facilities should be provided along outlying express bus rail lines and along major highways serving the city center
- Large suburban developments should be progressively redesigned to encourage more transit and pedestrian access and to limit the number of parking spaces. Management of parking supply and price is essential to stabilize or reduce congestion. Actions usually require restructuring of site plans and densities, and providing better transit and pedestrian access.

References

1. Turnbull KF, Pratt RH, Evans JE, Levinson HS (2004) TRCP Report 95, traveler response to transportation system changes, Chapter 3: park and ride/pool. Transportation Research Board, Washington, DC
2. Vaca E, Kuzmyak JR (2005) TRCP Report 95, traveler response to transportation system changes, Chapter 13: parking pricing and fees. Transportation Research Board, Washington, DC

3. Kuzmyak JK, Weinberger R, Pratt RH, Levinson HS (2003) TRCP Report 95, traveler response to transportation system changes, Chapter 18: parking management and supply. Transportation Research Board, Washington, DC
4. Levinson HS, Weant RA (1990) Parking. ENO Center for Transportation, Westport, CT
5. Levinson HS, Weant RA (1998) Parking and traffic congestion—changing perspectives. J Parking 1(1):17–25
6. Littman T (2010) Parking pricing and implementation guidelines. Victoria Transportation Policy Institute, Victoria, BC
7. Shoup D (2005) The high cost of free parking. American Planning Association, s.l
8. Morrall J, Bolger D (1996) The relationship between downtown parking supply and transit use. ITE J 66(2):17–18
9. Wilbur Smith and Associates, United States. Federal Highway Administration, Massachusetts. Department of Public Works (1974) An access oriented parking strategy for the Boston metropolitan area: summary report. s.l
10. Cervero R (1986) Suburban gridlock. Center for Urban Policy Research, Rutgers: The State University of New Jersey, New Brunswick, NJ
11. DiRenzo JF, Cima B, Barber E (1981) Parking management tactics, reference guide, vol 3. Sponsored by the Urban Planning Division of the Federal Highway Administration, Peat, Marwick, Mitchell and Co., Washington, DC
12. Transit Cooperative Research Program (TCRP) (2005) Report 95: Chapter 10—bus routing and coverage. Transportation Research Board, Washington, DC
13. Ferguson E (2000) Parking management and commuter rail—the case of northeastern illinois. J Public Transp 3(2):99–121
14. Transit Cooperative Research Program (TCRP) (2012) Report 153: guidelines for providing access to public transportation stations. s.l. Transportation Research Board
15. Levinson HS, Weant RA (2000) Transportation and parking demand analysis for bridgeport intermodal transportation center. Prepared for Wallace Floyd Design Group, Boston, MA, 30 Aug 2000
16. Keyani BI, Putnam ES (1976) Transportation system management: state of the art. s.l. Prepared for Urban Mass Transportation Administration US DOT
17. Hamer D, Currie GC (2010) Equity implications of parking taxes. Transp Res Rec 2319:21–29
18. Falcocchio JC, Darsin J, Prassas E (1995) An inquiry on the traffic congestion impacts of parking and pricing policies in the manhattan CBD. s.l. University Transportation Research Center, Region 2, and NYC Department of Transportation
19. Shoup D (1986) The effect of employer paid parking in downtown los Angeles—a study of office workers and their employers. s.l. Southern California Association of Governments, CA
20. Shoup D (1994) Cashing out employer-paid parking: an opportunity to reduce minimum parking requirements. University of California Working Paper, Berkeley, CA
21. Shoup D (1997) Evaluating the effects of parking cash out: eight case studies. Final report prepared for California air resources board, Research Division, Sacramento, CA
22. Willson R, Shoup D (1990) The effects of employer-paid parking in downtown Los Angeles: a study of office workers and their employers. Prepared for Southern California Association of Governments, Los Angeles, CA, pp 13–20

Chapter 23
Indirect Demand Strategies—Land Use, Transit, Alternative Modes

23.1 Introduction

This chapter describes strategies that reduce motor vehicle use in urban areas by (1) improving public transportation services (where appropriate) and (2) implementing land use policies that support travel mode alternatives to the automobile. These strategies are discussed in a single chapter because they are closely related and interdependent.

The chapter describes the relationship between development density and public transportation and how both affect highway travel demand and traffic congestion. It shows how both population and employment density influence modal choice and usage. It shows various public transit actions that can benefit existing riders, attract car drivers, and possibly support high density land development. It also identifies land development actions that can reduce the use of automobiles and can create pedestrian friendly communities. It also suggests future land development patterns that can foster compactness rather than spread and limit the growth of VMT as metropolitan areas expand.

23.2 Density, Transit, and Traffic Congestion

23.2.1 Land Use Density

Land use density is what distinguishes a suburban area from a city. In the suburbs traveling by car is a necessity because activities are separated by distances too long to walk, and effective transit service cannot be provided. This condition results in low density (uncongested) traffic on local streets but high density (congested) traffic on arterial roads that receive traffic from local roads to bring it to destinations far away.

J.C. Falcocchio and H.S. Levinson, *Road Traffic Congestion: A Concise Guide*, Springer Tracts on Transportation and Traffic 7,
DOI 10.1007/978-3-319-15165-6_23

In large cities, where many activities are located nearer to each other than they are in suburban areas, traveling options in addition to the car include walking, biking, and transit. However, lower per capita car use in cities does not translate to lower traffic congestion because cities with higher population and employment densities generate higher traffic densities on their street networks.

However, although city and suburban travelers are both impacted by traffic congestion, the severity of impact on travelers might not be the same:

- City streets provide access to more destination opportunities (per unit distance) than suburban streets
- Destination opportunities in cities can be reached by walking or biking while these modes usually are not feasible in suburban areas.

Therefore, the negative impact of traffic congestion on suburban travelers tends to be greater than on city travelers.

23.2.1.1 Population Density

Population densities reflect city size, city age, and physical features such as water bodies and hilly terrain that can constrain development to a limited supply of developable land. City age is also a factor. Older US cities that developed before the automobile era are more densely developed than cities that grew in the automobile era. The high rates of population growth and the availability and cost of land typically determine land use densities. Population densities typically increase with the cost of land.

23.2.1.2 Employment Density

Employment concentrations initially developed in city centers and in established suburban centers. They were created where major streets and transit lines crossed and converged, and along water bodies. Transportation technology has also played an important role in shaping densities and development. Until the 1929 Great Depression, rapid transit lines helped to concentrate employment in the city center. In the decades following World War II, major activity centers—largely auto dependent—emerged in the suburbs of large cities.

23.2.2 Population Density, Mode of Travel, and Traffic Density

The effects of population and employment density on person trip rates, travel modes, and per capita VMT have long been recognized.

A recent Transportation Research Board study requested by the US Congress, TRB Report 298 [1], examined the relationship between the built environment and motor vehicle travel. The study assessed whether petroleum use, and hence emissions of carbon dioxide (CO_2)—the primary greenhouse gas—could be reduced by more compact, mixed use development. The most reliable studies estimate that doubling residential density across a metropolitan area could lower household VMT by 5–12 %; and perhaps by as much as 25 % if coupled with higher employment concentrations, significant public transit improvements, mixed uses, and other supportive demand management measures.

The general effects of population density on daily person-trip rates, travel modes, and per capita VMT are shown in Table 23.1 [2]. As density increases above 10,000 persons per square mile, there is a pronounced decrease in daily per capita person-miles, vehicle miles, and auto trips. And there is a corresponding increase in the use of other modes.

However, while some compact land uses generate less VMT per capita, they can generate higher traffic density (VMT per square mile) resulting in higher traffic congestion. This is especially true for concentrations of office buildings in city centers. Therefore some of the most compact cities also tend to be the most congested. An illustration of this phenomenon is shown in Table 23.2, for the San Francisco Bay Area [3].

This table clearly shows why traffic congestion increases with population density: VMT decreases less than a corresponding increase in population density. Therefore, more compact (high density) developments do not necessarily lead to

Table 23.1 Daily trips per person in the US by population density and mode

Density (persons per square mile)	Daily person trips by mode							Daily person miles	Daily VMT per person
	Auto	Bus	Rail	Taxi	Walk and bike	Other	Total		
0–99	3.55	0.02	0.00	0.00	0.24	0.16	3.77	31.58	21.13
100–249	3.50	0.02	0.00	0.01	0.24	0.13	3.90	29.95	20.73
250–499	3.53	0.02	0.00	0.00	0.29	0.12	3.96	29.33	20.40
500–749	3.44	0.05	0.01	0.01	0.21	0.12	3.88	29.00	20.99
750–999	3.44	0.05	0.01	0.01	0.26	0.13	3.90	26.25	18.35
1,000–1,999	3.48	0.03	0.01	0.00	0.23	0.11	3.86	26.17	18.63
2,000–2,999	3.46	0.06	0.01	0.00	0.28	0.11	3.92	23.45	19.04
3,000–3,999	3.34	0.06	0.02	0.01	0.29	0.09	3.81	23.11	16.89
4,000–4,999	3.51	0.05	0.01	0.00	0.30	0.08	3.95	24.77	17.24
5,000–7,499	3.29	0.09	0.02	0.01	0.36	0.06	3.83	24.56	16.28
7,500–9,999	2.92	0.11	0.05	0.02	0.45	0.07	3.62	20.59	14.15
10,000–49,999	1.90	0.29	0.21	0.03	0.95	0.04	3.42	17.02	8.73
50,000<	0.59	0.42	0.61	0.16	1.55	0.07	3.40	12.55	2.31

Source Reference [2]

Table 23.2 Population density versus traffic density for the San Francisco bay are

Location	Population density (persons per acre)	VMT per capita (vehicle—miles per person per day)	Traffic density (daily VMT per acre)
Healdsburg	5	30	150
Berkeley	30	10	300
Downtown San Francisco	250	4	1,000

Source Created by Prof. Martin Wachs in Reference [3], p 15

less traffic congestion, but provide travelers with more travel options. Traveler benefits of high density include the availability of alternative modes to the automobile, and a higher number of destination opportunities accessible by alternative travel modes.

23.2.3 Densities for Public Transit

Many studies have shown that increasing population and employment densities produce higher shares of transit trips. As population density increases, the *number* of person trips made by public transportation (including pedestrian trips) increases. Conversely as densities decline, travel demands become more dispersed and transit becomes less effective in serving this demand. Public transit works best where travel is concentrated in time and space. It is well suited to serve high employment and population densities.

- The best transit markets are found where *both* employment and population densities are high
- Higher residential densities and lower auto ownership per household results in a higher proportion of regional and CBD trips by transit
- The proportion of CBD trips by public transport increases as employment density rises. For example, more than 90 % of all peak period person trips to the Manhattan CBD (NYC), where employment density approximates 800 persons per acre, use public transportation. In contrast, about 20 % arrive by transit in Denver (CO), where downtown employment approximates 150 persons per acre [4]
- Rapid transit and commuter rail relate closely to the number of jobs in the city center

The likelihood of commuting by public transit as a function of residential density and car ownership is shown in Fig. 23.1.

Mixed use and high-rise developments offer the highest likelihood of transit use, but use declines with increasing car ownership levels.

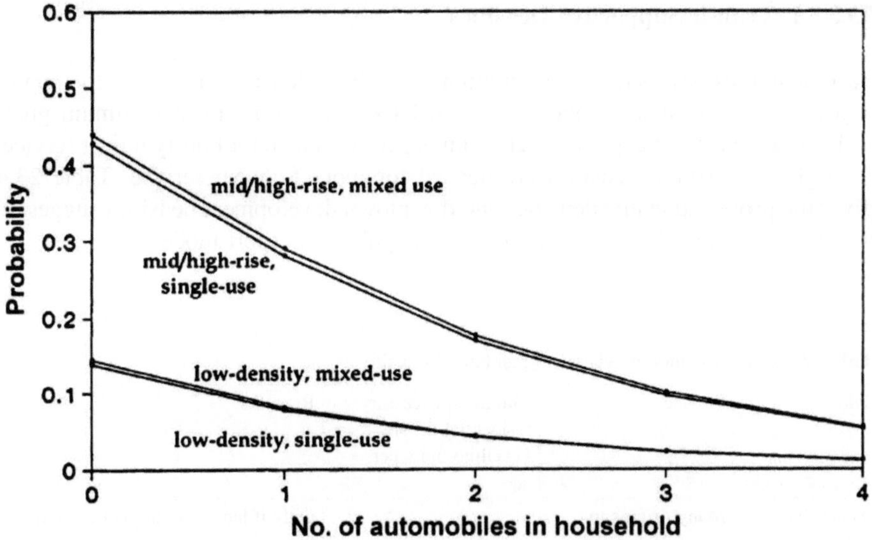

Fig. 23.1 Probability of commuting by transit as a function of auto ownership, for four land use scenarios. *Note* Based on modeling of survey results from the 11 metropolitan areas (MSAs or CMSAs) of Boston–Lawrence–Lowell, Dallas, Detroit, Los Angeles–Long Beach, Fort Worth–Arlington, Minneapolis–St. Paul, Philadelphia, Phoenix, San Francisco–Oakland, Tampa–St. Petersburg, and Washington, DC-MD-VA. *Source* Reference [5], pp 15–57. Figure 15.5 [2]

Employment and residential concentrations determine the potential transit demand. These relationships are conceptually shown in Table 23.3 for ranges of density.

- The best transit markets are created where *both* employment and residential densities are high—a condition that exists in major world cities. However, CBD employment density is generally more important than the residential density of within a catchment area of 45 min travel time to the CBD.
- Conversely, the weakest transit markets exist where both employment densities and residential densities are low. This condition exists in many smaller and medium sized US metropolitan areas.

Table 23.3 Generalized transit potential of CBD employment density and population density within 45 min of the CBD

Population density within 45 min of the CBD			
CBD employment density	High (>50,000/mi^2)	Medium (10,000–50,000/mi^2)	Low (<3,000–6,000/mi^2)
High (>150,000/mi^2)	+++++	++++	++
Medium (75,000–150,000/mi^2)	+++	++	+
Low (<75,000/mi^2)	+	–	–

Source Estimated

23.2.3.1 Transit Supportive Densities

Several studies have suggested minimum household densities necessary to support transit service. Pushkarev and Zupan [6] have suggested typical minimum gross residential density of approximately 4 units per gross acre for hourly transit service.

Higher residential densities are needed for more frequent service. Table 23.4 gives the gross residential densities and downtown development densities suggested by Pushkarev and Zupan to support various public transport modes [7].

Table 23.4 Transit modes related to residential density

Mode	Service	Minimum necessary residential density Dwelling units per acre	Remarks
Dial-a-bus	Many origins to many destinations	6	Only if labor cost are not more than twice those of taxis
Dial-a-bus	Fixed destination or subscription service	3.5–5	Lower figure if labor costs twice those of taxis: higher if thrice those of taxis
Local bus	“minimum” 1/2 mile route spacing,40 buses per day	4	
Local bus	“Intermediate” 1/2 mile route spacing, 20 buses per day	7	Average, varies as a function of a downtown size and distance from residential area to downtown
Local bus	“Frequent” 1/2 mile route spacing, 120 buses per day	15	
Express bus—reached on foot	Five buses during 2 h peak period	15 Average density over 2 mi^2 tributary area	From 10–15 miles away to largest downtowns only
Express bus —reached by auto	5−10 buses during 2 h peak period	3 Average density over 20 mi^2 tributary area	From 10 to 20 miles away to downtowns larger than 20 million ft^2 of nonresidential floor place
Light rail	5 min headway or better during peak hour	9 Average density for a corridor of 25–100 mi^2	To downtowns 20–50 million ft^2 of nonresidential floor place
Rapid transit	5 min headways or better during peak hour	12 Average density for a corridor of 100–150 mi^2	To downtowns larger than 50 million ft^2 of nonresidential floor place
Commuter rail	20 trains a day	1–2	Only the largest downtowns, if rail line exists

(*Source* Reference [7]). Courtesy of Indiana University Press, Bloomington.

Table 23.5 Transit modes related to residential density

Mode	<Minimum CBD employment
Rail rapid transit/commuter rail	100,000
Light rail/grade separated BRT	70,000
Express bus	35,000–50,000
Local bus	
10 min service	20,000–25,000
30 min service	7,500–10,000

Source Adapted from Reference [4] or [8]

Public transportation usage is higher in urban areas where:

- There are high residential densities and a large downtown or other cluster of non-residential activity
- Residential developments are located in close proximity to non-residential concentrations
- Transit service is convenient, reliable, and frequent
- Parking space is scarce and/or expensive

One set of guidelines [8] to assess the suitability of implementing rail rapid transit, light rail transit, and bus rapid transit is provided in Table 23.5.

23.3 Transit Improvements

23.3.1 Context

The importance of maintaining good transit for enhanced mobility—especially in major cities—is increasingly recognized in the US where many transit systems have been upgraded, and major new systems have been built. Bus rapid transit lines, used for many years in major world cities have emerged in the US and light and heavy rail lines have increased in number and extent. In addition, vehicle design has been advanced as well (for example, some buses now have doors on both sides).

The following sections show how transit operational treatments can enhance existing service, improve street system efficiency in moving people, reduce transit travel times, increase service reliability and reduce operating costs. They also describe the type of physical improvements that expand person capacity in major transit corridors to make service faster and more reliable, and provide a framework for achieving more compact land developments.

23.3.2 Transit Service Objectives

Transit service objectives should be established commensurate with the resources that are made available. The basic service objectives include providing [4]:

- High quality (convenient, comfortable, frequent, and reliable) network of transit services for residents and visitors
- Access to places of work, shopping, schools, and recreation. The amount of service supply should reflect both ridership demand and available resources
- Equitable and accessible services for the transit dependent—the elderly, the physically disabled, the young, and low income persons
- Expanded service to developing areas to capture emerging transit markets

A well designed transit system should serve major population, employment, retail, and medical centers. It should provide a simple, understandable route structure, convenient, frequent and reliable service, as well as affordable fares. It should provide coordinated transfers among bus and rapid transit lines. Where routes run, how well they are operated, and how effectively they tap major markets will determine their ability to retain existing riders and attract auto users to relieve traffic congestion.

23.4 Transit Operational Improvements

Many bus and rail service improvements can be implemented quickly at low cost. These short-range operational treatments address when, where, and how service should be provided in relation to land development, street system, travel patterns and fiscal resources. They can improve operating speeds and operating efficiency, improve service reliability, and retain/attract riders.

Service coverage can be expanded, service span and frequencies can be increased, and route structures can be simplified. Transit speeds can be increased by reducing the number and duration of stops. Timing traffic signals for transit, implementing transit signal priorities and installing transit lanes can further improve transit speeds and reliability.

Recommended practice and guidelines relating to transit improvements and their congestion mitigation impacts follow. Congestion reducing effects of improved transit will depend upon its ability to attract new riders *faster* than the growth in new auto users.

23.4.1 Expanding Coverage

Most urban transit systems provide good coverage of the population in "transit supportive" areas. However, many offices, shops, schools and multi-use developments are often located away from transit routes. While 80–90 % of an urban area's population is usually located within a quarter mile of a transit stop, transit's service area of the urban area's employment is smaller. Thus if 90 % of an area's population is served by transit, and 50 % of an area's employment is served by transit, this combines in 45 % coverage—resulting in fewer potential transit trips.

Increased transit coverage can be achieved by (1) modifying existing routes, (2) providing better pedestrian access to stops, and (3) locating future commercial developments near transit stops.

23.4.2 More Frequent Service

Bus service frequencies depend on transit demand—longer headways are found where demand is low, while short headways predominate when demand is high. Whenever practical, headways should be shorter than riding time. Where transit service frequency is less than 10–12 min, passengers arrive at transit stops at random, while longer headways require passengers to refer to schedules.

23.4.3 Route and Service Improvements

Transit routes and services are normally governed by transit agency policies and standards. Good operating practice follows these guidelines:

- There generally should be one route per arterial street. However, there can be more routes where streets and bus lines converge as they approach city centers. There also can be more services on a street where local, express, limited stop and bus rapid transit services run on the same street
- Fewer routes with shorter headways are preferable to many routes with long headways—operating on the same street.
- Excessively long routes can result in "bunching." Therefore, routes longer than about 1.5 h each way should be avoided
- Routes *through* the city center are preferred to looping routes. Looping routes can result in more turning movements and traffic conflicts. However, looped routes can be necessary where (1) ridership on each leg is not balanced, (2) the through routes would be too long, and (3) additional coverage is needed.
- Bus terminals are sometimes provided in city centers. They generally are desirable to serve long distance and low-frequency local services.

23.4.3.1 Reducing Delay from Transit Stops

The number of stops and the duration at transit stops should be minimized. This is because each stop delays transit vehicles, and sometimes motorists following buses or street cars. Time is also lost accelerating and decelerating to and from stops. Bus stops typically account for 25 % of bus travel time in city centers, 20 % within the city, and 15 % in the suburbs [9].

- Desirable minimum stop spacing for local buses are 300–600 ft in the city center, and 600–750 in other areas. Longer spacing (e.g., ½ to 2 miles) are normally provided for express, limited stop, and bus rapid transit (BRT) lines
- Stop spacing for LRT, rapid transit, and commuter lines depends on station access. Station spacing normally varies from about one half mile in densely developed area (where most people walk to and from stations) up to two miles or more in suburban areas where most riders drive to or from stations
- The duration of stops should be reduced wherever practical. Duration can be shortened by [1] providing off-vehicle fare collection, using several door channels for boarding and alighting and simplifying fare collection. The 2013 Transit Capacity and Quality of Service Manual [10] suggests the various service times in Table 23.6, for passengers boarding buses.

Values range from about 2.5 s per passenger for direct entry without payment to more than 4 s per passenger for complex fare payment. As more door channels are provided, passengers' service times are reduced. Where transit stops are removed from the travel lanes, delays to vehicle traffic are reduced.

23.4.3.2 Other Components of Transit Delay

Figure 23.2 breaks down total bus travel time into five components: (1) time while the bus is in motion, (2) time while the bus accepts or discharges passengers at bus stops, (3) time delay at traffic signals, (4) time delay for right turns, and (5) time spent in traffic congestion. It can be seen that as the total travel time rate increases, congestion increases, and the proportion of time spent in motion decreases.

The effect of area type on delays at bus stops, signals, and congestion, varies significantly—with the CBD causing the greatest amounts of delays and suburban areas the least. The most critical delay component is found at bus stops. As the

Table 23.6 Suggested bus passenger service times at bus stops as a function of fare payment

A. Boarding	
Type of fare payment	Seconds per passenger
Prepayment	2.5
Single ticket or token	3.5
Exact change	4.0
Swipe or dip card	4.2
Smart card	3.5
B. Alighting	
Front door	3.3
Rear door	2.1

(*Source* Reference [10])

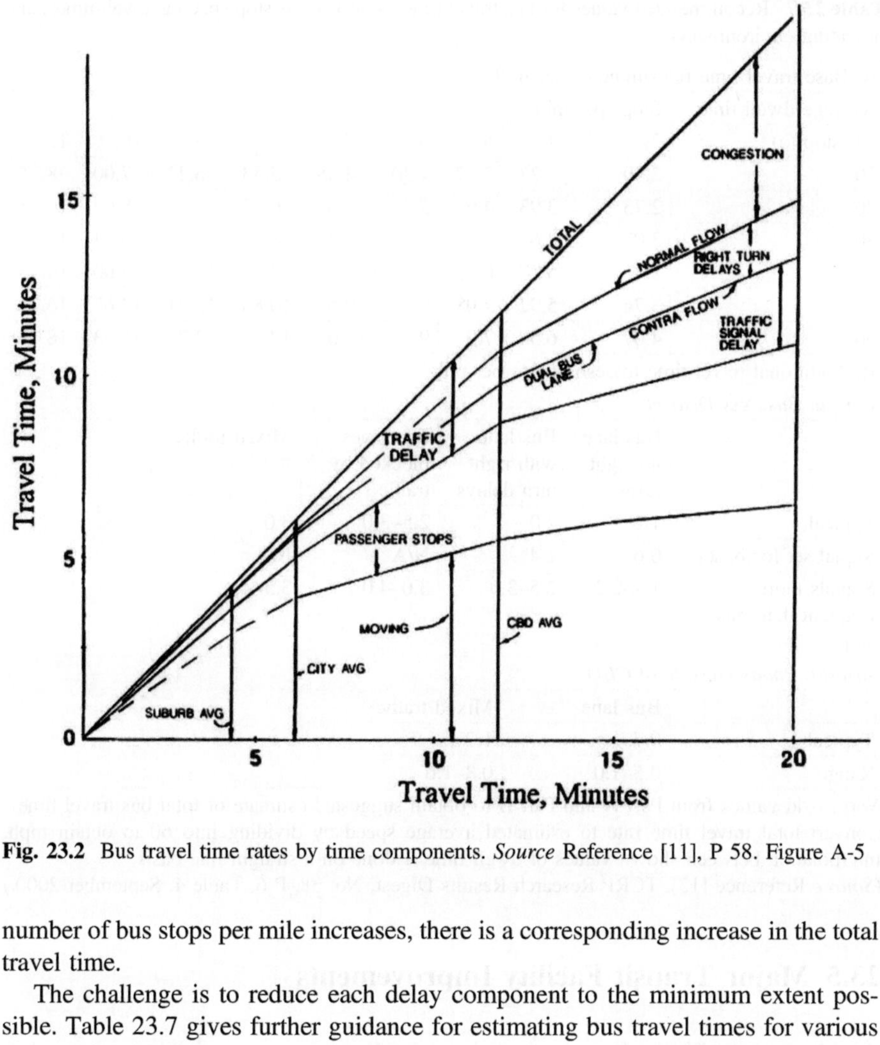

Fig. 23.2 Bus travel time rates by time components. *Source* Reference [11], P 58. Figure A-5

number of bus stops per mile increases, there is a corresponding increase in the total travel time.

The challenge is to reduce each delay component to the minimum extent possible. Table 23.7 gives further guidance for estimating bus travel times for various operating conditions.

- Base travel time rates reflect the number of stops per mile and the average dwell times per stop
- Additional transit travel times result from traffic congestion in city centers and along arterial streets. These times can be reduced by providing bus lanes, prohibiting right turns, and timing traffic signals for buses. For example, a bus route operating at 10 mph would save up to 1.0 min per mile with a bus lane, with a commonly used saving of 0.5 min per mile.

Table 23.7 Recommended values for bus travel times with various stop spacing, dwell times, and operating environments

A. Base travel time rates/minutes per mile

Average dwell time	Stops per mile								
Per stop (s)	2	4	5	6	7	8	9	10	12
10	2.40	3.27	3.77	4.30	4.88	5.53	6.23	7.00	8.75
20	2.73	3.93	4.6	5.3	6.04	6.87	7.73	8.67	10.75
30	3.07	4.6	5.43	6.3	7.2	8.2	9.21	10.33	12.75
40	3.4	5.27	6.26	7.3	8.35	9.53	10.71	12.00	14.75
50	3.74	5.92	7.08	8.3	9.52	10.88	12.21	13.67	16.75
60	4.07	6.58	7.9	9.3	10.67	12.21	13.7	15.33	18.75

B. Additional travel time losses/minutes per mile

Central Business District

	Bus lane no right turns	Bus lane with right turn delays	Bus lanes blocked by traffic	Mixed traffic flow
Typical	1.2	2.0	2.5–3.0	3.0
Signal set for buses	0.6	1.4	N/A	N/A
Signals more frequent than bus stops	1.7–2.2	2.5–3.0	3.0–4.0	3.5–4.0

Arterial Roads Outside of CBD

	Bus lane	Mixed traffic
Typical	0.7	1.2
Range	0.5–1.0	0.8–1.6

Notes Add values from Part A and Part B to obtain suggested estimate of total bus travel time Convert total travel time rate to estimated average speed by dividing into 60 to obtain mph. Interpolation between shown values of dwell time is done on a straight line basis
(*Source* Reference [12], TCRP Research Results Digest, No. 38, P 6, Table 4, September 2000.)

23.5 Major Transit Facility Improvements

Rapid transit facilities play a key role in expanding person carrying capacity and in providing and maintaining travel mobility in large metropolitan areas. These include commuter rail, heavy rail, light rail, and bus rapid transit. They are characterized as mainly operating in their own right of way and can they serve commuter trips without adding to the street and expressway congestion.

23.5.1 Objectives

The main reasons for building new transit infrastructure systems or improving existing systems are to enable person mobility and accessibility that is environmentally sustainable in densely populated areas, to provide transportation capacity for future growth, and to strengthen the viability of the city center and other mega centers.

Some specific benefits of major transit projects include:

- Expanding transportation capacity in a constrained environment/space. Transit facilities occupy 1/4 of the space needed for freeways.
- A transit "lane" can serve as much as 10–15 times the number of person trips served by a freeway lane.
- Allowing employment growth in the city center and other mega centers without increasing street traffic and parking requirements
- Simplifying surface transit routes by feeding bus lines to rapid transit stations and by reducing bus traffic in congested areas
- Providing more reliable trip times, since many causes of congestion are eliminated. Rail speeds are twice or more the speeds attained on surface transit routes
- Attracting auto users from heavily traveled congested corridors
- Helping to structure land development around both urban and suburban stations

These benefits could often offset the high transit development costs and environmental impacts normally associate with fixed guide-way transit facilities. Thus there should be a realistic balance between costs and benefits.

23.5.2 System Extensions

Market demand, political, physical, congestion factors and operating conditions determine how far transit line should extend or be extended. Basic guidelines include the following: (1) the length of the initial segment should be as short as possible to provide the desired service and to attract the needed ridership. Once the line is opened and its ridership is established, it then can be extended; (2) it should be long enough to provide a few good stations at its outer end that could develop the desired ridership; (3) the line should extend out far enough so that sufficient park-and-ride facilities can be provided at outer stations; and (4) it should serve existing riders and should capture new riders as well.

Ideally, line coverage and station spacing should capture *both* existing and future transit markets. New lines should extend beyond the limits of existing development and right of way should be preserved or assembled for subsequent extensions. Many successful lines and transit markets did not exist when some lines were initially built. This is apparent from the evolving population density profiles shown in Fig. 23.3.

Residential densities in New York, Chicago, Philadelphia, and Boston clearly reflect the result of rail transit development over the first half of the 20th century. In these cities the density patterns were similar. Rail transit lines had their greatest impact on development around stations located in undeveloped areas *farthest* from the city center, and not previously served by public transportation. Buildings were

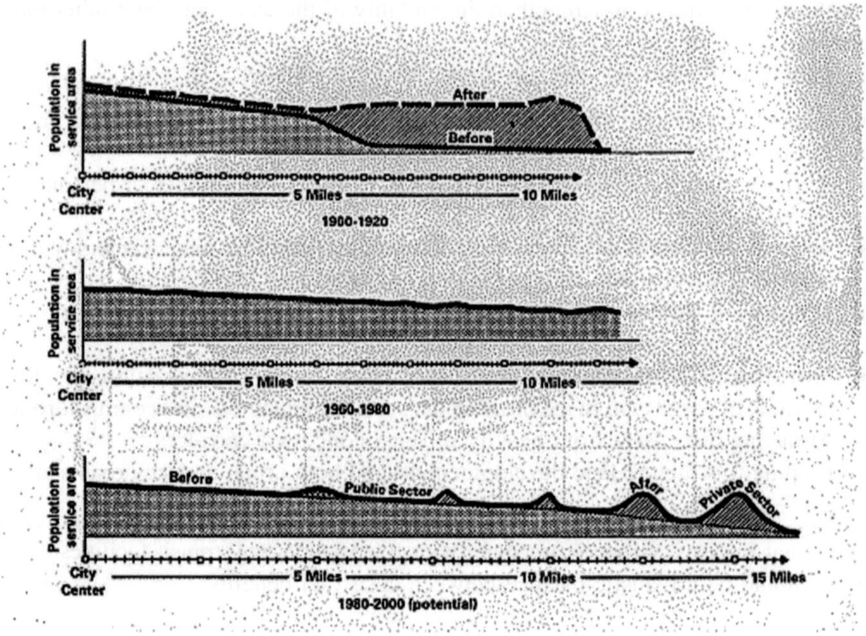

Fig. 23.3 Residential densities over time along a rapid transit line. *Source* Reference [13]

located as close to the station as land availability allowed. Rail transit impacts on developments in existing built-up areas were generally less strong [4].

23.5.3 Configuration and Design

Guidelines for rapid transit configuration, design and operations are as follows:

(a) Lines should radiate from the city center. Generally they should pass through the CBD rather than terminate there (commuter railroads are a possible exception).
(b) Generally, they should be grade-separated (e.g., subways in downtown areas). Cross town lines, when provided, should connect to the radial lines.

The CBD end of a rapid transit trip offers an excellent opportunity for travel time saving over a trip by automobile. It is essential, therefore, to maximize service convenience by placing routes through areas of heavy demand, and providing frequent stations, and interconnecting stations and mezzanines with major corridors for pedestrian movement. To the maximum extent possible, station facilities should serve as their own CBD distributors—thereby minimizing transfers to other transit vehicles or changes in travel mode.

(a) Lines should penetrate, rather than skirt, major market areas such as high-density residential neighborhoods, schools, medical centers, and outlying business areas. This practice makes it possible to increase ridership beyond the CBD market.
(b) A simplified route structure is essential. Therefore the number of branches should be minimized. Ideally heavy rail transit lines should have only one service per track, and never more than two.
(c) To maximize operating speeds, it is desirable to space stations far apart. Station spacing should reflect development densities and access modes. For walk access typical station spacing is 0.5 miles; for bus access is 0.5–1.0 miles; and when access is by car it is 1.5–3 miles.
(d) Convenient access to stations is essential. Good pedestrian access is important for all stations, especially those located in high density areas. Convenient bus-rapid transit interchange is necessary where bus lines serve or converge at stations. Escalators and elevators should provide necessary vertical transportation. Adequate park-and-ride facilities are essential at suburban stations since an inadequate supply of parking space can limit transit ridership.
(e) Transit-supportive development should be encouraged around stations that are currently surrounded by medium-to-high residential and commercial activities. This could enhance the station environs, increase transit ridership, and reduce automobile trips.

23.5.4 Congestion Implications of Rapid Transit Lines

Various rapid transit systems built before the automobile era, have had several congestion-related effects over time. They have made it possible for city centers to grow in density and they have induced residential development along their lines. Although the lines were grade-separated in the city centers, they induced employment growth resulting in major concentrations of activities that, in turn, increased the severity of street congestion—at least until traffic engineering was effectively applied in the downtown areas.

The congestion effects of major transit investments are somewhat different today. Urban dispersion has been more a phenomenon of the automobile. Longer travel distances have resulted in increased peak period freeway congestion. In contrast, grade separated transit provides improved mobility and can attract automobile travelers. Extensive park-and-ride facilities at outlying locations can reduce car trips destined to the CBD and street congestion. Even more significant, perhaps, is the ability of rapid transit lines to support concentrated developments around stations thereby reducing dependence on the automobile for travel mobility.

23.6 Estimating Ridership Response to Transit Service and Fare Changes

Passenger response to fare and service changes have been observed in many transit systems. Where service is improved, ridership has grown from two basic sources: (a) from existing riders who use transit more frequently, and (b) from new riders who previously used automobiles or walked. A convenient approach to estimating the impact of fare and service changes on transit ridership is the pivot-point elasticity method. Elasticity metrics indicate the responsiveness of ridership to changes in fare and/or service for a time period when all other factors that affect travel demand remain constant. They can provide realistic estimates of near term changes in ridership from changes in service.

The concept of transportation elasticity is adapted from the economist's measure of price elasticity of a particular product or service. Ridership elasticity is defined as the percentage change in ridership from a 1 % change in fares, service frequency, or transit-miles operated.

Two commonly used methods to calculate elasticity are the "Shrinkage Factor" and "Arc Elasticity".

The "shrinkage factor" has been commonly used to estimate a change in ridership resulting from fare increase. Hence the term shrinkage reflects the reduction in transit rides from a fare increase. However, the shrinkage factor is also applied to estimate ridership increase from a reduction in fare. It has been used as a "rule of thumb" in estimating the ridership effects of fare changes. It is the simplest method to use and gives a reasonable approximation for small fare changes. It is also used to estimate the change in ridership from changes in service (travel time, frequency, etc.).

Shrinkage Ratio Elasticity = [(Change in Ridership/Base Ridership)]/[(Change in Service Attribute (e.g. transit fare))/(Base Attribute (e.g., fare)]

Or

$$\mathbf{E} = [(\mathbf{R2} - \mathbf{R1})/\mathbf{R1})]/[(\mathbf{X2} - \mathbf{X1})/\mathbf{X1}] \qquad (23.1)$$

where:

E = elasticity
R1 = base ridership
R2 = calculated new ridership
X1 = value of base attribute (e.g., fare)
X2 = value of changed attribute (e.g., new fare)

Therefore:

$$\mathbf{R2} = \mathbf{R1} + [(\mathbf{E})(\mathbf{R1})] * [(\mathbf{X2} - \mathbf{X1})]/\mathbf{X1} \qquad (23.2)$$

Thus when the elasticity of ridership with respect to fares or service is known, using the above equations it is straight forward to calculate the change in ridership from a proposed change in fare and/or service.

A fare elasticity of –0.33 has often been used to estimate an average response of ridership change to a change in transit fares. Thus a 10 % fare *increase* would result in a 3.3 % *decrease* in ridership.

Arc Elasticity—A more accurate measure of elasticity is the "Arc Elasticity." This is similar to the Shrinkage Ratio, except that it uses the mid-point of fares or service, as the denominator, instead of their initial values.

$$\textbf{Arc Elasticity} = [(R2 - R1)/(R1 + R2)/2]/[(X2 - X1)/(X1 + X2)/2] \quad (23.3)$$

The variables are the same as those for Eq. 23.1

An American Public Transit Association study [14] used an "Integrated Moving Average" model to estimate fare elasticity. The following disaggregate values of fare elasticity were reported:

(a) Overall average = –0.40
(b) Systems in urbanized areas greater than one million = –0.36
(c) Systems in smaller cities = –0.43
(d) Average for peak hours = –0.23
(e) Average for off-peak = –0.42

It should be noted, however, that transit ridership is more responsive to *service* changes than to *fare* changes. Table 23.8 provides an example of cases where this difference is evident [15].

Elasticity estimates are also calculated for different types of service changes (service expansion or service frequency). It may be seen from Table 23.9 that ridership is more responsive to changes in service expansion (bus or train miles) than it is to travel time or transit frequency [15, 16].

23.7 Land Use for New Developments

Land development policies that improve livability and reduce/minimize VMT growth are desirable societal and environmental goals. Some promising land use design strategies that can be progressively implemented to mitigate traffic congestion are described below.

Table 23.8 Fare elasticities compared with service elasticities

Location	Fare elasticity	Service elasticity	Service measure used
Atlanta (1970–1972)	−0.15 to −0.20	+0.30	Bus miles
San Diego—all routes (1972–1975)	−0.51	+0.85	Bus miles
17 US Transit Operators (1960–1970)	(Deflated) −0.48	+0.76	Bus miles/capita
12 British Bus Operators (1960–1973)	−0.31	+0.62	Bus miles
30 British Towns (pre-1977)			
work trips	−0.19	+0.58	Bus miles/capita
non-work trips	−0.49	+0.76	Bus miles/capita
11 Spanish Towns/Cities (1980–1988)	(Deflated)		
Range (short term)	−0.16 to −0.44	+0.34 to +1.26	Bus kilometers
Average (short term)	−0.30	+0.71	Bus kilometers

(*Source* Reference [15], pp 10–12. Table 10.6)

Table 23.9 Typical arc elasticities for a range of transit service indicators and typical applications

Transit service indications	Travel time	Bus/train miles	Transit frequency
Typical application	New routes faster service	Service expansion	More frequent service on existing routes
Likely range	−0.3 to 0.5	0.4–1.3	0.13–0.5
Typical value	−0.4	0.9	0.4

(*Source* Reference [16], Page 3–19, Exhibit 3-19. And Ref. [15], pp 10-9, Table 10.3.)

23.7.1 Reducing VMT Through Land Use and Transportation Strategies

The two basic "smart growth" development objectives are to (1) make cities livable, and (2) preserve open spaces in the countryside adjacent to cities. While it is sometimes difficult to improve the built environment in cities, creative transportation and land use strategies can be applied to the exurban environment.

Coordinated approaches to land development and transportation are essential to manage congestion. Transportation and development agencies should work together in achieving this effect. It is imperative to anticipate the effects of new transportation facilities on development, and also the effects of new developments on transportation performance. Possible actions that can improve both livability and reduce congestion include the following:

1. Expand the concept of access management to include corridor management of *both* land development and road access. Consider "form based zoning" [17] to achieve both design and operations objectives.
2. Manage road access in the vicinity of freeway interchanges. Anticipate major developments at key interchanges along new or expanded freeways, and design interchanges by limiting access along major arterial roads to at least 1,000 ft or more from freeway interchanges.
3. Avoid excessive zoning of land for retail activities. Concentrate commercial developments at key junctions and discourage/limit strip development along major arterials.
4. Encourage transit-oriented development (TDO) near major transit stops and stations. These developments will improve accessibility, reduce parking requirements, and generate transit ridership.
5. Improve the continuity and connectivity of the local street system. The goals should be to maximize accessibility and to minimize use of arterial streets by local trips. Systems of streets that form a circuitous and discontinuous routing pattern increase trip lengths and discourage the use of modal alternatives to the automobile. Compared to cul-de sac neighborhoods, traditional neighborhoods built on grid systems combined with higher development (mixed use) densities, experience a lower VMT per capita and higher utilization rates of non-motorized modes [18]. On traditional grid networks, local streets provide an alternative to arterials for short trips and lessen the traffic demand on arterials.
6. Provide public transportation and pedestrian/bicycle access for new developments. Provide continuous sidewalks, bikeways, and storefronts along retail streets to encourage walking, biking, and transit riding. Figure 23.4 shows how building footprints can be transposed by locating buildings close to the streets, and placing parking in the rear.
   This transposition concept achieves several important objectives: (a) the rearrangement gives a "village" look to the development, (b) the building groups on each side of both streets are within close walk distance of each other, (c) buses operating along both streets can conveniently serve the various buildings, and (d) vehicle parking space is close to the rear of the buildings.
7. Extend rail or bus rapid transit to serve developing areas in large urban areas. Zone the land for commercial and residential densities that are supportive of transit service at selected suburban stations *in advance* of development.
8. Foster diverse land use activities in centers of developments to encourage multi-purpose trips, thus reducing the total vehicle trips generated.
9. Require large/major office and mixed use developments to be located near good public transportation (and also be accessible to nearby residential developments) as well as good highway access [20]. Large mega centers (usually over 500,000 ft^2 or more than 25,000 daily trip destinations) should have some form of rapid transit with stations within a few hundred feet of major employment concentrations. The need for multimodal access to mega centers was emphasized by Vuchik in his book, "Transportation for Livable Cities" [21].

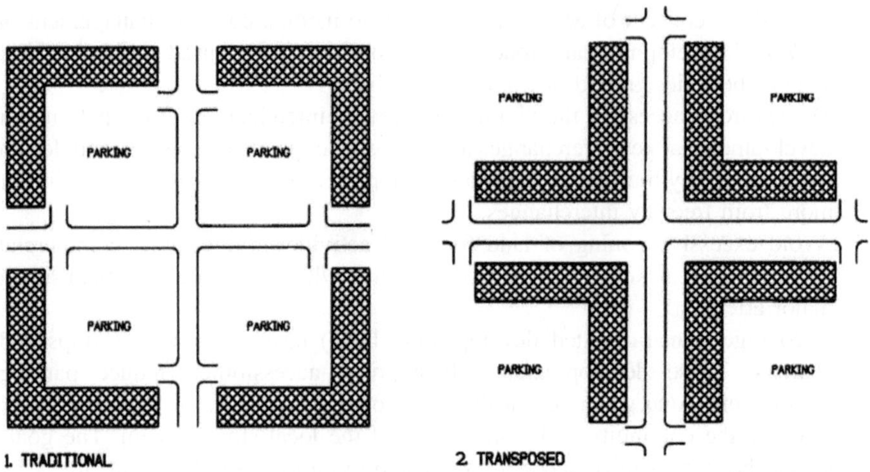

Fig. 23.4 Modifying strip development to support transit and pedestrian access. *Source* Reference [19], p 33. Figure 3.8

Figure 23.5 shows how a hypothetical large mixed-use activity center can become more transit and pedestrian friendly. A major off-street transit way with a centrally located station penetrates the center of the development. A system of pedestrian ways links the developments with the transit station, surrounding streets, and outlying commercial development along an arterial street. A landscape buffer separates the development from the nearby freeway.

To provide multimodal access to large new developments land use and transportation actions must be coordinated from the initial planning stages to the final site design stage.

10. Promote more densely developed and walkable areas within cities. Compared to low-density developments, households in developments with "twice the density, diversity of uses, accessible destinations (by modal alternatives to the private auto), and interconnected streets, drive about 33 % less" [21].
    Provide taxation policy incentives for high-density developments thereby reducing the per capita vehicle miles of travel. Also promote land use patterns that attract growth in infill areas of the city, allow for mixed use development, for higher densities, and for compact neighborhoods where walking/biking turn out to be a convenient mode choice.
11. Encourage transit oriented development at major transit stops and stations. These developments will improve walk-ability, reduce parking and traffic requirements, and increase transit ridership. Based on data from 25 sites. Table 23.10, based on data from twenty five sites, shows that transit–oriented development generate about half of the vehicle trips associated with typical suburban developments [22].

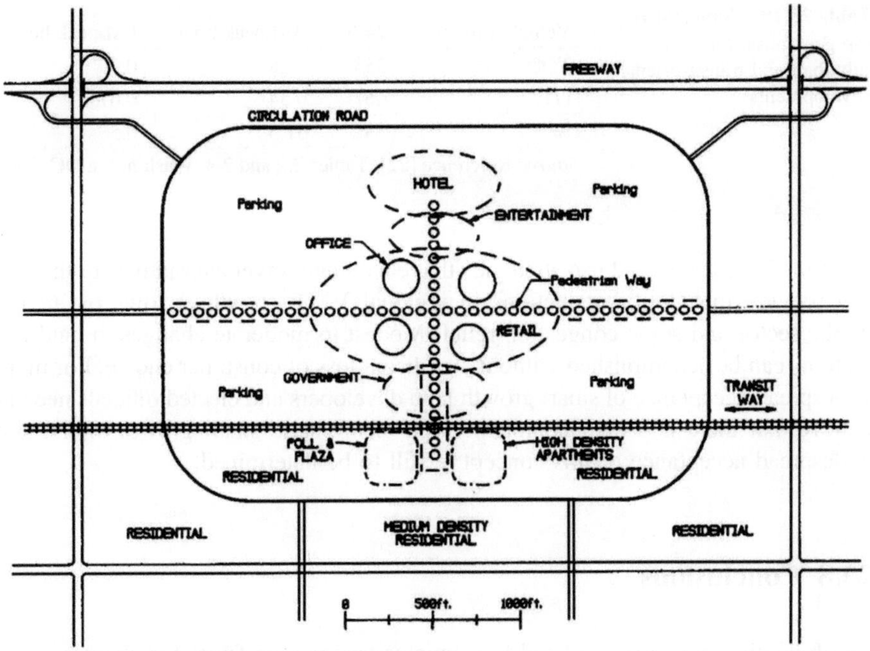

Fig. 23.5 Multimodal access for a major activity centers to support transit and pedestrian mobility. *Source* Reference [19], p 31. Figure 3.6

12. Achieve a better balance between jobs and housing. This has been a desirable planning goal for many years. To the extent that it can be realized, there can be corresponding reductions in peak period VMT as commuting trips become shorter and can be completed by waking, biking, or bus.
13. Consider value capture and financial incentives to attract and organize developments where they can be served by public transportation.

23.7.2 *Effects*

Although the above "smart growth" land use/transportation strategies are effective (in the near term) at the neighborhood scale, significant reductions in *regional* VMT impacts resulting from a change in land use patterns, however, takes a long time: "Even in rapidly growing urban areas, new urban developments and new land uses comprise only a fraction of all urban fabric. Thus, even dramatic changes to new development patterns would have to be maintained for decades before they could significantly reshape metropolitan land uses and, in turn, overall travel origins and destinations [3]."

Table 23.10 Comparative trip generation rates for suburban and transit-oriented developments

Vehicle trip rate	24 h	AM peak hour	PM peak hour
TOD	355	0.28	0.39
ITE	667	0.54	0.67
%	53.3	51.3	0.58

Source Reference [22], Tables 2.2 and 2.3, Washington DC, 2008

In the long term, land use strategies that encourage travel alternatives to the car, can provide meaningful reductions in regional VMT growth at low cost to the public sector and some congestion relief. Modest to moderate changes in land use patterns can be accomplished without significant loss of consumer choice. For more widespread acceptance of smart growth land developers and elected officials need to believe that there is a demand for the type of life style smart growth offers, and widespread acceptance of this concept is still to be determined.

23.8 Conclusions

The following congestion-related key conclusions emerge from this chapter.

1. Public transportation improvements can increase transit ridership, reduce car trips, and increase population and employment densities.
2. High employment and residential densities are desirable to minimize VMT and to maximize pedestrian and transit trips. But increasing densities may increase traffic congestion because compact land uses that generate less VMT per person generate higher traffic density (VMT/square mile)—the most compact cities also tend to be the most congested. An illustration of this phenomenon for the San Francisco Bay Area was shown in Table 23.2.
3. Land use generates travel demand. Its density and mix of uses determines the type of transportation system that best serves this demand. Putting people into fewer vehicles by increasing land use densities, encouraging walking and biking, and/or eliminate the need to travel (through telecommuting and teleshopping) reduces the VMT and can reduce freeway congestion.
4. Local zoning changes can contain and possibly reduce traffic congestion especially over the long run. Key actions include (1) downsizing zoning for commercial land, (2) discouraging strip developments, (3) coordinating access management with corridor development, and (4) requiring major activities to have good pedestrian and transit access.
5. VMT and congestion reductions from the various actions associated with public transportation and land development will become especially important in the long run as urban areas grow. These actions can provide the framework for creating more livable, sustainable, and accessible communities.

References

1. Transportation Research Board (2009) Driving and the built environment—the effects of compact development on motorized travel, energy use, and CO_2 emissions, special report 298. Transportation Research Board, Washington, DC
2. Dunphy RT, Fisher K (1996) Traffic congestion and density: new insights transportation research record 1552. 1990 NPTS Survey, as presented at Transportation Research Board, Washington, DC
3. Taylor B (2002) Rethinking traffic congestion. Access No. 21, Fall
4. Grey GE, Hoel LA (1991) Public transportation, 2nd edn. Prentice Hall, Englewood Cliffs
5. (2003) TCRP Report 95 traveler response to transportation system changes. Chapter 15—land use and site design. Transportation Research Board, Washington, DC
6. Pushkarev BS, Zupan JM (1976) Where transit works: urban densities for public transportation. Regional Plan News No. 99
7. Pushkarev B, Jeffrey M. Zupan, Regional Plan Association (1977) Public transportation and land use policy, a regional plan association book. Courtesy of Indiana University Press, Bloomington, New York. All rights reserved.
8. Levinson HS, Adams CL, Hoey WF (1975) NCHRP 155—bus use of highways: planning and design guidelines. Highway Research Board, National Research Council, Washington, DC
9. Levinson HS (1983) Analyzing transit travel time performance. Transportation Research Record No. 915
10. Kittelson and Associates, Inc. Kfh Group, Inc., Parsons Brinckerhoff Quade and Douglass, Inc. and Dr. Katherine Hunter-Zaworski (2003) TCRP Report 100: transit capacity and quality of service manual, 2nd edn. Transportation Research Board, Washington, DC
11. St. Jacques K, Levinson, HS (1997) TCRP Report 26: operational analysis of bus lanes on arterials. Transportation Research Board, Washington, DC
12. TCRP Research Results Digest, No.38. September 2000
13. Levinson HS Chapter 13: system and service planning. s.l. http://Ntl.Bts.Gov/Docs/11877/Chapter_13.html
14. Linsalata J, Pham LH (1991) Fare elasticity and its application to forecasting transit demand. American Public Transportation Association, Washington, DC
15. Pratt RH, Evans JC (2004) TCRP Report 95, traveler responses to transportation system changes, Chapter 10, bus routing and coverage. Transportation Research Board, Washington, DC
16. Kittelson and Associates, Levinson HS, Harris B (2007) TCRP Report 118 practitioner guide. Transportation Research Board, Washington, DC
17. http://en.wikipedia.org/wiki/Form-based_code
18. Ewing R, Cervero R (2010) Travel and the built environment. J Am Plann Assoc 76 (3):265–294
19. Koepke FJ, Levinson HS (1992) NCHRP Report 348: access management guidelines for activity centers. Transportation Research Board, Washington, DC
20. Levinson HS, Falcocchio JC (2011) Urban development and traffic congestion. In: Compendium of transportation and development institute congress 2011
21. Vuchik VR (1999) Transportation for livable cities. Center for Urban Policy Research, Rutgers the State University of New Jersey, New Brunswick
22. Arrington GB, Cervero R (2008) TCRP Report 128: effects of tod on housing, parking, and travel. s.l. Washington, DC

Part IV
Conclusions

Chapter 24
Recap and Concluding Observations

24.1 Introduction

This book has focused on metropolitan traffic congestion in the US and Canada. Its various chapters have described and analyzed the nature, causes, and consequences of traffic congestion in cities and suburbs, and they have set forth the various strategies and actions that can be taken to provide congestion relief. This concluding chapter summarizes the key findings, gives guidelines for congestion management and illustrates possible applications of congestion relief strategies in different settings.

24.2 Types of Congestion

Congestion can occur each day at the same time and location along a street or highway. This type of congestion is known as "recurring congestion." A second type of congestion is the "non-recurring" congestion that results from random events such as vehicle breakdowns, crashes, inclement weather, natural disasters or surges in travel demand. The US Federal Highway Administration indicates that non-recurring congestion accounts for about half of the total traffic delay in US urban areas. This recognition, coupled with the fact that relieving recurring congestion through added capacity has become increasingly difficult (high cost and environmental constraints to capacity expansion) to implement, have led transportation agencies to pay greater attention to reducing the delay impacts of non-recurring congestion through the application of ITS tools.

J.C. Falcocchio and H.S. Levinson, *Road Traffic Congestion: A Concise Guide*,
Springer Tracts on Transportation and Traffic 7,
DOI 10.1007/978-3-319-15165-6_24

24.3 Causes of Congestion

The major cause of traffic congestion is the imbalance between traffic demand and roadway capacity. This imbalance can occur at an intersection or along a major transportation corridor and its congestion impacts can propagate throughout an entire area. It can result each day at the same place and time (recurring congestion), or it can result at random at random places and times (non-recurring congestion). Specific causes of this demand—capacity imbalance include (1) population, employment, and motor vehicle growth, (2) concentrations of activities in space and time, (3) VMT growth from increasing travel distances between decentralized places or work locations and residence, (4) physical and operational deficiencies of streets and highways, (5) network capacity constrained by physical and topographic barriers, (6) inability of investments in highway transportation to keep pace with VMT growth, and (7) unexpected events (e.g., incidents, bad weather, work zones) that reduce the throughput capacity of roadways.

The intensity, extent, and duration of congestion generally increase as urban areas get larger and their economies expand. Therefore, larger cities generally are more congested than smaller cities; and dynamic, growing cities are more congested than cities facing an economic downturn.

24.4 Measuring Traffic Congestion Delay

Traffic congestion reflects the difference between the travel speed when a road is lightly traveled, and the travel speed during busy traffic periods. It is also expressed as the ratio of actual travel time to uncongested travel time or as the ratio of actual vs. uncongested travel time rates (e.g., min/mile). The three basic components of traffic congestion include intensity (amount), extent (area or network coverage), and duration (how long it lasts).

However, although the common practice in measuring congestion uses free-flow speed as the congestion threshold (e.g., see TTI's Annual Urban Mobility Reports), this practice can overstate the magnitude of rush-hour congestion in large urban areas.

Establishing how much congestion delay travelers are willing to tolerate has been a concern and a challenge for many years. Key considerations include trip length, city size and facility type.

- Longer trips are impacted more by congestion than shorter trips;
- Congestion is usually greater and lasts for longer periods in larger cities;
- In larger cities congestion is more tolerable than in smaller cities;
- Travelers expect to travel faster on freeways and suburban highways than on city streets.

It is vital, therefore, that standards of tolerable congestion delay reflect the size of the urban area and is developed from stakeholders' participation. When this is done the products of transportation professionals will have a better chance of influencing the decisions that provide congestion relief.

24.5 Consequences of Congestion

The consequences of congestion include longer and less reliable journey times, lower vehicle throughput, more crashes, reduced mobility and accessibility, and increased travel and environmental costs.

24.5.1 Trip Time

Longer trip times and slower trip speeds are the most perceptible user impacts. Because congested networks have a more adverse effect on longer trips, trip length should be considered in congestion analysis.

24.5.2 Mobility

Trip mobility varies with the door-to-door speed of travel, and can be defined as the number of trips taken and their distance (trip-miles) within the traveler's daily travel time and cost budgets. Lower traffic speeds resulting from congestion reduce the mobility of people who drive longer distances. The mobility of those who walk, bike or use local public transportation is **less impacted by congestion because their trip lengths are shorter**.

24.5.3 Accessibility

Accessibility is defined as the number of opportunities accessible from given location within an acceptable travel time and cost budgets. Typically, it is determined by (1) a traveler's mobility (the door-to-door distance one can cover within a travel time and cost budgets), (2) the number of desired opportunities located within this distance, and (3) the connectivity of the street network that determines the directness of travel between an origin and a desired destination. The impact of traffic congestion on accessibility is often determined by land use density patterns (density and mix) and design as it is by roadway network speed.

24.5.4 Traffic Productivity

Traffic congestion can reduce the capacity of freeways and other roads.

(1) Traffic volume on a roadway determines the traffic speed: as traffic volumes increase, speed drops. The lowest speed where the throughput volume reaches its maximum value is the *critical speed*—so called because when speed continued to drop below this value, the throughput volume of the roadway begins to drop as well.
(2) The critical speed for freeways approximates 50 mph. Lower speeds result in reduced throughput capacity.
(3) For arterial streets, critical speeds of about 10–15 mph have been reported.

Therefore, when traffic moves at below its critical speed there is a loss in throughput volume which also increases in the duration of congestion.

24.5.5 Crashes

Traffic congestion increases the density of vehicles occupying the roadway (vehicles per lane per mile of road increases). When vehicles follow each other at close spacing they tend to change lanes more frequently—merging into crowded lanes to exit or enter the roadway—increasing the risk and frequency of crashes.

24.5.6 Air Quality and Health

Traffic congestion degrades air quality with direct consequences to human health. Average emission rates are 2–3 times higher at speeds of less than 10 mph, than they are at speeds between 20 and 80 mph. Relieving traffic congestion is often cited as an air quality and sustainability improvement strategy.

24.5.7 Congestion Costs

The cost of congestion can be measured for both personal travel and goods movement, through the additional travel time, fuel consumption, and the additional crashes incurred.

- The personal travel hourly cost of congestion delay time is approximately valued at 50 % of the hourly wage rate, while the hourly cost of business travel approximates 100 % the hourly wage rate.

- Commercial vehicle travel time costs (expressed in 1995 dollars) by benefit category and vehicle type, have been developed by the Federal Highway Administration. Weighted hourly average costs range from about $14 for small autos to more than $30 for 4 and 6 axle combination trucks.

Congestion cost estimates are often developed using different congestion thresholds and different assumptions/methods. Therefore, when comparing the results of different studies it is vital to state the assumptions used in measuring congestion delay. The need to clarify the definition of congestion delay in calculating costs is essential.

The cost of traffic congestion in US urban areas is reported annually by the Urban Mobility Report (UMR) [1] and is widely quoted by the national press. The UMR defines the time and fuel costs (TFC) of congestion as:

$$\begin{aligned}\text{TFC} = {} & [(\text{actual travel time}) - (\text{free} - \text{flow travel time})] \times [\text{value of travel time}] \\ & + [(\text{fuel consumption in actual conditions}) - (\text{fuel consumption in free} \\ & - \text{flow conditions})] \times [\text{unit cost of fuel}]\end{aligned} \tag{24.1}$$

The 2012 UMR reported an annual congestion cost of about $120 billion in delay and fuel costs. Because these costs were calculated using free-flow speed, as the congestion threshold speed in all urban areas, the UMR report significantly overestimates the cost of congestion.

24.6 Congestion Relief Strategies

24.6.1 General Principles

Keeping congestion from adversely affecting a community's livability and economy is a key societal goal. A basic objective of congestion-relief actions in large urban areas is to reduce congestion to manageable levels since its complete elimination is usually neither practical nor cost-effective.

Transportation enables individuals, families, and businesses to achieve social, economic, and quality of life goals. It is a means to end (e.g., affordable housing costs, accessibility to desired destinations, etc.). Therefore, while important at the network level, reducing congestion should not be an end itself—rather we should ask "to what extent is congestion limiting our ability to reach desired destinations?" Therefore, the perspective on how best to deal with congestion should be broadened to include the goal of achieving vibrant, livable, and accessible communities.

Congestion relief strategies and related actions vary with (a) the type of congestion; (b) city size, structure, and street patterns; (c) the location, type and severity of specific problems, and (d) the likely future traffic growth. They are also influenced by agency and community support, and the availability of available resources.

24.6.2 Strategies that Relieve Nonrecurring Congestion

For nonrecurring events the strategies include shortening response and recovery times for incidents and real-time information to minimize adverse impacts on travelers. To accomplish these objectives two critical elements are necessary: (1) the application of ITS technologies (e.g., real-time information, fast computing algorithms, and communication) whose key function are to provide early detection of a random event(s) and inform responding agencies and travelers about its location, and travel alternatives, and (2) the coordination of functions among responding agencies.

Reducing the intensity, duration, and extent of congestion created by nonrecurring events involves the application of supply strategies (adaptation) and demand management strategies (mitigation) that are responsive to the type of event that reduces capacity or increases demand. Strategies suitable to manage the impacts of nonrecurring events involve the ability to detect the event as soon as it happens and to restore the roadway to full capacity as soon as possible; direct drivers to reduce speed during inclement weather or when the roadway is being repaired; mitigation strategies involve informing drivers about the location and times of special events that are likely to generate surge in traffic demand on the impacted roads so that they may plan changes in trip time, or route.

Strategies that reduce the impacts of non-recurring congestion delays involve the ability to detect the event as soon as it happens and to restore the roadway to full capacity as soon as possible, direct drivers to reduce speed during inclement weather or when the roadway is being repaired, or inform drivers about the location and times of special events that are likely to generate surge in traffic demand on the impacted roads. Safe and speedy evacuation is essential when major disasters occur.

24.6.3 Strategies that Relieve Recurring Congestion

Strategies that relieve the impacts of recurring congestion, include increasing the operational efficiency of existing road networks, creating new capacity, and managing (reducing) highway travel demand.

Strategies that reduce *recurring* congestion delays at physical bottlenecks involve grade separation of intersecting traffic streams; road widening at bottleneck locations; the addition of merging and turn lanes; and reconfiguration of entrance and exit ramps at freeways and expressways. However, to prevent new traffic attracted to the improved roadway from nullifying the travel time benefits of bottleneck removal, strategies that reduce bottleneck congestion in highly congested roads should be coupled with strategies that control traffic demand on these roads (e.g., ramp metering).

The Texas Transportation Institute Urban Mobility Report [1] recommends a balanced and diversified approach to reduce congestion—one that focuses on more of everything. The report states that "current investment levels have not kept up with the problems" and that population growth will require more systems, better operations and an increased number of travel alternatives. In addition, most urban regions have big problems now—more congestion, poorer pavement and bridge conditions, and less public transportation services than they would like to have. The report states that there will be a different mix of solutions in metro regions, cities, neighborhoods, job centers and shopping areas. Some areas might be more amenable to construction solutions, while other areas might use more travel options, productivity improvements, diversified land use patterns or redevelopment solutions. In all cases, the solutions need to work together to provide an interconnected network of services.

Various strategies to relieve recurring congestion, and where each works best, are summarized in Table 24. 1.

In smaller communities, congestion relief should focus on reducing the intensity, extent and duration of congestion. However, in larger cities priority generally should be given to reducing the duration of the congestion.

- Capacity expansion strategies generally are desirable in rapidly growing metropolitan areas to better balance roadway supply and demand. However, sometimes the increased capacity increases travel demand. For this reason, strategies that mitigate traffic demand should be combined with capacity expansion strategies.
- In very large urban areas (population more than 2 million) with strong city centers (employment more than 100,000) and rapid transit facilities, a combination of public transport improvements, managed freeway lanes, outlying park-and-ride facilities, and CBD parking ceilings can help relieve congestion.
- Effective coordination of land development and transportation investments are essential, especially in growing urban areas where transportation networks and land use design should facilitate the use of public transport and should be pedestrian and bike friendly.

24.6.4 Implementation Issues

Operational strategies that get the most use of the existing system by eliminating bottlenecks are desirable in all communities. However, they should not merely transfer the congestion from one location to another. Operational strategies are relatively easy to implement when they are the responsibility of one transportation agency.

Table 24.1 Congestion relief strategies related to urban area population

Relief strategy	Urban area population				
	1 Very small <100,000	2 Small 100,000–500,000	3 Medium 500,000–1.5 million	4 Large 1.5–3 million	5 Very large over 3 million
Roadway capacity enhancement					
Better use of existing streets and highways—traffic engineering and access management	X	X	X	X	X
Capacity expansion—new and improved freeways and arterials improved local connectivity	X	X	X	X	X
Managed lanes/variable road pricing		Very small	Limited	X	
Roadway demand management					
Parking management—park-and-ride CBD parking limits/pricing			Limited	X	X
Congestion pricing				X	X
Transit improvements (including new rapid transit lines)		Limited	X	X	X
Lane management		Limited	X	X	X
Incident event management	X	X	X	X	X

Source Estimated

Operational strategies are relatively easy to implement when they are the responsibility on a single agency, they are relatively easy to implement. Where the owners of the transportation infrastructure are agencies typically controlled by different units of government, coordination of congestion management strategies among these units may be time consuming and not always easily achievable—especially when the limited funding available for this purpose may not be transferrable between agencies.

Congestion relief actions are location specific. They should be keyed to the needs, opportunities, attitudes and resources of each urban area. Relevant considerations include:

- The location, type, duration and extent of congestion
- Likely future growth
- Physical and operational deficiencies of streets and highways
- The character of development patterns and their transportation connectivity
- The number, size, complexity and attitudes of the various governmental jurisdictions and public agencies involved

- Existing and future resources likely available for congestion relief
- Availability and use of public transportation

Agencies operating in large urbanized areas should coordinate their congestion management and relief actions. Arrangements will vary, depending on the size, complexity, and congestion problems of the area. Coordination of tasks among responding agencies is especially important to reduce response times to non-recurring congestion events.

24.7 Typical Application Scenarios

Typical applications of congestion relief strategies vary from simple cases (e.g., isolated intersections) to complex ones (e.g., city centers, transportation corridors).

24.7.1 Isolated Intersections

Intersection congestion can be reduced a number of ways—from simple retiming of signals to increasing roadway capacity to accommodate peak travel demands. Relief actions include (1) adding left-turn lanes, (2) improved traffic signal timings, and in some cases roadway widening. Complex signal phasing with long traffic signal cycles (e.g., greater than 120 s) should be avoided.

24.7.2 Suburban Areas

Suburban areas experience the largest share of metropolitan growth. In these areas travel demand growth may require a combination of the following congestion relief strategies: (1) increasing the traffic capacity of roadways as well as other modes of transportation, and (2) reducing private vehicle use by coordinating land use growth policies that reduce the need to drive with investments in alternative modes of transportation.

To be implementable, these strategies require effective coordination of land development and transportation decisions. New developments should promote walking and biking by providing a land use—street system designs that also support public transport use. These actions could include:

- Managing access on major roadways to minimize driveway conflicts and maintain good traffic signal coordination
- Cluster commercial developments and avoid commercial strips
- To facilitate walking and biking trips, provide continuous roads spaced at not more than 1/2 mile to 1 km intervals

- To minimize traffic delayed by large left-turning volumes at intersections, provide continuous collector/arterial roadways at 1/2 mile to 1 km spacing or less
- Provide sufficient residential densities to support bus service
- Provide rapid transit service to major activity and employment concentrations
- Provide local residential streets with sidewalks that in large urban areas can afford direct traveler access to bus service
- Provide local street patterns that can better accommodate emergency vehicles

24.7.3 Suburban Mega Centers

Mega centers in suburban areas are mainly accessible by private automobiles. These major concentrations of activities tend to generate heavy peak-period traffic demand that can exceed the capacities of the approach and boundary roads and result in severe traffic congestion on expressways and arterial roadways that provide access to the mega centers.

Congestion relief strategies on the roadways serving suburban mega centers involve proactive actions aimed at (1) increasing the capacities of the access modes to match the centers' traffic demand, or (2) constraining peak hour traffic loads by limiting the growth of these centers to the capacity of their access roads, and (3) coordinating the center's land use development/design to facilitate customer access by transit or as pedestrians.

Coordinating land use decisions (under local control), land developers' decisions (under private control) and transportation decisions (under State control) is a challenging task involving competing and diverse objectives (e.g., the financial interests of private developers, the importance of tax revenues to the local towns' operating budgets, and home rule in zoning decisions).

24.7.4 Central Business Districts (CBD) of Large Cities

Mega centers located in centers of large cities (e.g., Manhattan, Chicago, Boston, or Philadelphia central business districts) consist of high-density commercial and residential developments where rapid transit access predominates. There is a high volume of pedestrian trips, and a low per capita share of private vehicle use. Even where most trips are by public transport, these centers usually face extensive traffic congestion where CBD employment exceeds 100,000, and where off-street parking is expensive.

Congestion relief strategies for large city centers typically require a mix of strategies involving operational and physical improvements to increase multimodal transportation capacity, as well as multimodal travel demand management and parking management policies.

Strategies that reduce automobile use in peak periods, include

- transit service improvements,
- flexible work hours that spread the peak demand outside the peak hour,
- congestion pricing,
- downtown parking pricing and parking supply constraints, and
- providing outlying park-and-ride facilities connected to downtown by rapid transit lines (commuter rail, heavy, rail, light rail, BRT).

Parking pricing and road policies in city centers can relieve congestion by:

- Increasing the choice of public transport for accessing the area
- Increasing the availability of existing parking spaces in congested parts of the city center at different times of the day
- Reducing the incremental VMT added when searching for a parking space.

Construction of park-and-ride facilities at the periphery of the congested area and along outer stations of rapid transit lines can relieve congestion by:

- Reducing commuter VMT on radial express highways leading to the city center
- Reducing parking demands and needs in city centers
- Increasing transit ridership
- Extending the transit market to outlying areas.

24.7.5 Metropolitan Transportation Corridors

These corridors typically consist of freeways, arterial roads, and major transit lines connecting to centers of employment, commerce, and entertainment, as well as serving trips passing through the metropolitan area. Travelers are subject to recurrent congestion delays, every day, on the same roads, at the same time periods, resulting from physical bottlenecks. The effect of this recurring delay is frequently magnified by the additional delay from nonrecurring events such as incidents, weather, special events, and road maintenance.

Congestion strategies that reduce recurring congestion delays at physical bottlenecks involve road widening at bottleneck locations; the addition of merging and turn lanes; and reconfiguration of entrance and exit ramps at freeways and expressways, and grade separation of conflicting traffic streams. To prevent new traffic attracted to the improved roadway from nullifying the travel time benefits of bottleneck removal, strategies that reduce bottleneck congestion in highly congested roads could/should be coupled with strategies that control traffic demand on these roads (e.g., ramp metering and road pricing).

However, building new freeway capacity for general use is often inhibited by costs, environmental impacts and community opposition. Therefore, high occupancy vehicle (HOV) lanes are sometimes added to heavily traveled freeway lanes, to give travel time savings to motorists who car pool, and high occupancy toll

(HOT) lanes are added to freeway lanes so that motorists who car pool and pay tolls benefit from faster and more reliable travel.

However, the corridor congestion problem is often the result of the joint occurrence of recurring and nonrecurring events that can be effectively mitigated only through the application of real-time management tools applied to all modes of transportation service the corridor. This approach, known as Integrated Corridor Management (ICM) uses ITS tools and assets to manage traffic flow and influence traveler behavior to achieve operational objectives. Known as Active Transportation Demand Management (ATDM), this strategy consists of monitoring, controlling, and managing demand over the entire trip chain in the corridor [2].

24.8 Future Outlook

The ways that urban areas will grow and change in the future will have important bearing on when, how much, and where congestion will increase. Equally important will be peoples' preferences for living and work locations. Evolving technological breakthroughs in information, communications, and automation, will also influence the intensity, duration, and extent of traffic congestion. The ability of transportation system improvements and travel demand management programs to effectively relieve congestion in the future will also depend on the ability of independent public agencies to coordinate their transportation and land use decisions—a key requirement for effective congestion relief [3].

24.8.1 How Will Travel Demand Change in the Future?

In the last half of the twentieth century, highway travel outpaced population growth. More people living in suburban settings, driving more cars, led to increased vehicle miles of travel, declines in public transport use, increased land consumption, and increased traffic congestion.

In the twenty first Century, changes in public attitudes toward social responsibility for environmental preservation, recognition of the limits of auto mobility and increased investment in public transport have somewhat changed the perspective. Growth rates in the use of public transport, especially in larger metropolitan areas began to exceed the growth rate in VMT. "Managed" freeway lanes and road pricing have become more acceptable. Automated vehicles are becoming a reality and their increasing presence in the traffic stream is likely to increase the capacity throughput of freeways and expressways, as well as reducing crash rates. Expected changes in socio-demographic characteristics and household location decisions are likely to result in lower per capita VMT growth and lower future annual VMT growth rate [3].

24.8.1.1 Socio-Demographic Changes

The ways that urban areas will grow and change in the future will have important bearing on when, how much, and where congestion will increase. Equally important will be peoples' preferences for living and work locations. Evolving technological breakthroughs in information, communications, and automation, will also influence traffic congestion. However, the ability of transportation system improvements and travel demand management programs to effectively relieve congestion in the future will also depend on the ability of separate public agencies to work collaboratively and to coordinate their transportation and land use decisions—a key requirement for effective congestion relief [3] are:

- In the next 30–50 years the US population will grow more slowly with a corresponding reduction on VMT growth
- American work force is growing older, more female and more diverse. Resulting in a decrease in VMT per capita
- The difference between cities and suburbs will be less distinguishable, resulting in a decrease in VMT per capita, an increase in non-motorized trips and transit trips
- Mobile broadband will shape lifestyle choices—and possibly reduce car ownership, and VMT per capita for some trip purposes
- Increase in environmental concerns by the younger generation could result in lower car ownership, more transit, and non-motorized travel.

24.8.2 How Will the Transportation System Change in the Future?

Technology has played an important role in the shape and size of cities and metropolitan areas over the past 150 years. It has enabled cities to grow vertically and horizontally. It has brought about electric traction, steel framed skyscrapers, airplanes, automobiles, and big-box stores and it has also shaped the patterns of traffic congestion.

Future changes in communications technology could change life styles, travel behavior and development patterns, and could provide some congestion relief. Possibilities include better vehicle-to-vehicle communications, automated vehicles, and real-time information. Perhaps the automated highways proposed by Norman Bel Geddes in the Futurama Exhibit at the 1940 New York World's Fair will become a reality!?

24.8.3 How Will Regional Governance Change in the Future?

Congestion is everyone's business. Congestion relief strategies must have both agency and community acceptances and support. Sound governmental organization and management are essential to reduce both recurring and non-recurring congestion delay. Agencies must work together and provide the resources needed to relieve congestion.

Strategies for managing the supply and demand for roads (and parking spaces) and expanding public and active (walking and biking) transportation should be consistent with the vision of how urban areas should develop in future years. This is a fundamental requirement that entails coordinating the work of diverse agencies and private sector interests regarding of land-use decisions at the neighborhood level, and with transportation decisions at the regional and state levels.

A variety of coordinated congestion-reducing strategies are usually needed to address recurring and non-recurring congestion. There is no magic answer, no silver bullet, and change cannot be realized overnight. Getting more productivity from existing road and public transport systems is essential to reduce congestion and to improve travel time reliability. Businesses, employers and public policies could adopt various strategies to modify travel behaviors that reduce VMT. Variable road pricing in conjunction with transit improvements might become desirable strategies to reduce VMT growth. New developments can be designed to encourage more walking and transit trips. However, in many corridors, additional highway capacity and public transport capacity will be needed to move people and freight more rapidly and reliably.

24.8.3.1 Coordinating Land Use and Transport Decisions

Future urban growth could result in regional cities. How these regional cities are developed and designed will influence the growth and migration of traffic congestion. Better coordination of transport facilities and land development will be essential for both congestion relief and more livable communities. This includes:

- Locating different types of activities in the same zone with high density can reduce the per capita use of private motor vehicles and can enable more public transit use, walking, and bicycle trips, and
- Coordinating among the independent policies of local zoning boards, state, and federals transportation officials.

Urban land will need to be better managed to reduce VMT growth in future decades. This is a key strategy for long term congestion relief and livability in metropolitan areas.

24.9 Conclusions

This book has presented a comprehensive user-oriented account of traffic congestion. Its various chapters describe the characteristics, causes, consequences, and workable strategies to relieve both recurring and non-recurring congestion.

Long-term projections of traffic congestion are conditioned by many factors—many known, some unknown. Key unknown factors that have a bearing on future travel demand growth include the price of travel relative to disposable incomes, land use control, and the effect of information and communication technology on locational decisions of households and jobs.

However, effective congestion management strategies in the years ahead will require transportation agencies to apply coordinated, coherent, consistent, and continuing actions that adapt the transportation system to emerging new technologies and land development patterns. They will require land use regulations and selective pricing policies that can keep travel demands at manageable levels. Effective coordination of land development and transportation facility development will be essential especially where urban areas expand. In these areas, land use and network design elements should encourage non-motorized and public transportation use wherever practical.

References

1. Schrank B, Eisele B, Lomax T (2012) TTI's Urban mobility report powered by INRIX traffic data. s.l. texas A&M transportation institute, texas A&M university system
2. O'Connor, Arthur T (2014) Development of a framework for the deployment of intelligent transportation systems in a multi-agency environment. s.l. Ph.D. Dissertation, NYU Polytechnic School of Engineering
3. Transportation Research Board (2014) NCHRP report 750-strategic issues facing transportation. The effects of socio-demographics on future travel demand. Vol 6 Transportation Research Board, Washington, DC

MIX
Papier aus verantwortungsvollen Quellen
Paper from responsible sources
FSC® C105338

If you have any concerns about our products,
you can contact us on
ProductSafety@springernature.com

In case Publisher is established outside the EU,
the EU authorized representative is:
Springer Nature Customer Service Center GmbH
Europaplatz 3, 69115 Heidelberg, Germany

Printed by Libri Plureos GmbH
in Hamburg, Germany